D0710694

FINITE MATHEMATICS

FINITE MATHEMATICS

David G. Crowdis

Susanne M. Shelley

Brandon W. Wheeler

Sacramento City College

Rinehart Press / Holt, Rinehart and Winston

SAN FRANCISCO

Library of Congress Cataloging in Publication Data

Crowdis, David G.
　　Finite mathematics.

　　　1.　Mathematics—1961–　　　　　　　　I.　Shelley, Susanne,
1928–　　　　　　joint author.　　　　II.　Wheeler, Brandon W.,
joint author.　　　　　III.　Title.
QA39.2.C76　　　　　510　　　　73-18108
ISBN　0-03-002961-9

© 1974 by Rinehart Press
5643 Paradise Drive
Corte Madera, Calif. 94925

A division of Holt, Rinehart and Winston, Inc.

PRINTED IN THE UNITED STATES OF AMERICA

4 5 6 7　　038　　9 8 7 6 5 4 3 2 1

CONTENTS

PREFACE

Finite Mathematics represents one of the newest and fastest growing areas of mathematics applied to practical problems. The applications for the topics of *Finite Mathematics* are for the most part in business, economics, and the social sciences. This text is designed to show students from these subject areas the underlying mathematical concepts for such problem solving.

The topics are designed for students who have taken two years of high school algebra or the equivalent. There is sufficient material for a one-semester or two-quarter course. In addition, a number of special topics are included in the supplementary sections, including a computer programming language, BASIC, and a review of systems of equations. These topics are designed to extend the students' knowledge in special-interest areas or to strengthen weak points in the students' backgrounds.

Special care has been taken to make the material understandable to students with limited backgrounds in mathematics. Concepts are clearly presented and are coupled with numerous examples (over 300) chosen primarily from business and the social sciences. Each section is also supplied with a large set of graded exercises, including those especially chosen for students in business, economics, or the social sciences, and each chapter includes a set of summary problems to review the topics studied in the chapter. The vocabulary throughout the text was selected to be appropriate for such students.

The text developments present a balance between pure manipulation, theory, and applications. The format for any required proofs is informal but mathematically correct and consistent with the students' mathematical maturity. Whenever possible, developments point out how one topic can be combined with others to form new and powerful mathematical tools for problem solving.

The answers to the odd-numbered exercises and all problems in the chapter reviews are included in the text; the remaining answers are in a teacher's manual.

D. C., S. S., and B. W.

CHAPTER **1**

LOGIC

1.1 INTRODUCTION

Since this text starts with a discussion of logic, it is natural to ask two questions: What is logic? What does it have to do with finite mathematics? To answer the first question, a formal definition of logic might be:

Logic is the science of reasoning.

From this definition it is clear that a complete study of logic could be, and often is, a lifetime project. In this text we will consider one model of reasoning. It will be a *finite* model in the sense that only a finite number of cases will be allowed in analyzing reasoning patterns. It will be a *symbolic* model—that is, we will attempt to abstract patterns of valid or sound reasoning into symbolic patterns, much as a course in the algebra of real numbers abstracts the patterns of everyday arithmetic. When mathematicians first began to look at these abstract patterns, they did so because they needed logic as a tool in analyzing mathematical structures. Once they did this, they discovered that the patterns they found were exact models of other mathematical systems which on the surface had not appeared to be related to logic at all. Hence in this text we will view logic both as a tool with which to study mathematics and as a mathematical structure itself.

We have already answered the question of how logic relates to finite mathematics. It is a branch of finite mathematics. In general, finite mathematics deals with those parts of mathematics in which a finite number of cases or only finite values of a finite number of variables are involved. This contrasts with those topics examined in calculus and calculus-related studies, which tend to deal with the infinite.

The purpose of this chapter is to demonstrate how "valid conclusions" may be reached from given information, in the hope that the technique can be extended to apply to actual situations. In order to make this transfer easier, symbols will be used to replace actual statements in much the same manner that x's and y's are used in algebra. Therefore, the name *symbolic logic* is used to describe this study.

1.2 STATEMENTS OF LOGIC

Just as the basic structure of arithmetic consists of numbers, so the basic structure of logic consists of statements.

> **DEFINITION**
>
> **A statement in logic** is a sentence that can be assigned a single *truth value*.

The particular logic we are going to study is a *two-valued logic*; that is to say, there are two truth values: true or false. Thus a statement is a sentence that can be assigned the truth value *true* or the truth value *false* but not both at the same time.

In English grammar this type of sentence is called a *declarative sentence*. However, not all declarative sentences qualify as statements in logic.

Because the use of language to express statements often leads to semantic ambiguities, it is sometimes difficult to assign a single truth value to a statement, thus making it difficult to classify it as a logical statement. Consider, for example, statements such as "Zsa Zsa is a beautiful woman"; "Frank is a good man"; "Skiing is fun." It is not possible to arrive at a clear-cut true or false decision on the basis of subjective interpretations of "beautiful," "good," or similar value judgments.

EXAMPLE   Which of the following sentences are logical statements?

a. Today is Thursday.
b. I have $200 in my bank account.
c. $2 + 2 = 5$.
d. Friendly Mining is a good stock.
e. Don't buy glamour stocks.

Solution

a. This is a logical statement, since it can easily be determined if the statement is true or false. It must be one or the other.

b. Again, the truth or falsity of the statement can be verified, and (b) is a statement.

c. Yes. This is a good example of a false statement since it is known that $2 + 2 = 4$ and not 5.

d. Now the question arises: What is a good stock? There is no clear-cut decision as to true or false for this sentence unless some criteria for judging good or bad stocks have previously been defined. Thus (d) is *not* a statement in logic.

e. No. This sentence is not declarative and does not qualify as a statement.

Therefore, the logical statements are (a), (b), and (c).

The statements illustrated in the above example are **simple statements.** Often a statement contains more than one statement as component parts. For example, "Today is Thursday and I have \$200 in my checking account" is a statement consisting of statements (a) and (b) in the above example. This type of statement is called a **compound statement.** Examples of compound statements are:

a. If profits rise, then I will hire more employees.

b. The tenant will pay higher rent or he will move out.

c. The buyer selects line A and rejects line B.

The simple statements which are the components of (a) are

> Profits rise.
> I will hire more employees.

For (b) the component statements are

> The tenant will pay a higher rent.
> He (the tenant) will move out.

And for (c) the statements are

> The buyer selects line A.
> (The buyer) rejects line B.

If a compound statement is a logical statement, it also has to be either true or false but not both at the same time. The truth value of a compound statement will be studied in light of the truth values of its component statements.

Often a complicated statement can be analyzed easily if a symbol is used to replace the statement. Lowercase letters $p, q, r, s, \ldots$ are often the replacement choices. For instance, let p stand for the statement "Profits rise" and q stand for "I will hire more employees." Now the statement "If profits rise, then I will hire more employees" can be written "If p then q."

EXAMPLE Let p be the statement "This course is a degree requirement." Let q be the statement "John must pass this course." Translate each of the following symbolic statements:

    a. If p then q
    b. If q then p
    c. p or q
    d. p and q
    e. q only if p

Solution

    a. If this course is a degree requirement, then John must pass this course.
    b. If John must pass this course, then this course is a degree requirement.
    c. This course is a degree requirement or John must pass this course.
    d. This course is a degree requirement and John must pass this course.
    e. John must pass this course only if it is a degree requirement.

It should be evident that the words, *and*, *or*, and *if...then* were used to make compound statements from simple statements. These are not the only words which serve this purpose, but they are very important and will be carefully studied in subsequent sections of this chapter.

EXERCISES 1.2

Determine which of Exercises 1–10 are statements of logic.

1. The stock market crashed on Black Monday.

2. A fixed income is a disadvantage in times of inflation.

3. Real estate is a good business.

4. Either buy the coat or leave the store.

5. Professor Jones teaches business law and has a private law practice.

6. If you fly to Denver, you can attend the meeting.

7. The Los Angeles Rams won the football game and the San Francisco 49ers lost.

8. It is very pleasant here.

9. I'll fire you if you disagree with me.
10. If it rains on Tuesday and if the meeting is canceled, I will call you and we can go to lunch.

11. List the compound statements in Exercises 1–10.

Let p, q, and r be the following statements:

> *p: Government spending is increasing.*
> *q: Taxes are higher than ever.*
> *r: There is a freeze on wage increases.*

Translate each of the symbolic statements in Exercises 12–18.

12. *p* and *q*   13. If *p* then *q*
14. *q* and *r*   15. If *p* then *r*
16. *r* if *q*   17. If *q* and *r* then *p*
18. *p* only if *r*

19. Assume that *p* is the *true* statement "Jane is ill" and *q* is the *true* statement "Jane goes to work." What do you think the truth value of each of the following compound statements is? (An *actual* determination of the truth value of compound statements according to the rules of logic will be discussed in subsequent sections of this chapter.)
    a. *p* and *q*   b. *p* or *q*
    c. If *p* then *q*   d. If *q* then p

20. Assume that *p*: "Jane is ill" is a *true* statement and *q*: "Jane goes to work" is a *false* statement. What do you think the truth value of each of the following compound statements ought to be?
    a. *p* and *q*   b. *p* or *q*
    c. If *p* then *q*   d. If *q* then *p*

1.3 MODIFIERS, CONNECTIVES, AND TRUTH TABLES

Since a statement must be either true or false, the modification of a statement by the word *not* or the words *it is not the case that* changes the truth value of the statement.

Thus "Today is Thursday" must be either true or false. If it is true, then "Today is *not* Thursday" is false, and if it is false, then "Today is *not* Thursday" is true.

Symbolically, if *p* represents a statement, then *not p* is written $\sim p$, where $\sim$ is called a **modifier** of the statement *p*.

Truth tables are useful devices for testing truth values of statements. Use T to mean true and F to mean false.

p	$\sim p$
T	F
F	T

The table illustrates that if p is true, then not p ($\sim p$) is false, and if p is false, $\sim p$ is true.

Words that make a compound statement from simple statements are called **connectives**. From the discussion in the preceding section, the words *and*, *or*, and *if...then* serve such a purpose and are connectives.

The symbol $\wedge$ is used to mean *and*. If p and q are statements, then p *and* q is written $p \wedge q$.

The symbol $\vee$ is used to mean *or*. Thus p *or* q is written $p \vee q$.

EXAMPLE Consider statements p: "This car has a ninety-day warranty" and q: "Financing is available." Write a symbolic statement for each of the following:

a. This car has a ninety-day warranty, and financing is available.
b. Financing is available or this car has a ninety-day warranty.
c. This car does not have a ninety-day warranty and financing is not available.
d. It is not the case that this car has a ninety-day warranty or that financing is available.

Solution

a. $p \wedge q$
b. $q \vee p$
c. $\sim p \wedge \sim q$
d. $\sim(p \vee q)$ (The words *it is not the case that* usually modify the entire compound statement; therefore, the parentheses are used.)

Since a compound statement is also a logical statement, it must have a truth value, and this truth value in turn depends on the truth values of the component statements and on the definition of the connectives which form the compound statement.

In order to examine the truth value of a compound statement, we must consider all the logically possible arrangements of true and false of the component statements. For example, consider a compound statement with components p and q. The logical possibilities are: both p and q are true; both p and q are false; p is true and q is false; and p is false and q is true. Thus there are *four* possible entries in a truth table, as shown below.

$$\begin{array}{c|c} p & q \\ \hline T & T \\ T & F \\ F & T \\ F & F \end{array}$$

When three or more different statements are used to form a compound statement, more logical possibilities exist; this situation will be discussed in a subsequent section of this chapter.

We are now ready to consider the definitions of some compound statements. Why the statements are defined in the way they are is not easy to justify. Decisions for rules of games are made in such a way as to make the game interesting and to make the rules workable and reasonable, but two rules should not be in conflict with each other.

The first connective to be defined is *and*, $\land$. Consider the statement "Sunday the football team plays in Oakland and we will attend the game." If the game is played in Oakland and we attend, then both components are true, so let us consider the compound statement to be true. Thus if p is the statement "Sunday the football team plays in Oakland" and q is the statement "We will attend the game," we can symbolize the statement

$$\begin{array}{ccc} p & \land & q \\ T & T & T \end{array}$$

What happens if the team is playing in Oakland but we decide not to attend? Or worse yet, if we decide to attend but the team is playing elsewhere?

$$\begin{array}{ccc} p & \land & q \\ T & ? & F \\ F & ? & T \end{array}$$

A reasonable definition would assign the value *false* to such a combination and would certainly assign *false* to the event that neither p nor q is true:

$$\begin{array}{ccc} p & \land & q \\ F & F & F \end{array}$$

The above reasoning should make the following definition acceptable: We define $p \land q$ to be true only when p and q are *both* true. The truth table appears below.

p	q	$p \wedge q$
T	T	T
T	F	F
F	T	F
F	F	F

The connective $\wedge$ is also called a **conjunction**.

EXAMPLE   Let p be the *true* statement "This car has a ninety-day warranty" and q be the *false* statement "Financing is available." What is the truth value of the conjunction of p and q—that is, $p \wedge q$?

Solution

p	q	$p \wedge q$
T	F	F

Since $p \wedge q$ is true only when p and q are *both* true, $p \wedge q$ is false.

EXAMPLE   Make a truth table for the statement $\sim p \wedge q$.

Solution

(1)	(2)	(3)	(4)
p	$\sim p$	q	$\sim p \wedge q$
T	F	T	F
T	F	F	F
F	T	T	T
F	T	F	F

Column (2) has the opposite truth value of column (1) since it lists $\sim p$. Then columns (2) and (3) are considered for the conjunction in column (4).

The second connective to be defined is *or*, $\vee$. This connective is also called a **disjunction**. The truth table defining the disjunction appears below.

p	q	$p \vee q$
T	T	T
T	F	T
F	T	T
F	F	F

We define $p \vee q$ to be true as long as at least *one* of the component statements is true.

Does this definition of *or* seem to make sense? All an *or* statement really promises is that at least one of the components is true. For example, "I must deposit money in my account, or my check will bounce"; "In the presidential election of 1948, either Tom Dewey was elected or Harry Truman was elected"; "On Saturday evening I will go to a show, or I will watch the movie on television."

In the election example it would not be possible for both statements to be true simultaneously, whereas in the other two examples both statements could be true. Maybe I will go to an early show on Saturday and then come home and watch the TV movie.

A situation in which both components of a disjunction may be true is considered to be an **inclusive disjunction**. Thus the truth table for ∨ is the truth table for the inclusive disjunction. If it is not possible for both components to be true at the same time, the disjunction is said to be **exclusive**. The development of a truth table for the exclusive disjunction is left as an exercise.

EXAMPLE Let p be the *false* statement "The labor dispute has been settled" and q be the *true* statement "Everyone returns to work on Monday." Determine the truth value of each of the following:

a. $p \lor q$ b. $\sim p \lor q$
c. $p \lor \sim q$ d. $p \land q$

Solution

a.

p	q	$p \lor q$
F	T	T

b.

p	$\sim p$	q	$\sim p \lor q$
F	T	T	T

c.

p	q	$\sim q$	$p \lor \sim q$
F	T	F	F

d.

p	q	$p \land q$
F	T	F

EXAMPLE If p and q represent the same statements as in the above example but with undetermined truth values, make a truth table for the statement "It is not the case that the labor dispute has been settled or that everyone does *not* return to work on Monday."

Solution First, symbolize the statement: $\sim(p \lor \sim q)$. Next, list the possible cases:

p	q	$\sim q$
T	T	F
T	F	T
F	T	F
F	F	T

We are interested in $p \lor \sim q$ and then in the negation of that statement, so we test the possible truth values of the statement *inside* the parentheses first, and then we negate that statement. This procedure of working from the inside is helpful when working with grouped statements; it follows much the same pattern as removing parentheses in algebra, where the innermost statements are worked first, and so on.

p	$\sim q$	$p \lor \sim q$	$\sim(p \lor \sim q)$
T	F	T	F
T	T	T	F
F	F	F	T
F	T	T	F

Now, for example, if the dispute has been settled (p is *true*) and no one returns to work on Monday (q is *false*), the statement $\sim(p \lor \sim q)$ is *false*. Why? If p is true and q is false, then $\sim q$ is true. Thus line 2 of the above truth table is used. And $\sim(p \lor \sim q)$ is false in this entry.

   Note that the order of truth values for p and q is always written the same way:

p	q
T	T
T	F
F	T
F	F

This is done in order to facilitate the learning of the definitions and to make the checking of answers easier. We shall agree to this convention in this text.

   One more example is in order here. Suppose that you wish to make a truth table for the statement

$$(p \lor q) \land (p \land \sim q)$$

First, you must recognize that the compound statement is a *conjunction* of the compound statements $(p \lor q)$ and $(p \land \sim q)$. Thus the *final connective* is $\land$, and its truth values must be determined from the truth values of the two compound statements.

			(1)	(3)	(2)
p	q	$\sim q$	$(p \lor q)$	$\land$	$(p \land \sim q)$
T	T	F	T	F	F
T	F	T	T	T	T
F	T	F	T	F	F
F	F	T	F	F	F

Columns (1) and (2) were used to determine column (3).

EXERCISES 1.3

Let p be the statement "Logic is important" and q be the statement "I will pass this course." Write a verbal statement for each of Exercises 1–5.

1. $p \land q$ **2.** $p \lor q$ **3.** $\sim p \land q$

4. $p \lor \sim q$ **5.** $\sim(p \land q)$

Using the statements p and q from Exercises 1–5, write a symbolic statement for each of Exercises 6–10.

6. Logic is important or I will not pass this course.
7. Either I will pass this course or logic is not important.
8. It is not the case that logic is important and that I will not pass this course.
9. Either logic is important and I will pass this course, or logic is not important.
10. I will not pass this course and logic is not important.

Assume that p and q are statements. Make a complete truth table for each of Exercises 11–16.

11. $p \lor \sim q$ **12.** $\sim p \land q$ **13.** $\sim p \land \sim q$

14. $\sim(p \lor q)$ **15.** $\sim p \lor \sim q$ **16.** $\sim(p \land q)$

Assume that p and q are statements. Give the truth value for each of Exercises 17–23 under the given conditions.

17. p is true, q is false; $p \lor \sim q$
18. p is false, q is true; $p \land \sim q$
19. p is false, q is true; $\sim p \land q$

20. p is true, q is false; $\sim(p \wedge \sim q)$

21. p is true, q is true; $\sim(p \wedge \sim q)$

22. p is false, q is true; $\sim(\sim p \vee \sim q)$

23. p is true, q is false; $\sim p \vee \sim q$

24. If p and q are statements, make a truth table for $(p \wedge q) \vee (\sim p \wedge q)$. *Hint:* The *final* connective is $\vee$. You are concerned with a compound statement made up of component compound statements $(p \wedge q)$ and $(\sim p \wedge q)$. Therefore, write the truth tables for these statements first, and then formulate the disjunction.

25. Use the hint in Exercise 24 to make a truth table for the statement $(p \vee q) \wedge \sim p$.

1.4 MORE CONNECTIVES

One of the most important statements in mathematical reasoning, or for that matter, in any reasoning, is the *if...then* statement.

If $x + 2 = 5$, then $x = 3$.

If profits rise, then you will get a raise.

If I pass this course, then I will graduate.

Symbolically, if p and q are statements, then

"If p then q" is written $p \rightarrow q$

Another name for $p \rightarrow q$ is **conditional statement**. The *if* part of the statement is called the **hypothesis** or **antecedent**; the *then* part is the **conclusion** or **consequent**. Thus in the statement $p \rightarrow q$, p is the hypothesis and q is the conclusion, whereas in the statement $q \rightarrow p$, q is the hypothesis and p is the conclusion.

EXAMPLE In each of the following statements, name the hypothesis:

a. If prices drop, then I will buy a car.

b. I will buy a car if prices drop.

c. If I buy a car, then prices will drop.

Solution

a. Prices drop

b. Prices drop

c. I buy a car

EXAMPLE If p is the statement "Prices drop" and q is the statement "I buy a car," write each of the sentences in the preceding example symbolically.

Solution

a. $p \rightarrow q$
b. $p \rightarrow q$
c. $q \rightarrow p$

Note that (a) and (b) have the same symbolic representation, even though the English sentence appears to be different. But p is the hypothesis in both instances; therefore, the statement is $p \rightarrow q$.

How is the symbol $\rightarrow$ defined? At this time it should be emphasized that the definitions for the connectives are by convention or agreement and that they could have been defined differently. However, we must accept the definitions and recognize them as standard.

The table below is the accepted definition of the conditional statement.

p	q	$p \rightarrow q$
T	T	T
T	F	F
F	T	T
F	F	T

By this definition, a conditional statement is *false* only when the *hypothesis* is *true* and the *conclusion* is *false*. It is significant to note that by the definition a conditional statement is always true if the hypothesis is false, regardless of the truth value of the conclusion.

Does this definition seem reasonable? First, it should be stressed again that we are working with a two-valued logic and that there are no neutral statements—a statement is either true or it is false. If it is not true, it must be false.

Now let us look at the definition in light of an example: "If I am elected mayor, I will declare every Monday a holiday." The hypothesis, p, is "I am elected mayor," and the conclusion, q, is "I will declare every Monday a holiday." Clearly, if p and q are both true, the mayor has fulfilled his promise, and

$$p \rightarrow q$$
$$\text{T T T}$$

is acceptable.

If I am elected mayor and then welch on my holiday promise, the statement can reasonably be accepted as false; thus

$$p \rightarrow q$$
$$\text{T F F}$$

What happens if I lose the election? We *must* assign a truth value to the statement $p \to q$. It has been determined that since the hypothesis is not true, the initial condition has not been fulfilled; therefore, the outcome is relatively immaterial and a decision to call this situation *true* has been made. Thus

$$
\begin{array}{ccc}
p & \to & q \\
F & T & T \\
F & T & F
\end{array}
$$

EXAMPLE   Consider the statements p: "There is a wage freeze" and q: "There is a price freeze." Symbolize the statement "If there is a price freeze, then there will not be a wage freeze" and construct a truth table for this statement.

Solution   Symbolically, the statement reads

$$q \to \sim p$$

since the "price freeze" statement, q, is the hypothesis.

p	q	$\sim p$	$q \to \sim p$
T	T	F	F
T	F	F	T
F	T	T	T
F	F	T	T

$q \to \sim p$ is false only when q is true and $\sim p$ is false. This occurs in the first entry.

Another important statement which is closely connected to the conditional statement is the **biconditional.** If p and q are statements, the biconditional of p and q is

p if and only if q

Symbolically,

$$p \leftrightarrow q$$

This statement, as the name *bi*conditional implies, consists of *two* conditionals:

    1. $p \to q$ (If p then q.)
    2. $q \to p$ (If q then p.)

The truth table is developed from the truth table of the statement

$$(p \to q) \land (q \to p)$$

		(1)	(3)	(2)
p	q	$(p \rightarrow q)$	$\wedge$	$(q \rightarrow p)$
T	T	T	T	T
T	F	F	F	T
F	T	T	F	F
F	F	T	T	T

Column (1) is the truth table for $p \rightarrow q$; column (2) is the truth table for $q \rightarrow p$. Note that the conditional, $q \rightarrow p$, is false only in the third entry, since the hypothesis, q, is true and the conclusion, p, is false.

Column (3) is the conjunction of columns (1) and (2). The conjunction is true only when (1) and (2) are *both* true—that is, in the first entry and in the fourth entry.

From this truth table it is seen that $p \leftrightarrow q$ is true only when p and q have the *same* truth value—both true or both false.

p	q	$p \leftrightarrow q$
T	T	T
T	F	F
F	T	F
F	F	T

EXAMPLE   Let p and q be the same statements as in the preceding example.

    a. Symbolize the statement "There will be a wage freeze if and only if there is no price freeze."

    b. Construct a truth table for this statement.

Solution

a. Symbolically, $p \leftrightarrow \sim q$

b.

p	q	$\sim q$	p	$\leftrightarrow$	$\sim q$
T	T	F	T	F	F
T	F	T	T	T	T
F	T	F	F	T	F
F	F	T	F	F	T

Thus

$$p \leftrightarrow \sim q$$
$$\begin{array}{c} F \\ T \\ T \\ F \end{array}$$

For clarity, the truth values were repeated in the table under the letters p and $\sim q$ for $(p \leftrightarrow \sim q)$. The statement is only true when p and $\sim q$ have the same truth value—that is, in the second and third entries.

Recall that the biconditional $p \leftrightarrow q$ is defined as the conjunction $(p \rightarrow q) \wedge (q \rightarrow p)$.

But the verbal translation of $p \leftrightarrow q$ is "p if and only if q" or the two statements "p if q" and "p only if q." Symbolically, p if q is written $q \rightarrow p$ since q is the *if* part of the statement. Thus "p only if q" *must* be $(p \rightarrow q)$ in order to satisfy the definition. Thus "p only if q" is another way of saying "if p then q."

The following English translations of the statements $p \rightarrow q$ and $q \rightarrow p$ will be helpful:

$p \rightarrow q$	$q \rightarrow p$
If p then q	If q then p
q if p	p if q
p only if q	q only if p

EXERCISES 1.4

Construct truth tables for each of the statements in Exercises 1–10.

1. $\sim p \rightarrow q$ **2.** $q \rightarrow \sim p$

3. $\sim p \rightarrow \sim q$ **4.** $\sim q \leftrightarrow \sim p$

5. $(p \wedge q) \leftrightarrow \sim p$ **6.** $(p \vee q) \leftrightarrow \sim p$

7. $p \rightarrow (p \vee q)$ **8.** $p \rightarrow (p \wedge q)$

9. $p \rightarrow (\sim p \vee \sim q)$ **10.** $(p \vee q) \leftrightarrow \sim(\sim p \wedge \sim q)$

Let p be the statement "Two more men are hired" and q be the statement "Production is increased." Symbolize each of the statements in Exercises 11–15.

11. If two more men are hired, production is increased.

12. If production is not increased, then two more men are not hired.

13. Production is increased if two more men are hired.

14. Two more men are hired only if production is increased.

15. Production is increased if and only if two more men are hired.

*Suppose that in Exercises 11–15, p is a **true** statement and q is a **false** statement. Find the truth value of the statements in Exercises 16–20 as indicated.*

16. Exercise 11   **17.** Exercise 12
18. Exercise 13   **19.** Exercise 14
20. Exercise 15

Suppose that in Exercises 11–15, p is a **false** *statement and q is a* **true** *statement. Find the truth value of the statements in Exercises 21–25 as indicated.*

21. Exercise 11   **22.** Exercise 12
23. Exercise 13   **24.** Exercise 14
25. Exercise 15

26. Let the connective $\underline{\vee}$ be defined as the **exclusive or**—that is, $p \underline{\vee} q$ is true when exactly one of the component statements is true. Construct a truth table for $p \underline{\vee} q$.

27. What English words could be used as a good description of the exclusive or, $\underline{\vee}$?

28. Give an example of a sentence that contains the word *or* where it is exclusive.

1.5   TAUTOLOGIES AND CONTRADICTIONS

The following table summarizes the compound statements studied thus far, where each statement consists of two component statements:

p	q	(and) $p \wedge q$	(or) $p \vee q$	If . . . then $p \to q$	If and only if $p \leftrightarrow q$
T	T	T	T	T	T
T	F	F	T	F	F
F	T	F	T	T	F
F	F	F	F	T	T

What happens when three or more statements are involved? Let us consider the case of three statements p, q, and r. There are eight possible arrangements of true and false for this case. Each statement has two possible truth values: T or F. This choice occurs three times, so the number of possible arrangements is $2^3 = 2 \cdot 2 \cdot 2 = 8$. When four statements are involved, there are $2^4 = 2 \cdot 2 \cdot 2 \cdot 2 = 16$ possibilities to be considered, and in general, if n is the number of different statements, the truth table will have 2^n entries.

EXAMPLE   Construct a truth table for $(p \lor q) \to r$.

Solution   There are three *different* statements, p, q, and r. Therefore, the truth table will have eight entries. The final connective is $\to$, which will be composed of the truth values for $p \lor q$, and r.

			(1)	(2)	(1)
p	q	r	$(p \lor q)$	$\to$	r
T	T	T	T	T	T
T	T	F	T	F	F
T	F	T	T	T	T
T	F	F	T	F	F
F	T	T	T	T	T
F	T	F	T	F	F
F	F	T	F	T	T
F	F	F	F	T	F

Column (2) is the final truth table for $(p \lor q) \to r$.

Note that in the above table the general arrangement for the entries is: first four p statements true, last four p statements false; the q statements alternated, two true, two false, starting with true; and the r statements simply alternated true and false. This pattern is a simple arrangement to insure that all possibilities are accounted for. A similar arrangement can be used for the sixteen entries for four statements p, q, r, and s:

p: eight true, eight false
q: four true, four false, four true, four false
r: two true, two false,...
s: true, false, true, false,...

EXAMPLE   Construct a truth table for $(\sim p \land q) \lor \sim r$.

Solution   Again, the three different statements require eight entries:

				(1)	(2)	(1)
p	$\sim p$	q	r	$(\sim p \land q)$	$\lor$	$\sim r$
T	F	T	T	F	F	F
T	F	T	F	F	T	T
T	F	F	T	F	F	F
T	F	F	F	F	T	T
F	T	T	T	T	T	F
F	T	T	F	T	T	T
F	T	F	T	F	F	F
F	T	F	F	F	T	T

Column (2) is the final truth table, which is formed from the statements $(\sim p \land q)$ and $(\sim r)$

Sometimes a compound statement is such that *all* its final entries in a truth table are *true*. In other words, the compound statement is true regardless of the truth values of its component statements. Such a statement is called a **tautology**.

> **DEFINITION**
>
> A **tautology** is a compound statement which is always true regardless of the truth values of its component statements. A tautology is also called a **logically true statement**.

EXAMPLE   Construct a truth table for the statement $p \rightarrow (p \vee q)$.

Solution

		(1)	(2)	(1)
p	q	p	$\rightarrow$	$(p \vee q)$
T	T	T	T	T
T	F	T	T	T
F	T	F	T	T
F	F	F	T	F

Thus
$$p \rightarrow (p \vee q)$$
$$\text{T}$$
$$\text{T}$$
$$\text{T}$$
$$\text{T}$$

All entries for the conditional are true, and $p \rightarrow (p \vee q)$ is a tautology.

EXAMPLE   Let p be the statement "Personal income increases" and q be the statement "People eat more expensive food." Construct a truth table for the statement "If personal income increases, then people eat more expensive food, or personal income does not increase only if people do not eat more expensive food."

Solution   First, symbolize the statement:

$$(p \rightarrow q) \vee (\sim p \rightarrow \sim q)$$

Note that the final connective is $\vee$, so the compound statement is the disjunction of two conditionals. Only *two different* statements are involved; therefore, only four entries are needed:

				(1)	(2)	(1)
p	q	$\sim p$	$\sim q$	$(p \rightarrow q)$	$\vee$	$(\sim p \rightarrow \sim q)$
T	T	F	F	T	T	T
T	F	F	T	F	T	T
F	T	T	F	T	T	F
F	F	T	T	T	T	T

Since all entries in the final column, column (2), are T, the statement is a tautology.

It is also possible to have compound statements which are *never* true. Such statements are called **contradictions.**

DEFINITION

A **contradiction** is a compound statement which is always false, regardless of the truth values of the component statements. A contradiction is also called a **logically false statement.**

The simplest example of a contradiction is the statement $p \wedge \sim p$:

p	$\sim p$	$p \wedge \sim p$
T	F	F
F	T	F

EXAMPLE Classify the statement $[(p \rightarrow q) \wedge (q \rightarrow r)] \rightarrow (p \rightarrow r)$ as a tautology, a contradiction, or neither.

Solution Three different statements, p, q, and r, are involved. Therefore, there must be eight entries in the truth table:

			(1)	(3)	(2)	(5)	(4)
p	q	r	$[(p \rightarrow q)$	$\wedge$	$(q \rightarrow r)]$	$\rightarrow$	$(p \rightarrow r)$
T	T	T	T	T	T	T	T
T	T	F	T	F	F	T	F
T	F	T	F	F	T	T	T
T	F	F	F	F	T	T	F
F	T	T	T	T	T	T	T
F	T	F	T	F	F	T	T
F	F	T	T	T	T	T	T
F	F	F	T	T	T	T	T

Column (3) is the conjunction of columns (1) and (2). Column (5) is the conditional, with column (3) as the hypothesis and column (4) as the conclusion. The entries in column (5) are the final entries for the truth table. Since this column consists only of true entries, the statement is a tautology.

EXAMPLE   Let p be the statement "Only one in twenty Americans works in agriculture" and q be the statement "Agriculture is our largest single industry." Test the statement "Only one in twenty Americans works in agriculture and agriculture is our largest single industry, and if only one in twenty Americans works in agriculture, then agriculture is *not* our largest single industry" and determine if it is a tautology, a contradiction, or neither.

Solution   From the punctuation, the following symbolic statement is derived:

$$(p \wedge q) \wedge (p \to \sim q)$$

where the compound statement is the conjunction of $(p \wedge q)$ and $(p \to \sim q)$.

			(1)	(3)	(2)
p	q	$\sim q$	$(p \wedge q)$	$\wedge$	$(p \to \sim q)$
T	T	F	T	F	F
T	F	T	F	F	T
F	T	F	F	F	T
F	F	T	F	F	T

Column (3) is formed from the entries in columns (1) and (2). Since all entries in column (3) are false, the statement is a *contradiction*.

EXERCISES 1.5

Construct a truth table for each of Exercises 1–8.

1. $(p \wedge q) \vee r$ **2.** $(p \to \sim q) \wedge r$

3. $p \to (q \vee r)$ **4.** $p \to (q \wedge \sim r)$

5. $(p \vee q) \to (q \vee r)$ **6.** $(p \to q) \wedge (p \to r)$

7. $(p \leftrightarrow q) \vee (p \leftrightarrow r)$ **8.** $[(p \to q) \wedge (q \to r)] \to (p \to r)$

9. Which, if any, of the statements in Exercises 1–8 are tautologies?

Classify each of the statements in Exercises 10–15 as logically true (a tautology), logically false, (a contradiction), or neither.

10. $(p \to q) \leftrightarrow (\sim q \to \sim p)$

11. $(p \land q) \to p$

12. $(\sim p \lor \sim q) \land \sim p$

13. $(p \to q) \leftrightarrow (q \to p)$

14. $p \to (p \lor q)$

15. $[(p \land q) \to r] \leftrightarrow [p \to (q \to r)]$

In Exercises 16–18:

> *Let p be the* **true** *statement "Prices are up."*
> *Let q be the* **false** *statement "Taxes are up."*
> *Let r be the* **true** *statement "Net income is down."*

Symbolize each of the following compound statements and determine its truth value.

16. If prices are up and taxes are up, then net income is down.

17. Prices are up or net income is not down, and taxes are up.

18. Prices are up and net income is down if and only if taxes are up.

Let p be the **true** *statement "Real incomes in farming are below the national average" and let q be the* **false** *statement "The majority of Americans live in rural areas." Symbolize the compound statements in Exercises 19 and 20 and determine the truth value of each.*

19. It is not a fact that real incomes in farming are below the national average or that the majority of Americans live in rural areas.

20. Either real incomes in farming are below the national average, or the majority of Americans live in rural areas, or most Americans live in cities.

1.6 LOGICALLY EQUIVALENT STATEMENTS

The reader has probably observed by this time that apparently different statements have the same truth table. For example, consider $p \to q$ and $\sim q \to \sim p$.

p	q	$\sim q$	$\sim p$	$p \to q$	$\sim q \to \sim p$
T	T	F	F	T	T
T	F	T	F	F	F
F	T	F	T	T	T
F	F	T	T	T	T

DEFINITION

Two statements which consist of the same simple statements and which have identical truth tables are said to be **logically equivalent**. A statement may be substituted for any logically equivalent statement.

Recall also that the biconditional, $p \leftrightarrow q$, is true whenever p and q have the same truth value. It follows, therefore, that the biconditional of two logically equivalent statements is a tautology.

For example, examine the truth table for

$$(p \rightarrow q) \leftrightarrow (\sim q \rightarrow \sim p)$$

(From the above truth table, the truth values for $(p \rightarrow q)$ and $(\sim q \rightarrow \sim p)$ are obtained.)

$(p \rightarrow q)$	$\leftrightarrow$	$(\sim q \leftrightarrow \sim p)$
T	T	T
F	T	F
T	T	T
T	T	T

DEFINITION

A biconditional is an **equivalence** if and only if it is logically true.

That is, an equivalence is a biconditional which is true for all entries, regardless of the truth values of its component statements. Therefore, the component statements are logically equivalent.

There are many useful equivalences, some of which will be shown below.

EXAMPLE   Show that $\sim(p \wedge q)$ is logically equivalent to $\sim p \vee \sim q$.

Solution

p	q	$\sim p$	$\sim q$	$p \wedge q$	$\sim(p \wedge q)$	$\leftrightarrow$	$\sim p \vee \sim q$
T	T	F	F	T	F	T	F
T	F	F	T	F	T	T	T
F	T	T	F	F	T	T	T
F	F	T	T	F	T	T	T

Since the biconditional is a tautology, an equivalence has been established.

EXAMPLE   Prove that $\sim(p \to q)$ is logically equivalent to $(p \wedge \sim q)$.

Solution

p	q	$\sim q$	$p \to q$	$\sim(p \to q)$	$\leftrightarrow$	$(p \wedge \sim q)$
T	T	F	T	F	T	F
T	F	T	F	T	T	T
F	T	F	T	F	T	F
F	F	T	T	F	T	F

The following list of equivalences should be a useful guide to the reader. Some of them have already been proved as examples; the remaining proofs are left as exercises.

1. $\sim(\sim p) \leftrightarrow p$
2. $\sim(p \wedge q) \leftrightarrow (\sim p \vee \sim q)$
3. $\sim(p \vee q) \leftrightarrow (\sim p \wedge \sim q)$
4. $\sim(p \to q) \leftrightarrow (p \wedge \sim q)$
5. $(p \to q) \leftrightarrow (\sim p \vee q)$
6. $(p \to q) \leftrightarrow (\sim q \to \sim p)$
7. $(p \wedge q) \leftrightarrow (q \wedge p)$
8. $(p \vee q) \leftrightarrow (q \vee p)$

Through the use of equivalences, a statement may be expressed equivalently by using a different connective. This change is often a convenience.

EXAMPLE   Express $p \to \sim q$ as an equivalent statement using
a. $\wedge$ b. $\vee$

Solution

a. From the list of equivalences we note that

$$(p \to q) \leftrightarrow (\sim p \vee q)$$

or

$$(p \to \sim q) \leftrightarrow (\sim p \vee \sim q)$$

Also,

$$(\sim p \vee \sim q) \leftrightarrow \sim(p \wedge q)$$

Therefore,

$$(p \to \sim q) \leftrightarrow \sim(p \wedge q)$$

Use a truth table to test this equivalence.

p	q	$\sim q$	$p \wedge q$	$(p \to \sim q)$	$\leftrightarrow$	$\sim(p \wedge q)$
T	T	F	T	F	T	F
T	F	T	F	T	T	T
F	T	F	F	T	T	T
F	F	T	F	T	T	T

From the truth table it can be seen that $p \to \sim q$ can be stated equivalently in terms of $\wedge$ as $\sim(p \wedge q)$.

b. The second statement in solution (a) above states that

$$(p \to \sim q) \leftrightarrow (\sim p \vee \sim q)$$

The check of this equivalence is left to the reader.

It was stated earlier that the conditional is one of the very important compound statements. Three variations of the conditional are so frequently used that they are given special names.

Let $\qquad p \to q \qquad$ be a conditional statement

Then $\qquad \sim q \to \sim p \quad$ is its *contrapositive*

$\qquad\qquad q \to p \qquad$ is its *converse*

$\qquad\qquad \sim p \to \sim q \quad$ is its *inverse*

The first example established the conditional and its contrapositive as logically equivalent statements. Thus a conditional may be replaced by its contrapositive.

EXAMPLE   Let p be the statement "You are late for work" and let q be the statement "You will be fired."

a. Verbalize the symbolic statement $p \to q$.
b. Symbolize and verbalize the contrapositive of (a).
c. Symbolize and verbalize the converse of (a).
d. Use a truth table to check if $p \to q$ is equivalent to its converse.

Solution

a. $p \to q$ is the statement "If you are late for work, then you will be fired."
b. $\sim q \to \sim p$. "If you are not fired, then you are not late for work."
c. $q \to p$. "If you are fired, then you are late for work."
d.

p	q	$(p \to q)$	$\leftrightarrow$	$(q \to p)$
T	T	T	T	T
T	F	F	F	T
F	T	T	F	F
F	F	T	T	T

Since the biconditional is not a tautology, $p \to q$ is *not* equivalent to $q \to p$.

Since a conditional statement is equivalent to its contrapositive, the following conclusion concerning the converse and its inverse can be drawn.

Let $q \to p$ be the converse of $p \to q$:

a. $q \to p$ is a conditional statement.

b. $\sim p \to \sim q$, the inverse of $p \to q$, is the contrapositive of the converse, $q \to p$.

c. Therefore, the converse and the inverse are equivalent.

In summary, a conditional and its contrapositive are equivalent, and so are the converse and the inverse of the conditional:

$$(p \to q) \leftrightarrow (\sim q \to \sim p)$$
$$(q \to p) \leftrightarrow (\sim p \to \sim q)$$

EXERCISES 1.6

Prove that each of Exercises 1–6 is an equivalence.

1. $\sim(p \lor q) \leftrightarrow (\sim p \land \sim q)$
2. $(p \to q) \leftrightarrow (\sim p \lor q)$
3. $(p \land q) \leftrightarrow (q \land p)$
4. $(p \lor q) \leftrightarrow (q \lor p)$
5. $\sim(\sim p \lor q) \leftrightarrow (p \land \sim q)$
6. $\sim(p \land \sim q) \leftrightarrow (\sim p \lor q)$

7. Consider the statement "If efficiency in management is increased, profits will rise."
   a. Write the converse of this statement.
   b. Write the inverse of this statement.
   c. Write the contrapositive of this statement.
   d. Which of the above, if any, are logically equivalent?

8. Consider the statement "If fringe benefits are increased, the strike can be settled."
   a. Write the converse of this statement.
   b. Write the contrapositive of this statement.
   c. Write the inverse of this statement.
   d. Which of the above, if any, are logically equivalent?

By using equivalent statements, write each of Exercises 9–12 as a compound statement with the given connective. (It is permissible to use modifiers. For example, $p \lor q$ as an equivalent conjunction is $\sim p \land \sim q$.)

9. $p \to q$; $\lor$

10. $p \to q$; $\land$

11. $p \lor q$; $\to$

12. $p \land q$; $\to$

Consider the statement "I will purchase 100 shares of stock A only if there is a stock split." If p is the statement "I will purchase 100 shares of stock A" and q is the statement "There is a stock split,"

    a. *Symbolize each of Exercises 13–20.*
    b. *Determine if each statement is equivalent to the above statement.*

13. If I purchase 100 shares of stock A, then there is a stock split.
14. I will not purchase 100 shares of stock A, or there is a stock split.
15. If there is a stock split, then I will purchase 100 shares of stock A.
16. If there is no stock split, then I will not purchase 100 shares of stock A.
17. If I do not purchase 100 shares of stock A, then there is no stock split.
18. I will purchase 100 shares of stock A if and only if there is a stock split.
19. It is not the case that I will purchase 100 shares of stock A and that there is no stock split.
20. Only if there is a stock split will I purchase 100 shares of stock A.
21. The biconditional is the conjunction of a conditional with its converse. If p and q are statements, symbolize this definition of the biconditional using p and q.
22. True or false: The converse of the inverse of the statement $p \rightarrow q$ is the inverse of the converse. Justify your answer.
23. True or false: The contrapositive of the inverse is the converse of the conditional. Justify your answer.
24. Write a sentence which is logically equivalent to "If a is even, then a^2 is even."

1.7 IMPLICATION AND NEGATION

A biconditional tautology has been defined as an equivalence. A conditional tautology also has a special property which deserves a special name.

DEFINITION

In the conditional statement $p \rightarrow q$, if q is true whenever p is true, then $p \rightarrow q$ is called an **implication**, and we say that p **implies** q.

From the definition of $p \rightarrow q$, an implication must be a logically true conditional since it eliminates the case

$$p \text{ and } q$$
$$T \qquad F$$

$$
\begin{array}{c|c|c}
p & \to & q \\
\hline
T & T & T \\
F & T & T \\
F & T & F \\
\end{array}
$$

Proofs in algebra and geometry usually involve implications. For example, if we are asked to prove that the angles opposite the congruent sides of a triangle are congruent, the statement of the proof would go something like this:

Prove: If two sides of a triangle are congruent, then the angles opposite these sides are congruent. The hypothesis is *assumed to be true*, and if we can show that the conclusion is also true, then the *hypothesis implies the conclusion*.

We will come back to the importance of the implication when studying valid argument forms in the next section, but at this time the reader should be aware of the subtle difference between the *conditional* statement, which considers all possible outcomes for the truth value combinations of hypothesis and conclusion, and the *implication*, which considers only the case of the true hypothesis and the true conclusion.

EXAMPLE   Does the statement $(p \to q) \land (q \to r)$ imply $p \to r$?

Solution   Construct a truth table for $[(p \to q) \land (q \to r)] \to (p \to r)$

			(1)	(3)	(2)	(5)	(4)
p	q	r	$[(p \to q)$	$\land$	$(q \to r)]$	$\to$	$(p \to r)$
T	T	T	T	T	T	T	T
T	T	F	T	F	F	T	F
T	F	T	F	F	T	T	T
T	F	F	F	F	T	T	F
F	T	T	T	T	T	T	T
F	T	F	T	F	F	T	T
F	F	T	T	T	T	T	T
F	F	F	T	T	T	T	T

Column (5) indicates that the conditional statement is logically true; therefore,

$$[(p \to q) \land (q \to r)] \ implies \ (p \to r)$$

Some authors prefer to use a slightly different symbol to differentiate implication from conditional; often $\Rightarrow$ is used for implication and $\to$ for conditional. In this text we shall use $\to$ to mean either statement.

It is often handy to be able to recognize the *negation* of a compound statement without too much effort. The negation, or denial, of a statement is the modification of the statement by $\sim$.

EXAMPLE   Negate the statement $p \rightarrow q$ and express this negation as a conjunction.

Solution   The negation of $p \rightarrow q$ is $\sim(p \rightarrow q)$. From the established list of equivalences and previous examples,

$$(p \rightarrow q) \leftrightarrow (\sim p \vee q)$$

and

$$\sim(p \rightarrow q) \leftrightarrow \sim(\sim p \vee q)$$
$$\sim(\sim p \vee q) \leftrightarrow (p \wedge \sim q)$$

Thus

$$\sim(p \rightarrow q) \leftrightarrow (p \wedge \sim q)$$

Using a truth table to verify the equivalence,

p	q	$\sim q$	$p \rightarrow q$	$\sim(p \rightarrow q)$	$\leftrightarrow$	$(p \wedge \sim q)$
T	T	F	T	F	T	F
T	F	T	F	T	T	T
F	T	F	T	F	T	F
F	F	T	T	F	T	F

The equivalence has been verified.

EXAMPLE   Negate the statement $p \wedge q$ and write its equivalent as a disjunction.

Solution   From the list of equivalences it can be seen that

$$\sim(p \wedge q) \leftrightarrow (\sim p \vee \sim q)$$

The reader may verify this equivalence by checking the entries in a truth table.

A list of five useful negations of compound statements in terms of equivalent statements follows.

Statement	Negation
p	$\sim p$
$p \wedge q$	$\sim(p \wedge q) \leftrightarrow (\sim p \vee \sim q)$
$p \vee q$	$\sim(p \vee q) \leftrightarrow (\sim p \wedge \sim q)$
$p \rightarrow q$	$\sim(p \rightarrow q) \leftrightarrow (p \wedge \sim q)$
$p \leftrightarrow q$	$\sim(p \leftrightarrow q) \leftrightarrow [(p \wedge \sim q) \vee (q \wedge \sim p)]$

EXAMPLE   Negate the statement "The tax override passed and the bond issue failed."

Solution   Let p be the statement "The tax override passed" and q be the statement "The bond issued failed." Then the original statement is $p \wedge q$. The negation of $p \wedge q$ is $\sim(p \wedge q) \leftrightarrow (\sim p \vee \sim q)$.

Therefore, a negation of the statement is "The tax override did not pass or the bond issue passed."

EXERCISES 1.7

Which of Exercises 1–7 are implications? Justify your answers.

1. $(p \wedge \sim q) \rightarrow (q \rightarrow p)$
2. $(p \rightarrow q) \rightarrow (\sim p \wedge q)$
3. $(p \vee q) \rightarrow (\sim p \rightarrow q)$
4. $[(p \rightarrow q) \wedge q] \rightarrow p$
5. $[(p \rightarrow q) \wedge p] \rightarrow q$
6. $[(p \rightarrow q) \wedge \sim q] \rightarrow \sim p$
7. $[(p \rightarrow q) \wedge \sim p] \rightarrow \sim q$

Negate each of the statements in Exercises 8–15 and express each as a compound statement involving the suggested connectives.

8. $\sim p \rightarrow q$; $\wedge$ 9. $p \rightarrow \sim q$; $\wedge$
10. $p \wedge \sim q$; $\vee$ 11. $\sim p \vee q$; $\wedge$
12. $\sim p \wedge \sim q$; $\vee$ 13. $\sim p \vee \sim q$; $\wedge$
14. $(p \wedge q) \rightarrow r$; $\vee$, $\wedge$ 15. $(\sim p \vee q) \rightarrow r$; $\wedge$, $\wedge$

Write a negation for each of the statements in Exercises 16–20.

16. If John Bigmouth is elected, the country is in trouble.
17. The bank loan is approved and construction can be started.
18. Fringe benefits are increased if and only if no raise in salary is granted.
19. The Credit Union pays 6% interest or the Loan and Trust Company pays $6\frac{1}{4}$% interest.
20. If John gets a promotion and Bill quits, Mike will be hired.

Which of Exercises 21–25 represent a statement and its negation?

21. $\sim p \to q$; $\sim p \,\wedge\, \sim q$
22. $\sim p \,\wedge\, \sim q$; $p \,\wedge\, q$
23. $\sim p \,\vee\, q$; $p \,\wedge\, \sim q$
24. $\sim p \to \sim q$; $\sim p \,\wedge\, q$
25. $p \to q$; $\sim p \to \sim q$

1.8   VALID ARGUMENT FORMS

The discussion on statements of logic and the determination of the truth value of compound statements in the preceding sections serve as a basis for developing "correct reasoning" or "valid arguments."

The process of reasoning involves several statements, called **premises,** from which a **conclusion** may be drawn. If the conjunction of the premises *implies* the conclusion, the argument is said to be **valid.** If the conjunction of the premises does *not* imply the conclusion, the argument is said to be **invalid,** or a **fallacy.**

A standard way of writing arguments, either verbally or symbolically, is to list the premises on separate lines and then to separate the conclusion from the premises by a horizontal line.

For example, a pattern for an argument is

$$p \to q$$
$$\underline{p }$$
$$\therefore q$$

The first premise is $p \to q$.
The second premise is p.
The conclusion is q. The three dots ($\therefore$) mean *therefore.*

To test the validity of the above argument, it is not necessary to construct a complete truth table.

The premises must imply the conclusion; in other words, the conclusion must be true whenever the conjunction of the premises is true, and the conjunction is true only if every premise is true. (Clearly, if the conjunction of the premises is *false*, the argument is automatically valid, and almost any conclusion can be drawn!)

We therefore assume p to be true. Also, $p \to q$ is assumed to be true. Therefore, a partial entry in a truth table is

$$p \to q$$
$$\text{T T}$$

Clearly q *must* be true, because if q were false, $p \to q$ would be false.

Thus $[p \land (p \to q)]$ *implies* q, and the argument is valid.

Since the symbols p and q can stand for any statements, a valid argument pattern has been established, and any argument which can be expressed in the form

$$p \to q$$
$$\underline{p }$$
$$\therefore q$$

is a valid argument. Call this Pattern 1.

EXAMPLE   Test the validity of the argument:

If the boss wins the company golf tournament, all
employees will receive a bonus.
The boss wins the tournament.

Therefore, all employees will receive a bonus.

Solution   Let p be the statement "The boss wins the company golf tournament" and q be the statement "All employees will receive a bonus." Symbolically, the argument now reads

$$p \to q$$
$$\underline{p }$$
$$\therefore q$$

Since this argument follows Pattern 1, it is a valid argument.

Consider another argument form:

$$p \to q$$
$$\underline{\sim q }$$
$$\therefore \sim p$$

Recognizing that $p \to q$ is equivalent to its contrapositive, $\sim q \to \sim p$, this argument can be written

$$\sim q \to \sim p$$
$$\underline{\sim q }$$
$$\therefore \sim p$$

Now the argument fits Pattern 1 and is valid.

What about the following pattern, which we shall call Pattern 2?

$$p \to q$$
$$\underline{q }$$
$$\therefore p$$

This does *not* follow Pattern 1, so its validity must be tested.

Assume that q is true since it is a premise, and assume that $p \to q$ is true since it is also a premise. A partial entry for the truth table $p \to q$ with the given information is

$$p \to q$$
$$\text{T T}$$

The question is, *must p be true?* The answer is *no*; the statement, which has a true conclusion, is true regardless of the truth value of p, and

$$[(p \to q) \land q] \text{ does } not \text{ imply } p$$

Pattern 2 is a *fallacy.*

EXAMPLE   Test the argument:

If the boss wins the company golf tournament,
   everyone will get a raise.
The boss loses the tournament.

Therefore, someone will not get a raise.

Solution   Let p and q represent the same statements as in the previous example. Then the argument can be stated symbolically as

$$p \to q$$
$$\sim p$$
$$\therefore \ \sim q$$

Changing $p \to q$ to its equivalent contrapositive, $\sim q \to \sim p$, we can write

$$\sim q \to \sim p$$
$$\sim p$$
$$\therefore \ \sim q$$

Now the argument matches Pattern 2 and is a fallacy.

Whether or not an argument in its verbal form "sounds convincing" has little or nothing to do with its validity. Once the verbal statements are replaced by symbols and the subjective influence of a seemingly convincing argument is removed, the pattern can be tested objectively.

Often a series of conditional statements may lead to yet another conditional statement. For example, consider this set of promises made by Fred Runoften in a speech to voters.

If I am elected, I will reduce crime in the streets.
If I reduce crime in the streets, your children will be safe.
If your children are safe, you will be happy.

Therefore, if I am elected, you will be happy.

Let each of the statements be symbolized as follows:

> p: I am elected.
> q: I will reduce crime in the streets.
> r: Your children will be safe.
> s: You will be happy.

Thus the argument in symbolic form reads

$$p \to q$$
$$q \to r$$
$$r \to s$$
$$\therefore \; p \to s$$

It was shown in a preceding section that for the statement $[(p \to q) \land (q \to r)] \to (p \to r)$, $(p \to r)$ is true whenever $[(p \to q) \land (q \to r)]$ is true, so that $[(p \to q) \land (q \to r)]$ *implies* $(p \to r)$.
Thus

$$p \to q$$
$$q \to r$$
$$\therefore \; p \to r$$

is a valid argument pattern. Let us call this Pattern 3. Another name for this important pattern is the **law of syllogism,** and it is always valid.

Now through repeated applications this law may be extended to include more than two conditionals as premises:

$p \to q$	$p \to r$	$p \to s$
$q \to r$	$r \to s$	$s \to t$
$\therefore \; p \to r$	$\therefore \; p \to s$	$\therefore \; p \to t$, etc.

or

$$p \to q$$
$$q \to r$$
$$r \to s$$
$$s \to t$$
$$\therefore \; p \to t$$

It should be evident by now that Fred Runoften presented a valid argument in his campaign speech.

For the convenience of the reader, we state below the two valid argument patterns developed thus far and show some variations through the substitution of equivalent statements.

Valid Argument Patterns

PATTERN 1

$$p \rightarrow q$$
$$p$$
$$\therefore q$$

Variations

$$p \rightarrow q$$
$$\sim q$$
$$\therefore \sim p$$

$$\sim p \vee q$$
$$\sim q$$
$$\therefore \sim p$$

$$\sim q \rightarrow \sim p$$
$$\sim q$$
$$\therefore \sim p$$

$$\sim q \rightarrow \sim p$$
$$p$$
$$\therefore q$$

PATTERN 3
(Law of syllogism)

$$p \rightarrow q$$
$$q \rightarrow r$$
$$\therefore p \rightarrow r$$

Variation

$$p \rightarrow q$$
$$q \rightarrow r$$
$$r \rightarrow s$$
$$s \rightarrow t$$
$$\therefore p \rightarrow t$$

The reader may verify all the variations, although he has essentially already done this in previous exercises.

The following four fallacies occur often enough for special mention.

Common Fallacies

$p \rightarrow q$	$p \rightarrow q$	$p \vee q$	$p \rightarrow q$
q	$p \rightarrow r$	p	$\sim p$
$\therefore p$	$\therefore q \rightarrow r$	$\therefore \sim q$	$\therefore \sim q$

It is often necessary to substitute equivalent statements in order to simplify the testing of a lengthy argument.

EXAMPLE Test the validity of the following argument:

The new school can be built only if the tax override passes.
There must be at least a 60% voter turnout or the tax override will not pass.
If it rains, the voter turnout will be less than 60%.

Therefore, if the new school is built, it does not rain.

Solution Let the statements be symbolized as follows:

p: The new school can be built.
q: The tax override passes.
r: There must be at least a 60% voter turnout.
s: It rains.

Then the argument can be written

$$p \rightarrow q$$
$$r \vee \sim q$$
$$\underline{s \rightarrow \sim r}$$
$$\therefore p \rightarrow \sim s$$

The statements suggest a syllogism, but some changes in terms of equivalences must be made:

$r \vee \sim q$ is equivalent to $q \rightarrow r$
$s \rightarrow \sim r$ is equivalent to its contrapositive, $r \rightarrow \sim s$

Thus the argument is now written

$$p \rightarrow q$$
$$q \rightarrow r$$
$$\underline{r \rightarrow \sim s}$$
$$\therefore p \rightarrow \sim s$$

and it is valid.

EXERCISES 1.8

Determine the validity of each of the symbolic arguments in Exercises 1–10.

1. p
$$\underline{\sim q \rightarrow \sim p}$$
$$\therefore \sim q$$

2. $\sim p \wedge q$
$$\underline{\sim p}$$
$$\therefore q$$

3. $r \to t$

   $\sim t$

   ∴ $\sim r$

4. $p \to q$

   $r \lor s$

   $r \to p$

   ∴ $q \lor s$

5. $p \lor \sim q$

   q

   ∴ p

6. $p \to q$

   $\sim p$

   ∴ q

7. $p \to q$

   $q \to r$

   $q \to s$

   ∴ $r \to s$

8. $p \land q$

   $q \to r$

   r

   ∴ p

9. $p \to q$

   $\sim q \lor \sim r$

   $s \to r$

   ∴ $p \to \sim s$

10. $p \leftrightarrow q$

   q

   $p \to r$

   ∴ r

Symbolize each of the arguments in Exercises 11–20 and test their validity.

11. If the first of the month is on a Saturday, I must pay the bills on Friday. I paid the bills on Tuesday.

Therefore, the first of the month was not on Saturday.

12. You are entitled to a vacation only if you have been employed for at least six months.
You may start your vacation next Monday.

Therefore, you have been employed at least six months.

13. If two more men are hired, the project can be finished in three days.
If the project is finished in three days, there is no penalty.
If there is no penalty, the job makes a profit for the company.

Therefore, if two more men are hired, the job makes a profit for the company.

14. If peace negotiations are successful, stock prices rise.
The troops will return home only if peace negotiations are successful.
If peace negotiations are not successful, student demonstrations will occur.

Therefore, if stock prices do not rise, there will be student demonstrations.

15. A business must calculate its operating costs carefully, or it will not survive.
If it does not survive, then it must fire its employees.
It is not possible for a business to calculate its operating costs carefully and to fire its employees.

Therefore, the business does not calculate its operating costs carefully.

16. John is looking for work.
If John finds work, he can pay his bills.
He can stay out of jail if and only if he pays his bills.
He can pay his bills if and only if he looks for work.

Therefore, John finds work and stays out of jail.

17. If Proposition 14 passes, the tax structure for school support will be changed.
The sales tax will increase if Proposition 14 passes.
The sales tax was increased.

Therefore, Proposition 14 passed.

18. If taxes are reduced, consumer spending increases.
If spending increases, production will rise.
If profits decrease, production decreases.
An increase in profits implies higher employment.

Therefore, reduced taxes imply higher employment.

19. A degree in business administration is awarded only if two semesters of mathematics are on the transcript.
Calculus is a mathematics course.
If a student passes trigonometry, he can take calculus.
Mike passed trigonometry.

Therefore, Mike was eligible for a degree in business administration.

20. If I study logic, I learn how to reason correctly.
If I learn how to reason correctly, I can win arguments.
If I can win arguments, I will develop poise and self-assurance.
If I am poised and self-assured, I can be successful in business.
If I am successful in business, I can become a millionaire.

Therefore, if I study logic, I can become a millionaire.

1.9 INDIRECT ARGUMENT AND INFERENCE

A useful approach to testing the validity of an argument is the method of indirect proof.

If the conjunction of the premises is true, then the critical determinant of the validity of the argument is the truth value of the conclusion. If the conclusion is necessarily true, the argument is valid, and if the conclusion is false, the argument is invalid.

The method of indirect proof poses the question: Suppose the premises are *true* and the conclusion is *false*? In other words, $p \wedge \sim q$. Then since we are operating under a two-valued logic, one of two things will happen: (a) the assumption that the conclusion is false will lead to a *contradiction*, in which case the conclusion cannot be false and must therefore be true, or (b) there is no contradiction, in which case the conclusion is false and the argument is a fallacy.

As an example, consider a known valid argument form

$$p \rightarrow q$$
$$\underline{p}$$
$$\therefore q$$

Assume that the premises are true—that is,

$$p \rightarrow q \quad \text{and} \quad p$$
$$\text{T} \qquad\qquad \text{T}$$

and assume q to be false. Then the partial entry for $p \rightarrow q$ becomes

$$p \rightarrow q$$
$$\text{T} \textcircled{\text{T}} \text{F}$$

which is obviously incorrect. *A contradiction has been reached, q cannot be false, and the argument is valid.*

Another example of indirect proof is the following: Test the validity of the argument

$$p \rightarrow q$$
$$q \rightarrow r$$
$$\therefore r$$

We assume the following:

$$p \rightarrow q$$
$$\text{T}$$

$$q \rightarrow r$$
$$\text{T}$$

and

$$r$$
$$\text{F}$$

Now if r is false, q must be false in order to make $q \rightarrow r$ true, but if q is false, p must be false in order to make $p \rightarrow q$ true. Thus everything seems to be fine, *no* contradiction was reached, and the false conclusion makes the argument a fallacy.

Alternatively, the argument could have been tested as follows:

$$\left. \begin{array}{c} p \rightarrow q \\ q \rightarrow r \end{array} \right\} \quad p \rightarrow r \text{ by the law of syllogism}$$

Does $(p \rightarrow r)$ imply r? Using a truth table,

p	r	$(p \rightarrow r)$	r
T	T	T	T
T	F	F	F
F	T	T	T
F	F	T	F

Note that in the fourth entry $p \rightarrow r$ is true and r is false. Therefore, $(p \rightarrow r)$ does *not* imply r, and the argument is a fallacy.

EXAMPLE Use an indirect proof to test the validity of the argument:

If you work for the Jones Corporation, you are insured with Eternal Life Insurance Company.
You are not insured with Eternal Life.

Therefore, you don't work for the Jones Corporation.

Solution Let p and q be the following:

p: You work for the Jones Corporation.
q: You are insured with Eternal Life Insurance Company.

Then symbolically the argument is stated

$$\begin{array}{c} p \rightarrow q \\ \sim q \\ \hline \therefore \ \sim p \end{array}$$

We know this to be a valid argument pattern, but to test it indirectly, assume the following:

$$\begin{array}{c} p \rightarrow q \\ \text{T} \\ \sim q \text{ and } \sim p \\ \text{T} \qquad \text{F} \end{array}$$

Then q is false and

$$\begin{array}{c} p \rightarrow q \\ \text{T F} \end{array}$$

implies that p must be false. But we have assumed $\sim p$ to be false. Therefore, a *contradiction* is reached, and the argument is valid.

Of equal importance or perhaps of even greater importance than testing the validity of an argument is the problem of arriving at a conclusion from a given (assumed true) set of premises or hypotheses. The rules of this "game" are as follows:

1. Assume all premises to be true.
2. All premises must be used in order to arrive at
3. a conclusion, which must be true.
4. The premises together with the conclusion must form a valid argument.

EXAMPLE   What conclusion, if any, can you reach from the following premises?

$$p \rightarrow q$$
$$p \lor r$$
$$\sim r$$
$$\therefore \ ?$$

Solution   If $\sim r$ is true, then r is false. Thus

$$p \lor r$$
$$\text{T F}$$

implies that p must be true. But

$$p \rightarrow q$$
$$\text{T T}$$

implies that q must be true. Therefore, the conclusion, *which involved all the premises and is necessarily true*, is q.

$$p \rightarrow q$$
$$p \lor r$$
$$\sim r$$
$$\therefore \ q$$

EXAMPLE   What conclusion, if any, can be reached from the following premises?

If the sale starts on Monday, we must hire ten more salesmen.
We shall hire ten more salesmen only if we cannot handle the business.

Therefore,

Solution   Let p, q, and r be the following:
p: The sale starts on Monday.
q: We must hire ten more salesmen.
r: We can handle the business.

Then symbolically we write

$$p \rightarrow q$$
$$q \rightarrow \sim r$$
$$\therefore$$

By the law of syllogism we know that $[(p \rightarrow q) \wedge (q \rightarrow \sim r)]$ *implies*
$(p \rightarrow \sim r)$.

Therefore,

$$p \rightarrow q$$
$$q \rightarrow \sim r$$
$$\therefore p \rightarrow \sim r$$

Translating the symbols into their English equivalents, the conclusion is:

Therefore, if the sale starts on Monday, we cannot handle the business.

EXERCISES 1.9

Use an indirect proof to test the validity of each of the arguments in Exercises 1–12.

1. $p \vee q$
   $q \leftrightarrow r$
   $\sim p$
   $\therefore r$

2. $p \rightarrow \sim q$
   $\sim p \rightarrow r$
   q
   $\therefore r$

3. $p \vee q$
   $\sim r$
   $p \rightarrow s$
   $s \rightarrow r$
   $\therefore q$

4. $q \rightarrow p$
   $r \vee \sim p$
   $\therefore \sim r \rightarrow \sim q$

5. $\sim p \vee q$
   $q \rightarrow r$
   $r \vee \sim s$
   $\therefore p \rightarrow s$

6. $q \rightarrow \sim p$
   $p \vee r$
   q
   $\therefore r$

7. $p \leftrightarrow q$
   $q \vee \sim r$
   $s \rightarrow r$
   s
   $\therefore p$

8. $p \rightarrow \sim q$
   $r \rightarrow q$
   $\sim r \rightarrow s$
   s
   $\therefore \sim p$

9. If the treaty is signed on Tuesday, the embargo will be lifted immediately.
   If the embargo is lifted, stock prices will rise.
   The treaty is signed on Tuesday.

   ———

   Therefore, stock prices will rise.

10. The applicant must file a resumé with the placement office or he will be
   rejected.
   If the applicant is rejected, he cannot use the services of the placement
   office.
   The applicant used the services of the placement office.

   ———

   Therefore, he filed a resumé with the placement office.

11. A student can enter graduate school only if he takes the Graduate
   Records Examination.
   He can take the Graduate Records Examination or he must have a grade
   point average higher than 3.5.
   If he has a grade point average higher than 3.5, he will receive a fellow-
   ship.

   ———

   Therefore, if a student enters graduate school, he will receive a fellowship.

12. If you vote for Proposition 14, you vote to limit property taxes.
   If property taxes are limited, then the system for the financing of public
   education and social welfare services must be changed.
   Sales taxes will increase by 40% only if you vote for Proposition 14.
   You vote against Proposition 14 or all funds for community colleges will
   be eliminated.

   ———

   Therefore, if you vote for Proposition 14, all funds for community
   colleges will be eliminated.

*In Exercises 13–22, derive a conclusion from each of the sets using all the
premises, if possible.*

13. $p \rightarrow q$
   $q \rightarrow r$
   p
   ———
   $\therefore$

14. $p \leftrightarrow q$
   $q \leftrightarrow r$
   $r \vee s$
   $\sim s$
   ———
   $\therefore$

15. $p \rightarrow q$
   $q \leftrightarrow r$
   $r \vee s$
   $\sim s$
   ———
   $\therefore$

16. $p \rightarrow q$
   $\sim q \vee r$
   p
   ———
   $\therefore$

17. $p \rightarrow q$
   $\sim q \vee \sim s$
   $\sim s \vee t$
   _____
   $\therefore$

18. $\sim p \vee q$
   $q \rightarrow r$
   r
   _____
   $\therefore$

19. If a student passes Logic 24, then he is eligible for Business 36.
   But he is eligible for Business 36 only if he has sophomore standing.
   He does not have sophomore standing or he is not allowed to register for
      Logic 24.
   Sally passed Logic 24.

   Therefore,

20. John takes the swing shift on Thursday or Mike takes the swing shift on
      Thursday.
   Mike does not take the swing shift on Thursday or Bill takes Mike's place.
   Bill does not take Mike's place or Fred returns from vacation.
   Fred does not return from vacation.

   Therefore,

21. If Denver does not host the Winter Olympics, California will host them.
   California does not host the Winter Olympics or Montreal will bid for
      them.
   But Montreal fails to submit a bid.

   Therefore,

22. If students are bused to school, racial and ethnic imbalance will be
      prevented.
   A student in a rural area can get to school only if he is bused.
   If racial and ethnic imbalance are to be prevented, then parents must co-
      operate with school busing.
   But parents are not cooperating with school busing.

   Therefore,

REVIEW EXERCISES

Construct a truth table for each of the statements in Exercises 1–10

1. $p \wedge \sim q$

2. $(\sim p \vee q) \wedge p$

3. $\sim p \wedge p$

4. $(\sim p \vee q) \rightarrow q$

5. $(p \rightarrow q) \wedge r$

6. $(p \leftrightarrow q) \wedge (\sim p \rightarrow q)$

7. $(p \vee q) \rightarrow \sim q$

8. $\sim (p \rightarrow q) \leftrightarrow (p \wedge \sim q)$

9. $(p \vee q) \rightarrow (\sim p \rightarrow q)$

10. $(p \leftrightarrow q) \rightarrow (p \vee \sim q)$

Select the statements in Exercises 1–10 which fit each of the classifications in Exercises 11–13.

11. Tautology   **12.** Contradiction   **13.** Implication

*Assume p to be the **true** statement "Profits are rising," q to be the **false** statement "Government spending is decreasing," and r to be the **true** statement "Taxes always increase," Determine the truth value of each of the statements in Exercises 14–17.*

14. If profits rise, then government spending decreases.

15. Profits are not rising, and if government spending decreases, taxes always increase.

16. Taxes always increase if and only if government spending increases.

17. Government spending decreases only if profits rise, or taxes always increase.

Test the validity of each of the arguments in Exercises 18–22.

18. $p \lor q$
　　$q \to r$
　　$\sim r$
　　―――
　　$\therefore\ p$

19. $p \to q$
　　$q \leftrightarrow r$
　　$r \lor \sim s$
　　s
　　―――
　　$\therefore\ p$

20. $p \to q$
　　$q \leftrightarrow r$
　　$r \lor \sim s$
　　$\sim p$
　　―――
　　$\therefore\ \sim s$

21. If the office in Detroit opens, I can be transferred.
Either I am transferred or I quit.
I don't quit.

Therefore, the office in Detroit opens.

22. If the production of the new drill is not profitable, it will be discontinued.
If the promotion is successful, production of the drill will be continued.

Therefore, if production is discontinued, then the promotion is unsuccessful.

If possible, supply a conclusion for each of the sets of premises in Exercises 23–25.

23. $p \rightarrow q$

$\sim r \rightarrow \sim q$

$\underline{p}$

$\therefore$

24. $p \vee \sim q$

$q \rightarrow r$

$\sim r \leftrightarrow s$

$\underline{\sim s}$

$\therefore$

25. If inflation develops, there will be a price freeze.

If there is a price freeze, profits are limited.

Profits are limited or production is curtailed.

If production is curtailed, then inflation develops.

Therefore,

CHAPTER **2**

SETS

2.1 SETS, SUBSETS, AND EQUAL SETS

Basic to all branches of modern mathematics is the concept of sets. Many a cumbersome verbal statement in mathematics can be expressed neatly, concisely, and symbolically in terms of sets and set operations.

DEFINITION

A **set** is a well-defined collection of objects. The objects are called **members** or **elements** of the set.

Symbolically, capital letters $A, B, C, D, \ldots$ are usually used to designate sets, and lowercase letters usually designate elements, but this is not a hard-and-fast rule.

The significant words in the definition of set are *well defined*. This means that it must be unambiguously clear whether or not an element belongs in a given set. For instance, the set of stocks listed on the New York Stock Exchange is well defined. It can be verified whether or not a stock is so listed. The set of counting numbers, 1, 2, 3, . . ., is well defined. Clearly 5 belongs to the set, whereas $\frac{1}{2}$ does not, and neither does -3 or 0.

Sets are described in two ways.

1. *The listing method.* In this method, either all or a representative number of elements are listed within *braces* and separated by commas.

EXAMPLES

$$\text{Set } A = \{1, 2, 3, 4, 5\}$$

A verbal description of A is "A is the set of the first five counting numbers."

$$\text{Set } B = \{\text{president, vice-president, production manager,}$$
$$\text{sales manager, personnel manager}\}$$

A verbal description of set B might be "B is the set of executives of a certain company."

$$\text{Set } C = \{1, \tfrac{1}{2}, \tfrac{1}{4}, \tfrac{3}{8}, \ldots, \tfrac{1}{256}\}$$

The three dots, . . ., indicate that we continue the sequence as started, either until a stop is indicated, in this case $\tfrac{1}{256}$, or indefinitely.

The symbol $\in$ is used to denote the membership of an element in a set. For example, referring to the above sets, $2 \in A$, president $\in B$, $\tfrac{1}{16} \in C$, but $0 \notin A$. ($\notin$ means "is *not* an element of.")

2. *Rule or set-builder notation.* Whenever possible, set-builder notation is used to describe sets. It has the following form:

$$N = \{n \mid n \text{ is a counting number}\}$$

Read "N is the set of all elements n such that n is a counting number." The letter n is used here as a placeholder or sample element in the same manner in which x or y might be used in algebra; any letter or symbol may be used. The vertical bar, $\mid$, read "such that," separates the variable from the description of how it is used.

EXAMPLE   *List* the elements of the set N described above.

Solution

$$N = \{1, 2, 3, 4, \ldots\}$$

Note that the three dots indicate that we count without stopping. This type of a set is called an **infinite set.**

Even though finite mathematics deals with finite sets, we must include infinite sets relating to numbers.

DEFINITIONS

Let n be a counting number or zero. The set A is **finite** if and only if it has exactly n elements. A set which is not finite is **infinite**.

The set which has n elements and $n = 0$ is called the empty set.

DEFINITION

The set which contains *no elements* is called the **empty set**, or the **void set**, or the **null set**. It is symbolized by $\varnothing$ or $\{\ \}$.

EXAMPLE   List the elements in the set

$$R = \{x|\ x \text{ is a five-legged dog}\}$$

Solution R is well defined, but no dogs fit that description. Thus there are no elements in R. R is the empty set, $\varnothing$.

DEFINITION: Equal Sets

Two sets are said to be equal if and only if they contain exactly the same elements.

EXAMPLE   List the elements in the set

$$T = \{p|\ 2 < p < 6, p \text{ is a counting number}\}$$

Solution T is the set of counting numbers between 2 and 6; thus

$$T = \{3, 4, 5\}$$

EXAMPLE   List the elements in the set

$$S = \{p|\ 2 < p < 3, p \text{ is a counting number}\}$$

Solution   Now S is the set of counting numbers between 2 and 3.

There are none, so

$$S = \{\ \} \quad \text{or} \quad S = \varnothing$$

We often wish to discuss a set of objects contained in a given set or even in several such sets. We call the main set under discussion a *universal set* and the sets made from elements in this universal set *subsets*.

DEFINITION

Let $\mathcal{U}$ be a **universal set.** Then A is a **subset of** $\mathcal{U}$, written

$$A \subseteq \mathcal{U}$$

if and only if every element in A is also in $\mathcal{U}$.

Using the symbols of logic, we write

$$(A \subseteq \mathcal{U}) \leftrightarrow (\text{every } x \in A \to x \in \mathcal{U})$$

EXAMPLE Let $\mathcal{U} = \{2, 3, 4, 5, 6\}$. Which of the following are subsets of $\mathcal{U}$?

$A = \{2, 4\}$ $B = \{2, 4, 6\}$ $C = \{2, 3, 4, 5, 6\}$ $D = \varnothing$
$E = \{2, 5, 9\}$

Solution

$A \subseteq \mathcal{U}$, because $2 \in \mathcal{U}$ and $4 \in \mathcal{U}$.
$B \subseteq \mathcal{U}$, because $2 \in \mathcal{U}$, $4 \in \mathcal{U}$, and $6 \in \mathcal{U}$.
$C \subseteq \mathcal{U}$, because $2 \in \mathcal{U}$, $3 \in \mathcal{U}$, $4 \in \mathcal{U}$, $5 \in \mathcal{U}$, $6 \in \mathcal{U}$. As a matter of fact, C and $\mathcal{U}$ are actually the same set.
$D \subseteq \mathcal{U}$. Since D contains *no* elements, it does not contain any elements which are *not in* $\mathcal{U}$. We say that *the empty set is a subset of every set.*
$E \nsubseteq \mathcal{U}$, (read "E is *not* a subset of $\mathcal{U}$") because $9 \in E$ and $9 \notin \mathcal{U}$.
Note that A is also a subset of B, $A \subseteq B$, and $A \subseteq C$.

An alternate definition of equal sets can now be stated.

DEFINITION: Equal Sets

Let A and B be sets. Then

$$(A = B) \leftrightarrow (A \subseteq B \quad \text{and} \quad B \subseteq A)$$

In words, A equals B if and only if A is a subset of B and B is a subset of A.

Two important consequences of the definition of equal sets result:

1. The order in which the elements in a set are listed does not matter. If $A = \{3, 5, 6\}$ and $B = \{6, 3, 5\}$, then $A = B$, since $A \subseteq B$ and $B \subseteq A$.
2. Elements which are listed more than once need only be listed once. If $A = \{2, 2, 3, 3, 4\}$ and $B = \{2, 3, 4\}$, then $A = B$, because every element in A is also in B, and every element in B is also in A.

There are times when we wish to exclude the possibility that $A \subseteq B$ *and* $A = B$. In other words, we are interested in all subsets of B except set B itself. This situation is described in the following definition.

> **DEFINITION**
>
> A is a **proper subset** of B, written
>
> $$A \subset B$$
>
> if and only if $A \subseteq B$ and $A \neq B$.

In other words, A is contained in B, but there exists at least one element of B which is not an element of A.

EXAMPLE   Let $A = \{a, b, c\}$. List all *proper* subsets of A.

Solution   The proper subsets of A are:

$$\varnothing, \{a\}, \{b\}, \{c\}, \{a, b\}, \{a, c\}, \{b, c\}$$

EXAMPLE   Let the set of stocks listed on the New York Stock Exchange be a universal set. List some subsets of this set.

Solution   It is obvious that there are many such subsets, and here are some of these.

$A = \{x|\ x$ is a stock used to compute the Dow Jones averages$\}$
$B = \{$Comsat, McDonald's, CBS$\}$
$C = \{s|\ s$ is a stock listed in the "new highs" for November 10$\}$
$D = \{u|\ u$ is a utility$\}$

The description in B is quite specific, but A, C, and D need the additional information that $A \subseteq \mathscr{U}$, $B \subseteq \mathscr{U}$, and $C \subseteq \mathscr{U}$.

EXERCISES 2.1

Describe each of the listed sets in Exercises 1–5 by set-builder notation.

1. $A = \{1, 2, 3, 4, 5, 6, 7\}$
2. $B = \{$Alaska, Washington, Oregon, California$\}$
3. $C = \{$Washington, Adams, Jefferson, Monroe$\}$
4. $D = \{2, 4, 6, 8, 10, 12, \ldots\}$
5. $E = \{0, 1, -1, 2, -2, 3, -3, 4, -4, \ldots\}$

Describe each of the sets in Exercises 6–10 by the listing method.

6. $S = \{x|\ x$ is a counting number greater than 5 and less than 10$\}$
7. $S = \{n|\ n$ is a counting number greater than 5$\}$
8. $R = \{r|\ r$ is the capital of a New England state$\}$
9. $T = \{t|\ t$ is an office in the United States Cabinet$\}$
10. $A = \{x|\ x = x + 1, x$ is a real number$\}$

*Let $\mathcal{U} = \{$Joe, Dick, Sue, Dave, Brandon, Vivian$\}$. Classify each of the sets in Exercises 11–15 in relation to $\mathcal{U}$ as **one** of the following:* $\subset$, $=$, $\nsubseteq$.

11. $A = \{$Joe, Dick, Vivian$\}$
12. $B = \{x|\ x \in \mathcal{U}$ and x is male$\}$
13. $C = \{$Bill, Brandon$\}$
14. $D = \{$Sue, Vivian, Joe, Dick, Dave, Brandon$\}$
15. $E = \{x|\ (x \in \mathcal{U})$ and $(x$ is male or x is female$)\}$

The following entry appeared at the end of a given day in the local newspaper:

NEW YORK STOCK EXCHANGE

Stock	Sales	Closing Price	Net Change
Eastern Airlines	328,000	$23\frac{1}{8}$	$-\frac{7}{8}$
Phillips Petroleum	318,500	$36\frac{5}{8}$	$+\frac{3}{4}$
Comsat	4,400	59	$+\frac{1}{8}$
McDonald's	48,400	$60\frac{3}{8}$	$-1\frac{3}{4}$
TWA	155,400	$45\frac{3}{4}$	$\ldots$
Chrysler	189,800	$35\frac{1}{4}$	$+\frac{3}{8}$
IBM	57,800	$376\frac{1}{2}$	$-3\frac{1}{8}$

If the universal set is the set of listed stocks, list each of the subsets in Exercises 16–20 of this universal set.

16. $\{x|\ x$ had a net gain$\}$
17. $\{y|\ y$ closed above 60$\}$
18. The set of the four most actively traded stocks.
19. The set of stocks which showed a net loss of more than four points.
20. $\{s|\ s$ had fewer than 50,000 shares in sales$\}$

2.2 SET OPERATIONS

Subset, proper subset, and equality are *relations* on sets; they tell how two sets are related. We are now going to discuss some *operations* on sets. The first of these is the operation *union*.

DEFINITION

If A and B are sets, then the **union** of A and B, written

$$A \cup B$$

is the set of all elements in set A or in set B. Using the symbols of logic and of sets,

$$A \cup B = \{x \mid x \in A \ \lor \ x \in B\}$$

The union of two sets is again a set, and we have formed a new set from two sets. Since *union* is defined in terms of the *inclusive or*, the union set contains all elements in either set, including any elements which both sets may have in common. Also, from the definition of set equality, elements which both sets have in common need only be listed once in the union set.

EXAMPLE If $A = \{a, b, c\}$, $B = \{a, c, e, f\}$, $C = \{x, y\}$, and $D = \varnothing$, form the following sets:

a. $A \cup B$ b. $A \cup C$
c. $A \cup D$ d. $B \cup C$

Solution

a. $A \cup B = \{a, b, c, e, f\}$
b. $A \cup C = \{a, b, c, x, y\}$
c. $A \cup D = \{a, b, c\}$
d. $B \cup C = \{a, c, e, f, x, y\}$

Another operation on sets is *intersection*.

DEFINITION

If A and B are sets, then the **intersection** of A and B, written

$$A \cap B$$

is the set of all elements which A and B have in common. In the symbols of logic and sets,

$$A \cap B = \{x \mid x \in A \ \land \ x \in B\}$$

EXAMPLE   If $A = \{a, b, c\}$, $B = \{a, c, e, f\}$, $C = \{b\}$, and $D = \varnothing$, form the following sets:

a. $A \cap B$ b. $A \cap C$
c. $B \cap D$ d. $(A \cup B) \cap C$
e. $(B \cap C) \cup A$

Solution

a. $A \cap B = \{a, c\}$
b. $A \cap C = \{b\}$
c. $B \cap D = \varnothing$
d. Perform the operation within the parentheses first:

$$A \cup B = \{a, b, c, e, f\}$$

Now intersect this set with C:

$$\{a, b, c, e, f\} \cap \{b\} = \{b\} = C$$

e. Again, find $B \cap C$ first:

$$B \cap C = \varnothing$$
$$(B \cap C) \cup A = \varnothing \cup \{a, b, c\}$$
$$= \{a, b, c\}$$
$$= A$$

In the above example, it was seen that $B \cap C = \varnothing$. This indicated that B and C had no elements in common.

DEFINITION: Disjoint Sets

Sets A and B are said to be **disjoint** if and only if they have no elements in common, or symbolically, if and only if

$$A \cap B = \varnothing$$

Examples of disjoint sets are the even numbers and the odd numbers, men students and women students in a college, stocks which showed net gain on a given day and stocks which showed net loss, and so on.

The reader should have noted that union, $\cup$, was defined in terms of the logical connective $\vee$, and intersection, $\cap$, was defined in terms of the logical connective $\wedge$. In a subsequent section in this chapter, the parallel between sets and logic will be explored. The next set operation to be defined is analogous to the logical negation of a statement.

DEFINITION

Let $\mathcal{U}$ be a universal set and let $A \subseteq \mathcal{U}$. Then the **complement of** A, written

$$A'$$

is the set of all elements which are *not* in A but which are in the universal set. Symbolically,

$$A' = \{x \mid x \notin A \wedge x \in \mathcal{U}\}$$

EXAMPLE   Let $\mathcal{U} = \{1, 2, 3, 4, 5, 6, 7\}$
$A = \{1, 3, 5, 7\}$
$B = \{2, 4, 5\}$
$C = \{3\}$

Form each of the following sets:

a. A' b. B' c. C'
d. $(A \cup B)'$ e. $(A \cap B)'$ f. $(B \cap C)'$
g. $(A \cap C)' \cup B$ h. $A' \cap B$

Solution

a. $A' = \{2, 4, 6\}$
b. $B' = \{1, 3, 6, 7\}$
c. $C' = \{1, 2, 4, 5, 6, 7\}$
d. $A \cup B = \{1, 2, 3, 4, 5, 7\}$
$(A \cup B)' = \{6\}$
e. $A \cap B = \{5\}$
$(A \cap B)' = \{1, 2, 3, 4, 6, 7\}$
f. $B \cap C = \varnothing$
$(B \cap C)' = \{1, 2, 3, 4, 5, 6, 7\} = \mathcal{U}$
g. $A \cap C = \{3\}$
$(A \cap C)' = \{1, 2, 4, 5, 6, 7\}$
$(A \cap C)' \cup B = \{1, 2, 4, 5, 6, 7\}$
h. $A' = \{2, 4, 6\}$
$A' \cap B = \{2, 4, 6\} \cap \{2, 4, 5\} = \{2, 4\}$

EXERCISES 2.2

Let $\mathcal{U} = \{1, 2, 3, 4, 5, 6, 7, 8, 9\}$
$A = \{2, 4, 6\}$
$B = \{1, 8\}$
$C = \{x \mid x \in \mathcal{U} \text{ and } x \text{ is an even number}\}$
$D = \{y \mid y \in \mathcal{U} \text{ and } y \text{ is exactly divisible by 3}\}$

Form each of the sets in Exercises 1–20.

1. $A \cup B$

2. $B \cup C$

3. $B \cap C$

4. $D \cup A$

5. $D \cap A$

6. A'

7. $(A \cap D)'$

8. $A \cup B'$

9. $(B \cap C) \cup A$

10. $(B \cup C) \cap A$

11. $(D \cup A) \cap (B \cup C)$

12. $(A \cup B) \cap (A \cup C)$

13. $A \cup (B \cap C)$

14. $A' \cup (B \cap C)'$

15. $(D \cap C) \cup \mathscr{U}$

16. $\mathscr{U}'$

17. $(A \cup B)'$

18. $A' \cap B'$

19. $(A' \cup B')'$

20. $((A \cap D)')'$

Suppose A is the set of all business administration majors in a college, B the set of economics majors, and C the set of psychology majors. The chart below illustrates the numbers of students who are enrolled in statistics (S), English (E), and data processing (D).

	S	E	D
A	128	68	216
B	140	94	65
C	85	110	160

If we wish to find out how many students are in the set $A \cap E$, we read 68, because $A \cap E$ is the set of business administration majors who are taking English. Find the number of students in each of the sets in Exercises 21–28. Assume that no student takes more than one of these classes.

21. $(A \cup B) \cap D$

22. $(B \cap E) \cup (C \cap D)$

23. A'

24. $(B \cap S)'$

25. $C \cup S$

26. $C \cap S$

27. E'

28. $D' \cap B'$

In Exercises 29–36, write a verbal description of each of the sets in the designated exercise.

29. Exercise 21

30. Exercise 22

31. Exercise 23

32. Exercise 24

33. Exercise 25

34. Exercise 26

35. Exercise 27

36. Exercise 28

37. If $A \cup B = B$, what, if anything, can be said about the relation between A and B?

38. If $A \cap B = A$, what, if anything, can be said about the relation between A and B?

39. If $A \cap B = \varnothing$, what can be said about A and B?

2.3   VENN DIAGRAMS

A graphic description of sets, relations on sets, and set operations is achieved in terms of **Venn diagrams.**

Let a rectangle represent a specific universal set, and let circles inside the rectangle represent subsets of the universal set.

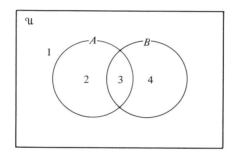

FIGURE 2.3.1   A Venn diagram.

In Figure 2.3.1, $\mathcal{U}$ is the universal set, and A and B are subsets of $\mathcal{U}$. The numerals 1, 2, 3, 4 designate four specific regions or subsets of $\mathcal{U}$. Clearly, region 3 designates $A \cap B$. This set can be shaded, as in Figure 2.3.2.

Figure 2.3.3 illustrates $A \cup B$, which is also designated by regions 2, 3, and 4.

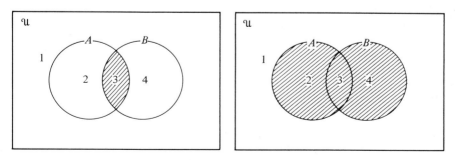

FIGURE 2.3.2 $A \cap B$ is shaded.   **FIGURE 2.3.3** $A \cup B$ is shaded.

This leaves region 1 (Figure 2.3.4). Since region 1 represents everything which is *not* in $A \cup B$, region 1 is the complement of $A \cup B$, $(A \cup B)'$.

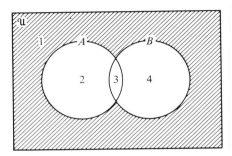

FIGURE 2.3.4 $(A \cup B)'$ is shaded.

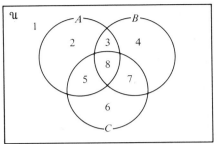

FIGURE 2.3.5 A Venn diagram showing three subsets of $\mathcal{U}$.

A Venn diagram illustrating three subsets of $\mathcal{U}$ is shown in Figure 2.3.5. Now there are *eight* regions to be identified as follows:

Region 1: $(A \cup B \cup C)'$
Region 2: $A \cap (B \cup C)'$
Region 3: $(A \cap B) \cap C'$
Region 4: $B \cap (A \cup C)'$
Region 5: $(A \cap C) \cap B'$
Region 6: $C \cap (A \cup B)'$
Region 7: $(B \cap C) \cap A'$
Region 8: $A \cap B \cap C$

These set designations of the indicated regions are not unique—there are different but equivalent descriptions, and some of these are left as exercises.

EXAMPLE   Use a Venn diagram to illustrate $(A \cup B) \cap C$.

Solution   First shade $A \cup B$ (regions 2, 3, 4, 5, 7, 8). Then use different shading for C (regions 5, 6, 7, 8). (See Figure 2.3.6.) Since we are interested in the intersection of $A \cup B$ with C, the *doubled shaded* region (5, 7, 8) is the desired set, as shown in Figure 2.3.7.

Set equations may be verified or disproved by means of Venn diagrams. If two Venn diagrams illustrating the same sets have identical shading, then the set statements involved are equivalent.

EXAMPLE   Use Venn diagrams to verify that

$$A \cup (B \cap C) = (A \cup B) \cap (A \cup C)$$

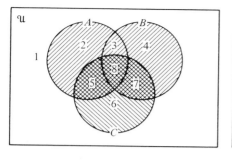

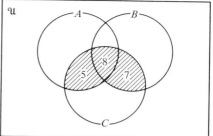

FIGURE 2.3.6

FIGURE 2.3.7 $(A \cup B) \cap C$ is the shaded region.

Solution Draw two Venn diagrams, as shown in Figures 2.3.8 and 2.3.9:

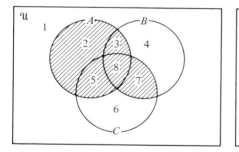

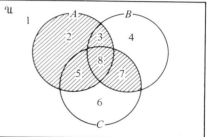

FIGURE 2.3.8 $A \cup (B \cap C)$ is shaded.

FIGURE 2.3.9 $(A \cup B) \cap (A \cup C)$ is shaded.

$A \cup (B \cap C)$ is shown as the shaded regions 2, 3, 5, 7, 8 in Figure 2.3.8.
$A \cup B$ is represented by regions 2, 3, 4, 5, 7, 8.
$A \cup C$ is represented by regions 2, 3, 5, 6, 7, 8.
Therefore, the intersection of these sets,

$$(A \cup B) \cap (A \cup C) = \text{regions 2, 3, 5, 7, 8}$$

Since Figures 2.3.8 and 2.3.9 have identical shading, we have verified the conjecture that

$$A \cup (B \cap C) = (A \cup B) \cap (A \cup C)$$

We can use Venn diagrams to verify certain useful theorems concerning set operations. The theorems are stated below, and their verification is left as an exercise.

Theorems

1. $A \cup B = B \cup A$	$\cup$ is commutative.
2. $A \cap B = B \cap A$	$\cap$ is commutative.
3. $(A \cup B) \cup C = A \cup (B \cup C)$	$\cup$ is associative.
4. $(A \cap B) \cap C = A \cap (B \cap C)$	$\cap$ is associative.
5. $A \cup (B \cap C) = (A \cup B) \cap (A \cup C)$	$\cup$ distributes over $\cap$.
6. $A \cap (B \cup C) = (A \cap B) \cup (A \cap C)$	$\cap$ distributes over $\cup$.
7. $(A \cup B)' = A' \cap B'$	The complement of the union of two sets is the intersection of their complements.
8. $(A \cap B)' = A' \cup B'$	The complement of the intersection of two sets is the union of their complements.
9. $A' \cup A = \mathscr{U}$	The universal set is the union of a set and its complement.
10. $(A')' = A$	The complement of the complement of a set is the set itself.

In the next section we shall use Venn diagrams to illustrate the relationship between sets and logic.

EXERCISES 2.3

Use a shaded Venn diagram to illustrate each of Exercises 1–10.

1. $A \cap B'$	2. $A' \cup B$	3. $(A \cup B) \cap C$
4. $A' \cap C'$	5. $(A \cap B)' \cup C$	6. $(A \cup B) \cap C'$
7. $A' \cup B' \cup C'$	8. $(A \cap B \cap C)'$	9. $(A \cup C) \cap B$
10. $A \cup (C \cap B)$		

Consider the regions designated in Figure 2.3.5. Describe each of the sets in Exercises 11–15. (For example, $\{5, 3, 8\} = (A \cap B) \cup (A \cap C)$.) Note that there is more than one correct answer for each problem.

11. $\{2, 3, 5, 7, 8\}$	12. $\{4, 7, 6\}$
13. $\{1, 2, 4, 6\}$	14. $\{5, 7, 8\}$
15. $\{1, 8\}$	

Use Venn diagrams to verify the theorems as listed in Exercises 16–23.

16. Theorem 1   **17.** Theorem 2
18. Theorem 3   **19.** Theorem 4
20. Theorem 5   **21.** Theorem 6
22. Theorem 7   **23.** Theorem 8

The set descriptions of regions 1–8 in this section are not unique. Use alternate forms of set descriptions from those listed to describe each of the regions in Exercises 24–26 as illustrated in Figure 2.3.5.

24. Region 3
25. Region 6
26. Region 7

2.4   SETS AND LOGIC

An interesting and useful relationship between statements of logic and their connectives, and sets and operations on sets can be established.

For any given simple statement there are two logical possibilities: true and false. For a compound statement involving two simple statements there are four logical possibilities, and for a compound statement involving three simple statements there are eight logical possibilities. These possibilities were discussed in the chapter on logic.

Consider a universal set, $\mathcal{U}$, which consists of all the logical possibilities of a statement. If p is a statement whose logical possibilities are considered in $\mathcal{U}$, then let A be the subset of $\mathcal{U}$ for which p is true. Call this subset A the *truth set* of p. Clearly, A' contains the logical possibilities for which p is false or for which $\sim p$ is true (see Figure 2.4.1).

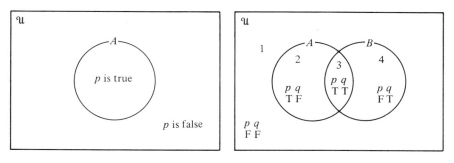

FIGURE 2.4.1   **FIGURE 2.4.2**

Now consider two statements p and q. Let A be the truth set of p, and let B be the truth set of q, the set of all logical possibilities for which q is true. Then there are four logical possibilities, corresponding to regions 1, 2, 3, and 4 as shown in Figure 2.4.2.

> *Region 1: p* false   *q* false
> *Region 2: p* true   *q* false
> *Region 3: p* true   *q* true
> *Region 4: p* false   *q* true

Now note that region 3 corresponds to the case when $p \land q$ is true:

$$p \land q$$
$$\text{T T T}$$

It also corresponds to the set statement $A \cap B$.

Regions 2, 3, and 4 correspond to the case when $p \lor q$ is true:

p	$\lor$	q
T	T	T
T	T	F
F	T	T

But regions 2, 3, and 4 also correspond to the set statement $A \cup B$.

Thus a parallel between the set operation $\cap$ and the logical connective $\land$ has been established, and a similar parallel has been established between $\cup$ and $\lor$.

What about the logical connective $\rightarrow$? Let us examine the truth table to determine the regions for which $p \rightarrow q$ is true.

	p	$\rightarrow$	q
Region 3:	T	T	T
Region 2:	T	F	F
Region 4:	F	T	T
Region 1:	F	T	F

The shaded diagram, Figure 2.4.3, illustrates the subsets of true logical possibilities for $p \rightarrow q$. Call this set the truth set of $p \rightarrow q$. What set description can be used for this set? Region 2 can be described as

$$A \cap B'$$

We are interested in the *complement* of region 2, or

$$(A \cap B')'$$

Using set operation Theorem 8 from the preceding section,

$$(A \cap B')' = A' \cup (B')'$$

But $(B')' = B$; therefore,

$$p \to q \text{ corresponds to } A' \cup B$$

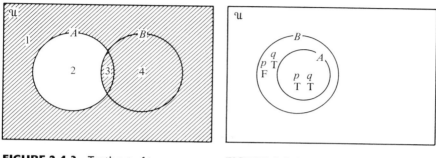

FIGURE 2.4.3   Truth set of $p \to q =$ A′ ∪ B.

FIGURE 2.4.4 $A \subseteq B$; p implies q.

What if p *implies* q? This requires the case where q is true whenever p is true. From Figure 2.4.2 we see that this occurs only in region 3. Actually, if $A \subseteq B$, then p implies q, because the entire truth set of p is included in the truth set of q, as illustrated in Figure 2.4.4.

A chart summarizing the correspondences between sets and some statements of logic is given below.

Language of logic	*Language of sets*
Statement p	Set A, truth set of p
Statement q	Set B, truth set of q
$\sim p$	A'
$p \wedge q$	$A \cap B$
$p \vee q$	$A \cup B$
$p \to q$	$A' \cup B$
p implies q	$A \subseteq B$

Venn diagrams were used to verify set statements. These diagrams can also be used to verify statements of logic.

EXAMPLE   Use a Venn diagram to verify that $p \vee (\sim p \vee q)$ is a tautology—that is, show that the statement is logically true.

Solution   If the statement is logically true, it must be true for all cases, and the entire set of logical possibilities must be used. Thus for a logically true statement, all of $\mathcal{U}$ must be shaded, since $\mathcal{U}$ is the truth set of the statement.

Using the set language for $p \vee (\sim p \vee q)$, we write $A \cup (A' \cup B)$ and shade the appropriate regions in the Venn diagram (see Figure 2.4.5).

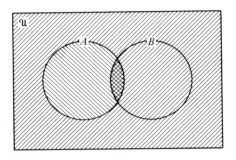

FIGURE 2.4.5 $A \cup (A' \cup B)$ or $p \vee (\sim p \vee q)$.

First shade $A' \cup B$; then shade A. The union is the entire shaded region, which is the entire universal set. Therefore, $p \vee (\sim p \vee q)$ is logically true.

EXAMPLE   Use a Venn diagram to verify that

$$p \to q$$
$$\sim q$$
$$\overline{\therefore \quad \sim p}$$

is a valid argument form.

Solution   For an argument to be valid, the conjunction of the premises must imply the conclusion. In other words, the argument is valid if $[(p \to q) \wedge (\sim q)]$ implies $\sim p$, or the truth set of the conjunction of the premises must be a *subset* of the truth set of the conclusion. Let A be the truth set of p and B be the truth set of q.

Then $p \to q$ corresponds to $A' \cup B$
                                   $\sim q$ corresponds to B'
and $\sim p$ corresponds to A'

We wish to show that

$$[(A' \cup B) \cap B'] \subseteq A'$$

Figure 2.4.6 is the Venn diagram of the statement

$$(A' \cup B) \cap B'$$

Clearly, $[(A' \cup B) \cap B'] \subseteq A'$

therefore, $[(p \rightarrow q) \land \sim q]$ *implies* $\sim p$

and the argument form is valid.

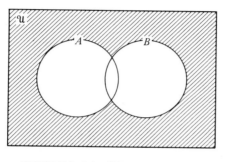

FIGURE 2.4.6 $[(A' \cup B) \cap B'] \subseteq A'$.

EXAMPLE   If A and B are the truth sets of statements p and q, respectively, what statement of logic corresponds to the diagram in Figure 2.4.7?

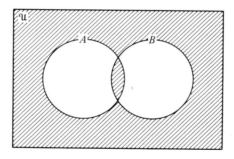

FIGURE 2.4.7

Solution   The shading in the diagram indicates

$$(A \cap B) \cup (A \cup B)'$$

Taking this as it stands, the logical parallel is

$$(p \land q) \lor [\sim (p \lor q)]$$

We simplify $\sim (p \lor q)$ to read $\sim p \land \sim q$ and obtain

$$(p \land q) \lor (\sim p \land \sim q)$$

By using the diagram, we also see that p and q are either both true or both false, and an alternative, equivalent, and simpler statement of logic is

$$p \leftrightarrow q$$

EXERCISES 2.4

1. If a logically true statement is shown as a Venn diagram with the entire universal set shaded, how is a logically false statement related to sets?

Use Venn diagrams to classify each of the statements in Exercises 2–6 as logically true, logically false, or neither. (Let A be the truth set of statement p and B be the truth set of statement q.)

2. $p \vee \sim p$

3. $p \wedge \sim p$

4. $(p \wedge q) \vee (\sim p \wedge q)$

5. $(p \vee q) \wedge (\sim p \vee q)$

6. $\sim (p \rightarrow q) \wedge q$

7. Determine a set statement for the biconditional, $p \leftrightarrow q$. Draw a Venn diagram to illustrate this concept.

If Venn diagrams can be used to verify statements of logic, then truth tables can be used to determine set relationships. Use truth tables to determine which of the sets in Exercises 8–10 are empty. (Hint: A logically false statement corresponds to the empty set.)

8. $(A \cap B) \cap B'$

9. $(A \cup B') \cap (A' \cup B)$

10. $B \subseteq (A \cap A')$

*Recall that if A and B are the truth sets of statements p and q, then p **implies** q if and only if $A \subseteq B$. Thus if a valid argument is such that the conjunction of the premises **implies** the conclusion, the intersection of the truth sets of the premises must be a subset of the conclusion. Use Venn diagrams to test the validity of each of the arguments in Exercises 11–15.*

11. $p \rightarrow q$
$\underline{p}$
$\therefore q$

12. $\sim p \rightarrow \sim q$
$\underline{\sim p}$
$\therefore \sim q$

13. $p \vee q$
$\underline{p}$
$\therefore q$

14. If the advertising campaign is a success, Mr. Jones will get a promotion. Mr. Jones was promoted.

Therefore, the advertising campaign was a success.

15. The cost of production of model *B* has increased or the sales of model *B* have fallen off. Model *B* had its highest sales record this month.

Therefore, the cost of production of model *B* has increased.

2.5 SWITCHING CIRCUITS

An interesting application of the laws of sets and logic is in the realm of switching circuits or simple switching networks.

Consider an electrical switch which can be either on or off. When the switch is in the on position, current can flow through it, and when the switch is in the off position, no current flows through. Figures 2.5.1 and 2.5.2 illustrate this concept. In Figure 2.5.1, current flows through the switch from terminal *A* to terminal *B*, whereas Figure 2.5.2, illustrating the switch in the off position, shows no current flowing from *A* to *B*.

FIGURE 2.5.1 Switch, S, in on posi-   **FIGURE 2.5.2** Switch, S, in off position; current flows from A to B.   tion; current cannot flow from A to B.

Suppose now that two switches *S* and *T* are connected in *series* between *A* and *B*, as shown in Figure 2.5.3.

FIGURE 2.5.3 Two switches connected in series.

It seems reasonably obvious that the only way current can travel from *A* to *B* is for both switches to be on. There are, however, four logical possibilities for *S* and *T*.

S	T
ON	ON
ON	OFF
OFF	ON
OFF	OFF

The analogy between two statements s and t and the truth values of the conjunction $s \wedge t$ are immediately obvious if we let S correspond to s, T correspond to t, and on and off correspond to true and false, respectively.

Thus if two switches are connected in series, the only way current will flow through these switches is when

$$s \wedge t \text{ is true}$$

Circuits may also be wired in *parallel*, as shown in Figure 2.5.4.

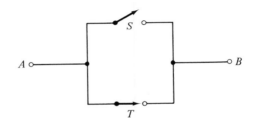

FIGURE 2.5.4 Two circuits connected in parallel.

Current will flow from A to B as long as at least one of the switches is in the on position. Again, if S and T correspond to statements s and t, with s true when S is on and t true when T is on, then two circuits in parallel can transport current from A to B if and only if

$$s \vee t \text{ is a true statement}$$

This simple notion can be extended to more complicated network arrangements.

EXAMPLE Construct a switching circuit for three switches R, S, and T such that R and S are connected in series and T is parallel to R and S. What true logical statement will insure the flow of current from A to B?

Solution Figure 2.5.5 illustrates the desired arrangement. Let r, s, and t correspond to switches R, S, and T, respectively, and let each statement be true for its respective switch in the on position. Then current will flow from A to B if and only if $r \wedge s$ is true or t is true.

Symbolically, we write

$$(r \land s) \lor t$$

and we are interested only in the cases when this disjunction is true.

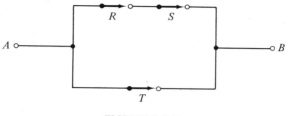

FIGURE 2.5.5

EXAMPLE   List the logical possibilities of the statement $(r \land s) \lor t$ to be true by constructing a tree diagram.

Solution

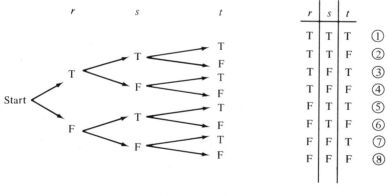

FIGURE 2.5.6

From the tree diagram in Figure 2.5.6 we see that $r \land s$ is true only in cases 1 and 2.

   For $(r \land s) \lor t$ to be true, either $r \land s$ must be true or t must be true. t is true in cases 1, 3, 5, and 7.

   Thus $(r \land s) \lor t$ is true in cases 1, 2, 3, 5, and 7, and there are five logical possibilities.

   It is possible to have switches in a network arranged so that several switches are off or on simultaneously; in other words, if a given switch is on, then certain other switches are automatically on also. We designate such switches by the same letter and say these switches are equivalent. Similarly, if two networks are such that their current flow is operated under identical conditions, the networks are said to be equivalent.

EXAMPLE   Show that the networks in Figures 2.5.7 and 2.5.8 are equivalent.

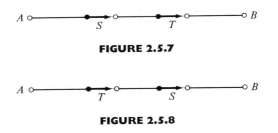

FIGURE 2.5.7

FIGURE 2.5.8

Solution   Switches S and T are connected in series in both figures. Current will flow from A to B in Figure 2.5.7 if and only if the associated statement

$$s \wedge t \text{ is true}$$

Similarly, in Figure 2.5.8, for current to flow from A to B, the associated statement

$$t \wedge s \text{ must be true}$$

But $$(s \wedge t) \leftrightarrow (t \wedge s)$$

or $$s \wedge t \quad \text{and} \quad t \wedge s$$

are logically equivalent; therefore, the two networks are equivalent.

Similarly, a network may be arranged so that if one switch is on, another is automatically off. This may be interpreted as a *negation*, and if S designates the on switch, then S' designates the off switch. Thus s corresponds to S and $\sim s$ corresponds to S'.

It seems reasonable from the above example that in order to verify the equivalence of two switching circuits, one need only to verify the equivalence of their associated logical statements.

The parallel which was developed in the preceding section between sets and logic can be applied here to simplify switching circuits.

EXAMPLE   Consider the switching circuit illustrated in Figure 2.5.9. Use set theory to simplify this circuit and draw an equivalent simpler switching circuit.

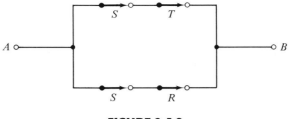

FIGURE 2.5.9

Solution   Let sets R, S, and T correspond to switches R, S, and T, respectively. Then the circuit corresponds to the set sentence

$$(S \cap T) \cup (S \cap R)$$

Shade a Venn diagram for this statement and observe that

$$[(S \cap T) \cup (S \cap R)] = S \cap (T \cup R)$$

Figure 2.5.10 illustrates the corresponding switching circuit.

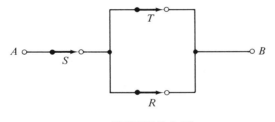

FIGURE 2.5.10

This rather simplified idea of switching circuits is applicable to computers, which contain what is known as logic circuitry.

EXERCISES 2.5

If statements s, t, and r correspond to switches S, T, and R in the same manner as established in this section, design switching circuits to represent each of Exercises 1–4.

1. $s \lor (t \land r)$ **2.** $r \land s \land t$

3. $(r \land s) \lor (r \land t)$ **4.** $r \lor (\sim s \land \sim t)$

Translate each of the switching networks in Exercises 5–8 into statements of logic.

5.

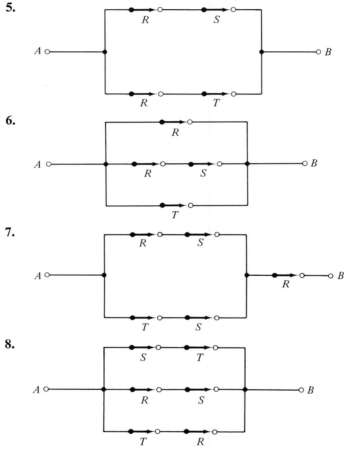

6.

7.

8.

9. In a committee of three members, a majority vote of two carries an issue. However, the chairman of the committee may veto the issue. Thus the issue must have two yes votes, one of which must be the chairman's. If each member of the committee has a button to press to indicate his yes vote, design a circuit which lights up if and only if an issue passes.

10. Design a circuit for a committee of four, with a majority vote of three to pass and veto power for the chairman of the committee for the conditions set up in Exercise 9.

11. Draw a switching network to correspond to the statement

$$\{[(r \wedge s) \wedge \sim s] \vee [\sim r \wedge s]\} \wedge \sim s$$

12. Find an equivalent but simpler network for Exercise 11.

13. A common switching arrangement in houses is to have two switches that turn on or off a single light. Construct a switching circuit to do this and write out the corresponding logical statement.

2.6 CARTESIAN PRODUCTS AND TREE DIAGRAMS

We have discussed the set operations of complement, union, and intersection. Another operation on sets is the operation Cartesian (or cross) product.

DEFINITION: Cartesian Product

Let A and B be sets. Then A $\times$ B (read "A cross B") is the set

$$\{(a, b)| \; a \in A \text{ and } b \in B\}$$

EXAMPLE Let $A = \{p, q, r\}$ and $B = \{1, 2\}$. List the elements in $A \times B$ and in $B \times A$.

Solution For $A \times B$, the elements in the form (a, b) will be such that a is selected from set A and b is selected from set B:

$$A \times B = \{(p, 1), (p, 2), (q, 1), (q, 2), (r, 1), (r, 2)\}$$

$B \times A$ is such that the first object is selected from B and the second object is selected from A:

$$B \times A = \{(1, p), (2, p), (1, q), (2, q), (1, r), (2, r)\}$$

The fact that the elements in $B \times A$ are like those from $A \times B$ except in *reverse order* is very significant.

We call the objects or elements in a Cartesian product, (a, b), **ordered pairs** because the order of the listing is important.

DEFINITION

An **ordered pair** is an element of the Cartesian product of two sets. Two ordered pairs (a, b) and (c, d) are equal if and only if a = c and b = d.

It was shown in a previous section that $A \cup B = B \cup A$ and $A \cap B = B \cap A$—that is, $\cup$ and $\cap$ are commutative operations. Since the ordered pair (a, b) is not the same as the ordered pair (b, a) unless $a = b$, it is clear that

$$A \times B \neq B \times A$$

and the operation Cartesian product is *not* commutative in general.

In order to insure that all the possible ordered pairs in a Cartesian product are listed, the following table is helpful. For $A \times B$ from the above example, list set A vertically and set B horizontally as shown, and tabulate the ordered pairs.

$\times$	1	2
p	$(p, 1)$	$(p, 2)$
q	$(q, 1)$	$(q, 2)$
r	$(r, 1)$	$(r, 2)$

Then $B \times A$ can be tabulated as follows:

$\times$	p	q	r
1	$(1, p)$	$(1, q)$	$(1, r)$
2	$(2, p)$	$(2, q)$	$(2, r)$

It is also possible to find the Cartesian product of a set with itself.

EXAMPLE   Let $A = \{a, b, c\}$. Find $A \times A$.

Solution   Make a table as in the above illustration.

$\times$	a	b	c
a	(a, a)	(a, b)	(a, c)
b	(b, a)	(b, b)	(b, c)
c	(c, a)	(c, b)	(c, c)

The ordered pairs in the table are the elements of $A \times A$.

EXAMPLE   An insurance firm moves into a new suite of offices, and the branch manager (b), personnel manager (p), and account executive (a) must choose their offices. The choice is as follows: office with large window but facing busy street (A), office with no view but quiet location (B), and office with nice view and quiet location but near water cooler (C). What are all the possible executive-office arrangements which could occur?

Solution $S = \{b, p, a\}$, the set of executives, and $T = \{A, B, C\}$, the set of offices. Since we are interested in the executive-office choices, let us tabulate $S \times T$.

$$T$$

$\times$	A	B	C
b	(b, A)	(b, B)	(b, C)
p	(p, A)	(p, B)	(p, C)
a	(a, A)	(a, B)	(a, C)

S on the left bracing b, p, a.

We have seen that an ordered pair is an element in a Cartesian product of two sets and consists of two components, say a and b, whose order is significant. Thus $(a, b) = (b, a)$ if and only if $a = b$.

(a, b, c) is an **ordered triple** if the order in which a, b, and c appear is significant.

DEFINITION

An **ordered triple** is an element of the cross product of three sets:

$$A \times B \times C = \{(a, b, c) \mid a \in A, b \in B, c \in C\}$$

A **tree diagram** is a useful device for listing ordered triples. Suppose that $A = \{a, b\}$, $B = \{1, 2, 3\}$, and $C = \{p, q\}$.

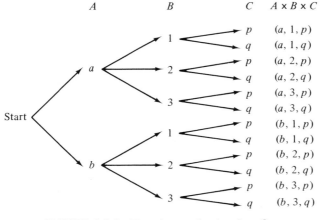

FIGURE 2.6.1   Tree diagram for $A \times B \times C$.

Figure 2.6.1 is read as follows: Begin at *start* and follow the arrows to the outermost "branch" of the tree, thus obtaining all possible ordered triples.

EXAMPLE Use a tree diagram to obtain the logical possibilities of three statements p, q, and r and compare the result with the listing in the appropriate section in the chapter on logic.

Solution The logical possibilities for each statement are $\{T, F\}$.

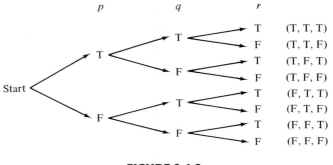

FIGURE 2.6.2

Figure 2.6.2 verifies the eight logical possibilities.

It is, of course, possible to have product sets involving more than three sets. The ordered elements in these sets are called ordered **n-tuples**.

DEFINITION

$(a_1, a_2, a_3, \ldots, a_n)$, where n is a natural number, is called an **n-tuple**. If the order in which the components appear is significant, then the n-tuple is an **ordered n-tuple**, and

$$(a_1, a_2, a_3, \ldots, a_n) = (b_1, b_2, b_3, \ldots, b_n)$$

if and only if

$$a_1 = b_1, a_2 = b_2, a_3 = b_3, \ldots, a_n = b_n$$

Thus if $n = 2$, we have an ordered pair, if $n = 3$, an ordered triple, if $n = 6$, an ordered six-tuple, and so on.

The reader should note that this definition follows the standard practice of extending a concept (in this case ordered pairs) in such a way that previous definitions and results are unchanged in both their meaning and application.

EXERCISES 2.6

Let $A = \{1, 2, 3\}$, $B = \{a, b\}$, *and* $C = \{x, y, z\}$. *List the ordered n-tuples in each of the sets in Exercises 1–8.*

1. $A \times B$ **2.** $B \times A$ **3.** $A \times C$

4. $A \times A$ **5.** $B \times B$ **6.** $A \times B \times C$

7. $A \times B \times B$ **8.** $C \times A \times B$

9. At an amateur-pro golf tournament, the following amateurs are scheduled to play: Bill Jones, Joe Smith, Frank Davis, and Jim Franklin. The pros are Jack Nicklaus, Sam Snead, Lee Trevino, and Arnold Palmer. List all possible amateur-pro pairings.

10. A publishing house decides to produce a certain book in three different editions: hardbound, paperback, and book club. The book is available in English, German, French, and Italian. List all available book-language options.

11. An automobile manufacturer makes three models of cars: four-door sedan (F), two-door sports coupe (T), and station wagon (S). Each model is available in four colors: gold (g), red (r), blue (b), and white (w). Each color can be ordered plain (p) or metallic (m). Use a tree diagram to list all possible options of car-color-finish.

12. The elements of the Cartesian product of two sets are ordered pairs, of three sets are ordered triples, and of four sets are ordered four-tuples. Write a definition for the cross product of four sets, $A \times B \times C \times D$.

13. If $A = \{a, b\}$, $B = \{1, 2, 3\}$, $C = \{p, q\}$, and $D = \{x, y\}$, make a tree diagram to list the four-tuples of $A \times B \times C \times D$.

14. If A is a nonempty set and $B = \varnothing$, what set specifies $A \times B$?

15. Use a tree diagram to list all logical possibilities of a compound statement involving four different simple statements.

16. Records for patients at the Speedy Recovery Hospital are cross-filed under sex (M and F), religion (c: Catholic, p: Protestant, j: Jewish, o: other), and age group (U: under 18, Y: 18 to 35, M: 36 to 60, O: over 60). Use a tree diagram to list all possible classifications of patients at this hospital.

17. The Quickie Drive-In Restaurant lists its "$1.89 special" as follows:

    Choice of soup or salad
    Choice of hamburger, roast beef sandwich, or ham sandwich
    Choice of jello or ice cream
    Choice of coffee, tea, or cola

List all possible complete selections available at this special price. (*Hint:* Use a tree diagram.)

2.7 CARDINALITY AND EQUIVALENT SETS

A finite set has been defined as a set which has exactly n elements, where n is a counting number or zero. This number, n, is said to be the **cardinality** of the set, and the numbers 0, 1, 2, 3, 4, . . . are also called **cardinal numbers.** We denote the cardinality of a set S by the notation

$$n(S) \text{ (read "} n \text{ of } S\text{")}$$

Thus if $S = \{a, b, c, d\}$, then $n(S) = 4$, and if $T = \varnothing$, then $n(T) = 0$.

It is important to recognize that S is a set, whereas $n(S)$ is a number. The operations on sets, such as union, intersection, and cross product, are applicable to sets only, whereas the operations addition, subtraction, multiplication, and division are applicable to numbers such as $n(S)$.

Suppose that $A = \{1, 2, 3\}$ and $B = \{2, 5, 9, 7\}$. What is the cardinality of $A \cup B$ if $n(A) = 3$ and $n(B) = 4$?

$$A \cup B = \{1, 2, 3, 5, 9, 7\}$$

Thus
$$n(A \cup B) = 6$$

But since $2 \in A$ and $2 \in B$, $A \cap B = \{2\}$ and $n(A \cap B) = 1$. Therefore,

$$n(A \cup B) = n(A) + n(B) - n(A \cap B)$$
$$6 = 3 + 4 - 1$$

This introduces the following theorem (verified in next chapter) for $n(A \cup B)$.

THEOREM

If A and B are finite sets, then

$$n(A \cup B) = n(A) + n(B) - n(A \cap B)$$

If A and B are disjoint sets—that is, $A \cap B = \varnothing$—then

$$n(A \cup B) = n(A) + n(B)$$

This, of course, is consistent with the above theorem, since $n(\varnothing) = 0$.

What about $n(A \times B)$? The names *Cartesian product* and *cross product* suggest how one derives $n(A \times B)$, and the following theorem (presented without proof) provides a formula.

THEOREM

If A and B are finite sets, then $n(A \times B) = n(A) \cdot n(B)$, the product of $n(A)$ and $n(B)$.

EXAMPLE If $A = \{a, b, c, d\}$ and $B = \{2, 3\}$, find $n(A \times B)$ and $n(B \times A)$.

Solution $n(A) = 4, n(B) = 2$

$$n(A \times B) = n(A) \cdot n(B) = 4 \cdot 2 = 8$$

Also, $n(B \times A) = n(B) \cdot n(A) = 2 \cdot 4 = 8.$

Since multiplication is a commutative number operation, $n(A) \cdot n(B) = n(B) \cdot n(A)$ for all cardinal numbers. Thus

$$n(A \times B) = n(B \times A)$$

even when $A \times B \neq B \times A$

This is reasonable because the only difference between $A \times B$ and $B \times A$ is the change in the order of the components of the ordered pairs, *not* the number of ordered pairs.

There is a special name for sets which have the same number of elements.

DEFINITION: Equivalent Sets

Two sets A and B are equivalent if and only if $n(A) = n(B)$.
In symbols, $A \sim B$, where $\sim$ is read "is equivalent to."

Clearly, all equal sets are equivalent. But not all equivalent sets are equal, since the only requirement for equivalence is that the sets have the same *number* of elements, not that the elements be the same.

Thus $(A \times B) \sim (B \times A)$ even if $A \times B \neq B \times A$.

If two sets A and B are equivalent, then their elements can be put in *one-to-one correspondence*—that is, every element in A can be paired with exactly one element in B, and every element in B can be paired with exactly one element in A.

EXAMPLE Establish two possible one-to-one correspondences between set $A = \{$Joe, Bill, Fred$\}$ and set $B = \{$Jane, Gina, Karen$\}$.

Solution One possibility is: Joe, Jane
                          Bill, Gina
                          Fred, Karen

Another possibility is: Joe, Karen
                          Bill, Jane
                          Fred, Gina

Note that in each possibility both sets are completely used up.

Let us examine $A \times B$ to see how these one-to-one pairings compare with $n(A \times B)$. That is, are the number of one-to-one pairing possibilities the same as $n(A \times B)$?

$\times$	Jane	Gina	Karen
Joe	(Joe, Jane)	(Joe, Gina)	(Joe, Karen)
A Bill	(Bill, Jane)	(Bill, Gina)	(Bill, Karen)
Fred	(Fred, Jane)	(Fred, Gina)	(Fred, Karen)

We know that $n(A \times B) = 9$. From examining the chart, we see that the possible pairings are

1. (Joe, Jane), (Bill, Gina), (Fred, Karen)
2. (Joe, Jane), (Fred, Gina), (Bill, Karen)
3. (Bill, Jane), (Joe, Gina), (Fred, Karen)
4. (Bill, Jane), (Fred, Gina), (Joe, Karen)
5. (Fred, Jane), (Bill, Gina), (Joe, Karen)
6. (Fred, Jane), (Joe, Gina), (Bill, Karen)

Since the possible pairings are now exhausted, there are six possibilities, and $6 \neq n(A \times B)$. The mathematical formula for determining the number of one-to-one pairings is left for Chapter 3, "Partitions and Counting."

EXERCISES 2.7

In Exercises 1–10:

$$Let \ \mathcal{U} = \{1, 2, 3, 4, 5, 6, 7, 8, 9\}$$
$$A = \{x \mid x = 2n, n \in \mathcal{U}\}$$
$$B = \{x \mid x \ is \ exactly \ divisible \ by \ three \ and \ x \in \mathcal{U}\}$$
$$C = \{2, 4, 7, 9\}$$
$$D = \{2, 4, 8\}$$

1. List the elements of A.
2. List the elements of B.
3. Find the cardinality of A, B, C, and D.
4. Find $n(A \cup B)$.
5. Find $n(A \cup C)$.

6. Find $n(C \cup D)$.

7. Find $n(A \times B)$.

8. Find $n(A \times C)$.

9. Find $n[(A \times B) \cup (C \times D)]$.

10. Find $n(A' \cup D)$.

11. If A, B, and C are finite sets, how should $n(A \times B \times C)$ be defined?

12. Use a tree diagram to determine $n(A \times B \times C)$ if $A = \{1, 2, 3\}$, $B = \{a, b\}$, $C = \{T, F\}$.

13. Are any of the sets, A, B, C, D as defined for Exercises 1–10 equivalent? If so, name these sets.

14. In an election for president and vice-president of an organization, the candidates for president were Joe, Joan, Harvey, and Sandra. The vice-presidential candidates were Betty, Harold, Ted, and Alice. Since the set of presidential candidates is equivalent to the set of vice-presidential candidates, a one-to-one correspondence can be established between these sets. List three possible one-to-one correspondences of presidents and their vice-presidential running mates.

15. For the election in Exercise 14, how many possible outcomes exist?

16. A catalog lists a two-piece woman's costume as being a jacket with either a skirt or slacks, available in white, navy, brown, or black, in wool or synthetic fiber. How many complete outfits can be ordered for this costume?

17. Verify your answer for Exercise 16 by making a tree diagram for the possible selections.

A magazine publisher wished to determine the reader reaction to his magazine by the educational background of his readers. One thousand readers responded to the questionnaire in the following manner:

	(L) Liked	(H) Disliked	(N) No opinion
(A) Formal education through 8th grade	30	150	20
(B) High school diploma	45	20	130
(C) Two-year college degree	220	15	10
(D) Four-year college degree	112	20	46
(E) Education beyond four-year college degree	94	48	40

Find the number of people who belonged to each of the sets in Exercises 18–23.

18. $A; B; C; D; E$ **19.** $L; H; N$
20. $A \cap L$ **21.** $(C \cup D \cup E) \cap L$
22. $(C \cup D) \cap (H \cup N)$ **23.** $(D \cup E) \cap L$

24. From the data for Exercises 18–23, can a reasonable conclusion be drawn about the educational background of readers who are most likely to renew the magazine subscription?

In Exercises 25 and 26:

A neighborhood door-to-door survey was conducted by a baby food manufacturer. Each respondent was asked three questions:

    a. Do you have children under five years of age?
    b. Do you use our brand of baby food?
    c. Would you accept a free sample of our brand?

Three hundred doorbells were rung by the survey team with the following results:

    a. 185 responded yes
    b. 125 responded yes
    c. 165 responded yes
    a. and b. 85 responded yes
    b. and c. 115 responded yes
    a. and c. 115 responded yes
    a., b., and c. 75 responded yes

25. If A, B, and C are the truth sets of (a), (b), and (c), respectively, determine the cardinality of each of the following:

    a. $n(A \cup B)$ b. $n(B \cup C)$
    c. $n(A \cup C)$ d. $n[(A \cup B \cup C)']$

26. Construct a Venn diagram for this data, and determine:

a. How many people either did not respond or responded no to all the questions asked?
b. How many people used the manufacturer's brand but were unwilling to accept a free sample?
c. How many people had children under five years of age but were unwilling to accept a free sample?
d. Of those who had children under five years of age, how many were not using the manufacturer's product?

REVIEW EXERCISES

Let $\mathcal{U} = \{0, 1, 2, 3, 4, 5, 6, 7, 8, 9\}$. Classify each of the sets in Exercises 1–4 in relation to $\mathcal{U}$ as **one** *of the following:* $\subset, =, \not\subseteq$.

1. $A = \{2, 4, 6\}$
2. $B = \{x\mid x = \frac{1}{2}n, n \in \mathcal{U}\}$
3. $C = \{n\mid n \text{ is a cardinal number}\}$
4. $D = \{d\mid d \text{ is a cardinal number less than 10}\}$

$\mathcal{U}$ is the set of personnel employed by a certain company with subsets designated as follows:

$$A = \{x\mid x \text{ is a secretary}\}$$
$$B = \{p\mid p \text{ is an executive}\}$$
$$C = \{f\mid f \text{ is a female})$$
$$D = \{m\mid m \text{ is a male}\}$$
$$G = \{c\mid c \text{ has a custodial position}\}$$
$$H = \{h\mid h \text{ is a part-time employee}\}$$
$$K = \{s\mid s \text{ is a salesperson}\}$$

Describe each of the sets in Exercises 5–10.

5. A' 6. $B \cap C$ 7. $G \cup H$
8. $G \cap H$ 9. $A \cap D$ 10. $(D \cap H) \cap K$

Let $A = \{1, 5, 7\}$, $B = \{2, 4\}$, and $C = \{1, 4\}$. Find each of Exercises 11–20.

11. $A \times B$ 12. $B \times A$
13. $A \times B \times C$ 14. $C \times B \times A$
15. $n(A \times B)$ 16. $n(A \times B \times C)$
17. $n(A \cup B)$ 18. $n(A \cup B) + n(A \cup C)$
19. $n[(A \times B) \cup (A \times C)]$ 20. $n(A \cup B \cup C)$

21. If $S = \{e, t, a, m\}$, establish all possible one-to-one correspondences of S with itself, such that the correspondence forms a proper English word.
22. If $A = \{1, 2\}$ and $B = \{1, 2, 4\}$, find a set S such that

$$A \cup S = B$$

$$A \cap S = \varnothing$$

23. If $A \cup B = B$, what relationship *must* exist between A and B?

Draw an appropriately shaded Venn diagram to illustrate each of Exercises 24–27.

24. $(A \cap B) \cup C$ **25.** $(A' \cap B) \cup C'$

26. $(A \cup B)' \cup C$ **27.** $(A' \cup B') \cup C'$

Use Venn diagrams to classify each of the statements in Exercises 28–31 as logically true, logically false, neither, or an implication.

28. $(p \wedge q) \vee (\sim p \vee \sim q)$

29. $(p \wedge q) \rightarrow (\sim p \vee \sim q)$

30. $(\sim p \wedge \sim q) \wedge (p \vee q)$

31. $[(p \vee q) \wedge r] \wedge \sim p$

32. An advertisement for a job opening reads as follows:

> Wanted: applicants male or female, two years of college work completed or three years of experience or one year of college and two years of experience; able to speak French, German, or Japanese in addition to English, for public relations work with growing export-import company.

Make a tree diagram to list the qualifications of all possible applicants.

33. For the applicants for the position listed in Exercise 32, how many logical possibilities are there for qualified female applicants with two years of college work completed?

CHAPTER **3**

PARTITIONS AND COUNTING

The objective of this chapter is to develop techniques for counting the elements of a set. Of course, these techniques will be much more sophisticated than simple one-to-one matching with the cardinal numbers 1, 2, 3, ... For instance, we will see how to *calculate* quickly the number of different "diamond flushes" (five cards, all diamonds) that could be created from a deck of poker cards or the number of committees of three that could be formed of United States senators. "Counting" will be shown to have many uses in its own right, and in the next chapter we will see that it is a cornerstone of probability theory.

3.1 PARTITIONS

Theorems for counting are best introduced in terms of set theory—specifically, partitions. To partition a room means to divide it into distinct smaller rooms. To partition a set has an analogous meaning. The partition of a set is made in terms of distinct subsets of the set. That is, it is made up of subsets such that no two have a common element. In other words, they are *disjoint*. A partition of a set is also such that the union of all subsets is the set being partitioned.

DEFINITION

Let $S_1, S_2, \ldots, S_n$ be subsets of A. Then the collection of sets $(S_1, S_2, S_3, \ldots, S_n)$ is a **partition of set** A if and only if $S_i \cap S_j = \varnothing$ for $i \neq j$ and $S_1 \cup S_2 \cup S_3 \cup \cdots \cup S_n = A$.

The sets $S_1, S_2, \ldots, S_n$ are called the **cells** of the partition. The property $S_i \cap S_j = \varnothing$ for $i \neq j$ is sometimes described by saying the sets are **mutually exclusive**, while $S_1 \cup S_2 \cup S_3 \ldots \cup S_n = A$ is denoted by saying the cells of a partition are **exhaustive**.

EXAMPLE   Three partitions of $A = \{1, 2, 3, 4\}$ are:

1. $(\{1\}, \{2\}, \{3\}, \{4\})$
2. $(\{1, 2\}, \{3, 4\})$
3. $(\{1, 2, 3\}, \varnothing, \{4\})$

Notice that each of the partitions in the above example is such that the member sets are disjoint and that their union is A.

EXAMPLE   The collection of sets $(\{1, 2, 3\}, \{3, 4\}, \varnothing)$ is *not* a partition of $A = \{1, 2, 3, 4\}$ because $\{1, 2, 3\}$ and $\{3, 4\}$ are not disjoint.

If a set has two different partitions, then a refined partition can be formed by taking all possible intersections of the sets from the different partitions.

DEFINITION

If $(S_1, S_2, S_3, \ldots, S_n)$ and $(R_1, R_2, R_3, \ldots, R_m)$ are partitions of A, the following is the **cross partition** of the given partitions:

$$(S_1 \cap R_1, S_1 \cap R_2, S_1 \cap R_3, \ldots, S_1 \cap R_m$$
$$S_2 \cap R_1, S_2 \cap R_2, S_2 \cap R_3, \ldots, S_2 \cap R_m$$
$$S_3 \cap R_1, S_3 \cap R_2, S_3 \cap R_3, \ldots, S_3 \cap R_m$$

$$\cdot \qquad \cdot \qquad \cdot \qquad \qquad \cdot$$
$$\cdot \qquad \cdot \qquad \cdot \qquad \qquad \cdot$$
$$\cdot \qquad \cdot \qquad \cdot \qquad \qquad \cdot$$

$$S_n \cap R_1, S_n \cap R_2, S_n \cap R_3, \ldots, S_n \cap R_m)$$

The above cross partition can be shown to be in fact a partition by using the algebra of sets and mathematical induction.*

EXAMPLE   Two partitions of $A = \{1, 2, 3, 4\}$ are

$$(\{1, 2\}, \{3\}, \{4\})   \text{ and }   (\{1, 3\}, \{2, 4\})$$

Form their cross partition.

Solution   The cross partition is formed by creating all possible intersections of a set from the first partition with a set from the second:

$$\{1, 2\} \cap \{1, 3\} = \{1\}, \{1, 2\} \cap \{2, 4\} = \{2\}, \{3\} \cap \{1, 3\} = \{3\}$$
$$\{3\} \cap \{2, 4\} = \varnothing, \{4\} \cap \{1, 3\} = \varnothing, \{4\} \cap \{2, 4\} = \{4\}$$

Therefore, the cross partition is:

$$(\{1\}, \{2\}, \{3\}, \varnothing, \{4\})$$

EXAMPLE   From the Venn diagram in Figure 3.1.1, two partitions of $\mathcal{U}$ are (A, A') and (B, B'). Form the cross partition.

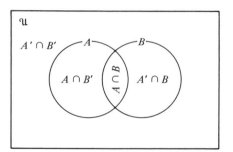

FIGURE 3.1.1   The cross partition of (A, A') and (B, B').

Solution   The cross partition consists of all intersections of pairs, one from each partition. That is:

$$(A \cap B, A \cap B', A' \cap B, A' \cap B')$$

Note that the cross partition defines disjoint subsets of $\mathcal{U}$ whose union is $\mathcal{U}$.

Cross partitions are used in problems of classification or identification. A bacterial culture is identified by assigning it to a particular cell of a partition that is formed by cross-partitioning many partitions. One partition is (A, A'), where A is the set of all bacteria that can survive in air and A' is the set of bacteria that cannot survive in air.

* For a brief discussion on mathematical induction, see Chapter A in the back of this book.

A second partition for the set of bacteria is (F, F'), where F is the set of all bacteria that have flagella (taillike appendages used for locomotion) and F' is the set of all bacteria that do not. The cross partition

$$(A \cap F, A \cap F', A' \cap F, A' \cap F')$$

is a finer partition than either of the originals. If a biologist can assign an individual culture to one of the cells of the cross partition, he is a long way toward discovering exactly which bacteria he has. The final identification would be completed by using more partitions and cross partitions of the set of all bacteria. Other partitions are created using width and length of individuals and the effects of certain stains on slides of the culture. Eventually the test culture is assigned to a cell with only one member and is thus classified.

Another example of cross-partitioning may be found in the following classification situation.

A large Japanese manufacturing company recently discovered that 80% of its employees were not married. Believing that marriage is important to the general well-being of its employees and that it thus increases productivity, the company management is attempting to promote "romance" by matching suitable male-female pairs. One partition is (A, A'), where A is the set of all single employees. A second partition is (B, B'), where B is the set of all male employees. The cross partition

$$(A \cap B, A \cap B', A' \cap B, A' \cap B')$$

is a further partition, where $A \cap B$ represents the set of single male employees and $A \cap B'$ is the set of all single female employees. In order to match age groups, another partition may be made by (C, C'), where C is the set of all employees under thirty years of age, and so forth. By making more partitions and creating further cross partitions, a "dating center" can be established, and the management hopes that the resulting marriages will increase productivity and profits for the company.

The number of elements in pairs of subsets of a universal set is easily counted if the subsets are members of a partition. The cells of a partition are mutually exclusive. If $n(X)$ denotes the number of elements in set X, then if R and S are cells of the same partition of $\mathscr{U}$, $n(R \cup S) = n(R) + n(S)$ because R and S are disjoint. It is more difficult to find the number of elements in the union of two sets when they are not disjoint.

For instance, suppose that a manufacturer polls known users of his product to discover if they like situation comedies or medical dramas on television. The results are that 180 like medical dramas and 570 like situation comedies; everyone polled had an opinion. Does this imply that $180 + 570 = 750$ people were polled? The answer is no, since some may have liked both types of program. If it is also known that 75 individuals liked both types of program, then the number polled was $180 + 570 - 75 = 675$. The

75 is subtracted from the sum because it is being counted twice—once with the group of 180 who liked dramas and again with the group of 570 who liked comedies.

THEOREM

If two sets R and S are not disjoint, then the number of elements in their union is:

$$n(R \cup S) = n(R) + n(S) - n(R \cap S) \tag{1}$$

The above theorem can be verified from Figure 3.1.2. Set R contains $R \cap S$ and so does S. Therefore, $n(R)$ includes the elements of $R \cap S$, as does $n(S)$. Thus $n(R) + n(S)$ counts the elements of $R \cap S$ twice. By subtracting one of two numbers, $n(R \cap S)$, in $n(R) + n(S)$, equation (1) counts the elements of $R \cup S$ only once.

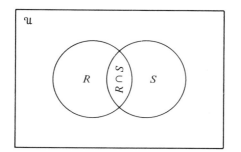

FIGURE 3.1.2 $n(R \cup S) = n(R) + n(S) - n(R \cap S)$.

Formulas can be found for the number of elements in the union of three or more nondisjoint sets. The equation for three sets is:

$$n(A \cup B \cup C) = n(A) + n(B) + n(C) - n(A \cap B) - n(A \cap C)$$
$$- n(B \cap C) + n(A \cap B \cap C) \tag{2}$$

The rationale behind equation (2) can be seen by examining the Venn diagram of Figure 3.1.3. Note that:

$n(A)$ includes elements from cells 2, 3, 5, and 6
$n(B)$ includes elements from cells 3, 4, 6, and 7
$n(C)$ includes elements from cells 5, 6, 7, and 8

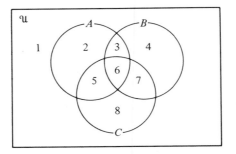

FIGURE 3.1.3 $n(A \cup B \cup C) = n(A) + n(B) + n(C) - n(A \cap B) -$
$n(A \cap C) - n(B \cap C) + n(A \cap B \cap C)$.

Thus if $n(A) + n(B) + n(C)$ was taken as the number of elements in $n(A \cup B \cup C)$, the elements in cells 3, 5, and 7 would be counted twice and the elements of cell 6 would be counted three times. This repeated use of elements can be corrected by subtracting the number of elements in the intersections of the sets—that is, by subtracting $n(A \cap B)$, $n(A \cap C)$, and $n(B \cap C)$. However, the indicated subtractions completely eliminate the elements of cell 6 because it is in all three intersections of pairs of sets:

$$A \cap B \text{ contains cells 3 and 6}$$
$$A \cap C \text{ contains cells 5 and 6}$$
$$B \cap C \text{ contains cells 7 and 6}$$

Therefore, to count the elements of $A \cup B \cup C$, we add the number of elements in A, B, and C, then subtract the number of elements in the intersections of A and B, A and C, and B and C. The subtractions eliminate entirely the elements of cell 6, which is $A \cap B \cap C$. Thus after subtracting, $n(A \cap B \cap C)$ must be added so that every element in $A \cup B \cup C$ is counted, as it is in equation (2).

Other equations can be created for use in the analysis of the numbers of elements in sets that are not elements of a partition. However, Venn diagrams are often the best tool for such an analysis.

EXAMPLE The Industrial Fabricating Company manufactures a variety of small engines. These engines are confined by their design to uses in electrical generators, small boats, and tractors. The company produces seven models. Three models can be used exclusively in generators, tractors, and boats, respectively. A fourth can be used in boats or generators only, a fifth in boats or tractors only, and a sixth model can be used in generators or tractors only. The seventh model is adaptable to boat, tractor, or generator use. The company's warehouse records keep the numbers of engines on hand by their possible use. The records show the following types of engines are in the warehouse:

Generator	450
Boat	700
Tractor	290
Boat or generator	100
Boat or tractor	200
Generator or tractor	150
Generator or boat or tractor	60

The sales manager wishes to know how many engines are in the warehouse and how many engines there are for tractors only.

Solution We cannot simply add the numbers given to answer the first question because the sets are not a partition. The 450 generator engines, for instance, must include the 60 engines that can be used in generators, tractors, or boats.

Figure 3.1.4 is a Venn diagram with the G set representing the set of generator engines, B representing boat engines, and T tractor engines. The numbers are placed in the cells of Figure 3.1.4 by using the given data in

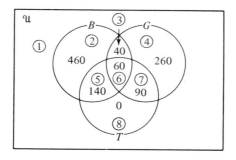

FIGURE 3.1.4

reverse order and by assigning value to the innermost cells of the partition first. Now 60 is the number of elements in the intersection of B, G, and T, cell 6. From the given data, cells 6 and 7 must contain 150 engines. Since there are 60 in 6, cell 7 must contain 90. We are given that there are 200 engines for boats or tractors. This means that cells 5 and 6 must have a total of 200 engines. Since cell 6 has 60, cell 5 must have 140. Cell 3 contains 40 engines because there must be a total of 100 in cells 3 and 6. Cell 8 must have engines enough so that tractor engines number 290. However, cells 5, 6, and 7 total 290, so cell 8 must be empty. There are no engines for tractor use exclusively, which was a question to be answered. Cell 2 is found to contain

460 since the total for boat engines is 700 and cells 3, 6, and 5 account for 240 engines. Cell 4 contains 260 engines. The 260 is derived by adding the elements in cells 3, 6, and 7—190 engines—and subtracting from 450, the total for generator engines. Figure 3.1.4 now shows a partition of $\mathcal{U}$ with the number of elements in each cell. The $n(B \cup G \cup T)$ is found by adding the elements of all the cells. There is a total of 1,050 engines.

Venn diagram analysis will yield the total number of elements in three nondisjoint sets as in the last example, but it has an added advantage over an equation. Venn diagrams give the "whole picture." For instance, in Figure 3.1.4, it is easily seen that there are 460 engines that must be used exclusively in boats. Such a conclusion could not be seen as quickly using equation (2).

EXERCISES 3.1

Form two different partitions of the sets in Exercises 1 and 2.

1. $\{a, b, c\}$ **2.** $\{a, b\}$

Form all possible partitions of the sets in Exercises 3 and 4 such that no cell is empty.

3. $\{a, b, c\}$ **4.** $\{a, b, c, d\}$

Find the cross partitions for the partitions in Exercises 5–9.

5. $(\{1, 2\}, \{3, 5\}, \{4\}); (\{1, 4\}, \{2, 3, 5\})$
6. $(\{a, b\}, \{c, d\}); (\{a\}, \{b\}, \{c\}, \{d\})$
7. $(P, P'); (Q, Q')$
8. Find the cross partition of (R, R') with the cross partition of Exercise 7.
9. $(\{a, b\}, \{c, d\}, \{e, f\}); (\{a, b, c\}, \{d, e, f\}); (\{a, f\}, \{b, c\}, \{d, e\})$

10. In Exercise 6 above, the second partition is a refinement of the first. Generalize about what will happen with the cross partition in all such cases, where a "unit" partition is crossed with a second partition. (A unit partition is a partition which contains a single element.)
11. In the bacteria example of this section a number of partitions of the set of all bacteria were given. These partitions could then be used to cross-partition the set of all bacteria. Give four partitions for the set of all animals.
12. Repeat Exercise 11 for the set of all plants.

13. Draw a Venn diagram and give the cross partition for (P, P'), (Q, Q') if $Q \subset P$.

14. Draw a Venn diagram and form the cross partition of (P, P'), (Q, Q'), (R, R') if $R \subset Q$.

15. A survey of students in a small community college found that 40 students were taking French and 50 students were taking German. How many students took only French if:

    a. No students took both French and German?
    b. Eight students took both classes?

16. In a political poll 100 people were interviewed. Seventy liked the incumbent candidate and 45 liked the challenger. Thirty liked both the incumbent and the challenger. How many liked neither candidate?

Exercises 17 and 18 are based on the following information, which was received from a survey of students:

> *27 were taking economics*
> *45 were taking business administration*
> *27 were taking public administration*
> *12 were taking economics and business administration*
> *13 were taking economics and public administration*
> *10 were taking business administration and public administration*
> *5 were taking all three courses*
> *15 were taking none of the three courses mentioned*

17. How many students were surveyed?

18. How many students in the survey were taking just one of the mentioned courses?

Exercises 19 and 20 are based on a survey of 400 people, which showed that:

> *110 liked football*
> *75 liked basketball*
> *180 liked baseball*
> *40 liked baseball and basketball*
> *45 liked baseball and football*
> *50 liked football and basketball*
> *25 liked all three sports*

19. How many people surveyed like none of the sports?

20. How many of those surveyed liked football but not baseball?

21. Analyze the following data, which were collected to determine career preferences among a national sample of 1,000 individuals:

> 650 liked economics
> 800 liked business
> 500 liked data processing
> 325 liked economics and business
> 420 liked economics and data processing
> 470 liked data processing and business
> 330 liked all three career fields

22. In a survey of 100 college graduates it was discovered that:

> 30 read *Time* Magazine
> 35 read *Newsweek*
> 27 read *U.S. News & World Report*
> 14 read *Time* Magazine and *Newsweek*
> 16 read *Time* Magazine and *U.S. News & World Report*
> 12 read *U.S. News & World Report* and *Newsweek*
> 7 read all three magazines

How many in the survey read *Time* Magazine if and only if they read *Newsweek*?

23. Human blood classification or "typing" is based on the presence or absence of three antigens, known as A, B, and rh. If a specimen of blood has all three, it is called AB positive. If it has antigen A but not B antigen or rh antigen, it is called A negative. In general, blood types are identified using the letters *A* or *B* or both for the A and B antigens and the words or symbols for *positive* or *negative* to show the presence or absence of the rh antigen. If a blood sample contains no A or B antigen, it is called type O. The Pine Wood Hospital has 400 patients that belong to the following blood types:

> rh +,   220 patients
> O +,   80 patients
> O −,   70 patients
> A +,   75 patients
> A −,   30 patients
> B +,   25 patients
> B −,   55 patients

How many patients have AB − blood and how many are rh − ?

3.2 PERMUTATIONS AND THE FUNDAMENTAL THEOREM OF COUNTING

This section will develop techniques for counting the elements in sets known as *permutations*. Permutations are arrangements. The idea of a permutation is quite simple.

EXAMPLE Suppose that Bill, Donna, and Dick are to have their picture taken together. How many different pictures are possible, assuming that the three will simply be standing next to one another? That is, a "different picture" will mean a different arrangement only.

Solution Since there are only three people involved, we can list the possible arrangements as they line up from left to right:

1. Donna, Bill, Dick
2. Donna, Dick, Bill
3. Bill, Donna, Dick
4. Bill, Dick, Donna
5. Dick, Donna, Bill
6. Dick, Bill, Donna

Thus there are six different pictures possible of Donna, Dick, and Bill standing side by side.

DEFINITION

An ordering of (or an arrangement in a specific order of) *r* objects where no two of the objects are identical is called a **permutation of the *r* objects.**

In the above example, arrangement 4—Bill, Dick, and Donna—places the three persons in a special order. It is a permutation of the three people. Each of the six listings in the example is a permutation.

If *r* is small, a listing is a satisfactory method for finding the number of permutations of *r* objects. If *r* is large, other methods which use the Fundamental Theorem of Counting are required. The statement of the theorem uses the word *event*. In this context an *event* will be any action whose outcomes can be counted. For example, there are two possible outcomes for the event "flipping a coin." There are three possible outcomes for the event "placing a person on the left" in the example above. There are nine possible outcomes for the event "choosing the first digit for a house number."

THE FUNDAMENTAL THEOREM OF COUNTING

If event A has m possible outcomes, and after event A happens event B has n possible outcomes, and after event B happens event C has p possible outcomes, and so on, then the sequence of events, one from A, then one from B, then one from $C, \ldots$ can occur in $m \cdot n \cdot p \ldots$ possible ways.

An example of the theorem follows.

EXAMPLE The previous example—finding the number of possible pictures of Donna, Bill, and Dick—is easily computed using the Fundamental Theorem of Counting. Let event A be "placing a person on the left." Let event B be "placing a person in the center of the picture." Let event C be "placing a person on the right in the picture." Event A can be performed three ways, since any of the three people could be placed on the left. After event A happens, event B could happen two ways, since after the first person is placed on the left, two remain to go in the center. After event B, there is only one person to go on the right, so event C can happen in only one way. That is event A has three possible outcomes, and after it, event B has two possible outcomes, and after B, C can happen only one way.

Therefore, by the Fundamental Theorem of Counting, the sequence of events can happen in $3 \cdot 2 \cdot 1 = 6$ ways, which is the number of arrangements possible for the three people.

The Fundamental Theorem of Counting may be justified in the following manner. Let the set $\{a_1, a_2, a_3, \ldots, a_m\}$ be all the individual outcomes possible for event A and let the set $\{b_1, b_2, b_3, \ldots, b_n\}$ be all the individual outcomes for event B. Then the sequence of events, one from A and then one from B, can be represented by an ordered pair. One such ordered pair is (a_1, b_5). Then the number of the possible sequence of events, one from A and one from B, is the number of ordered pairs in $A \times B$, the cross product of A and B. But

$$n(A \times B) = n(A) \cdot n(B) = m \cdot n$$

and the theorem is proved for two events. It can be shown by mathematical induction that the Fundamental Theorem of Counting is also true for any finite number of events: $A, B, C, \ldots$.

EXAMPLE A builder of tract houses designs his homes by connecting predrawn architectural blueprints of garages, family living areas, and bedroom wings. For any house, he has a choice of five different garages, eight different family living areas, and six different bedroom wings. How many different houses can be built from these plans?

Solution The Fundamental Theorem of Counting applies. The number of different houses is the number of possible sequences of events that can occur, where the first event is choosing a garage, the second is choosing a family living area, and the third is choosing a bedroom wing. The possibilities are:

$$5 \cdot 8 \cdot 6 = 240$$

Therefore, 240 houses can be built with no two exactly alike.

EXAMPLE A doctor wishes to put serial numbers on his files. Each case is to be identified by two letters followed by three digits. How many files can be identified in this manner?

Solution The first letter may be any of 26, as may the second letter. There are 10 choices for each of the digits. Therefore, using the Fundamental Theorem, the number of unique identification tags of the type described is:

$$26 \cdot 26 \cdot 10 \cdot 10 \cdot 10 = 676{,}000$$

The fundamental Theorem of Counting is the main tool used to compute permutations.

EXAMPLE How many permutations are possible using all four of the letters A, B, C, and D once?

Solution The problem can be viewed as one of placing the letters next to one another starting from the left. The first letter placed may be any one of the four. After the first is in place, there are three letters remaining, one of which will go next to the first placed letter. After the first two are in place, there are two remaining, one of which will go next to the second letter placed. Finally, there is one letter remaining to be placed on the right side of the permutation. Therefore, the number of permutations possible is:

$$4 \cdot 3 \cdot 2 \cdot 1 = 24$$

In the above example, the product $4 \cdot 3 \cdot 2 \cdot 1$ gives the number of permutations. Products like this one—products of all the natural numbers from 1 to n—will occur often in this and the next chapter. A notation for these products will allow later formulas to be concisely given. The notation used is $n!$ (read "n factorial").

DEFINITION: *n* **Factorial**

If n is a natural number and greater than 1, then:

$$n! = n\,(n-1)(n-2)(n-3) \cdots (4)(3)(2)(1)$$

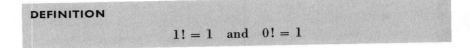

DEFINITION

$$1! = 1 \quad \text{and} \quad 0! = 1$$

The definition of 1! and 0! is necessary so that the factorial notation can be used in formulas that will be given later.

EXAMPLES

$$
\begin{array}{ll}
1! = 1 & 4! = 4\cdot3\cdot2\cdot1 = 24 \\
2! = 2\cdot1 = 2 & 5! = 5\cdot4\cdot3\cdot2\cdot1 = 120 \\
3! = 3\cdot2\cdot1 = 6 & 6! = 6\cdot5\cdot4\cdot3\cdot2\cdot1 = 720
\end{array}
$$

The number of permutations of n objects all used once and only once in each arrangement is denoted by $_nP_n$, read "the permutation of n objects taken n at a time." The formula for $_nP_n$ is

$$_nP_n = n!$$

The above formula is derived by letting an ordered n-tuple, $(p_1, p_2, p_3, \ldots, p_n)$, represent a single permutation of n objects and by using the Fundamental Theorem of Counting to discover how many such n-tuples can be created. The first element, p_1, can be chosen n ways. After p_1 is chosen, p_2 can be selected $(n - 1)$ ways. After p_2 is chosen, p_3 can be selected from $(n - 2)$ elements. With each subsequent selection of elements p_1 through p_n, the number of unused elements is reduced by 1. Only 1 element will remain when it comes time to choose an element for p_n. Therefore,

$$_nP_n = (n)(n - 1)(n - 2)(n - 3)\cdots(3)(2)(1) = n!$$

EXAMPLE Find the number of permutations of five objects when all five are used in each arrangement.

Solution The number is $_5P_5$, which equals

$$_5P_5 = 5! = 5\cdot4\cdot3\cdot2\cdot1 = 120$$

Permutations are often created by taking arrangements of fewer than the total number in a set.

EXAMPLE How many permutations of A, B, C, D, E are possible if each arrangement is to contain three objects, each used once?

Solution If the ordered triple (p_1, p_2, p_3) represents such an arrangement, then there are five choices for p_1, four choices for p_2, and finally three choices for p_3. The number of permutations is

$$5\cdot4\cdot3 = 60$$

In the above example, we were discussing permutations of three objects from a set of five. The notation for the number of such permutations is $_5P_3$. The symbol $_5P_3$ is read "the number of permutations of five objects taken three at a time," and as was shown in the example, $_5P_3 = 60$. The symbol $_nP_r$ represents the number of permutations from a set of n objects with r objects in each arrangement. The formula for $_nP_r$ is derived by examining the ordered r-tuple, $(p_1, p_2, \ldots, p_r)$, which represents one arrangement of r objects taken from a set of n objects. The first element, p_1, can be chosen from n objects. After it is in place, the next element of the permutation, p_2, is chosen from $(n - 1)$ objects. Each successive p_i is chosen from a set of objects that numbers 1 less than the set from which p_{i-1} was chosen. The following list shows the number of elements from which each p_i is chosen:

$$p_1 \qquad p_2 \qquad p_3 \qquad p_4 \qquad p_5 \quad \cdots \quad p_r$$

$$n - 0 \quad n - 1 \quad n - 2 \quad n - 3 \quad n - 4 \qquad n - r + 1$$

The key to the following formula is the number of elements in the set from which p_r is chosen. Note that the number of elements in the sets from which p_1, p_2, p_3, p_4, and p_5 are chosen is n minus a number that is 1 less than the subscript of p. Therefore, p_r is chosen from a set that numbers $n - r + 1$ elements, and

$$_nP_r = n(n - 1)(n - 2)\cdots(n - r + 1) \tag{1}$$

Above it was shown that $_5P_3 = 60$. Using equation (1),

$$_5P_3 = 5\cdot4\cdot(5 - 3 + 1) = 5\cdot4\cdot3 = 60$$

Equation (1) is not usually used for finding $_nP_r$ because it is somewhat difficult to remember and clumsy to write. A more compact equation is created by multiplying and dividing equation (1) by $(n - r)!$, which gives

$$_nP_r = n(n - 1)(n - 2)\cdots(n - r + 1)\left(\frac{(n - r)!}{(n - r)!}\right)$$

Now $(n - r)! = (n - r)(n - r - 1)(n - r - 2)\cdots(3)(2)(1)$

so $_nP_r$ can be written

$$_nP_r =$$

$$\frac{n(n - 1)(n - 2)\cdots(n - r + 1)(n - r)(n - r - 1)(n - r - 2)\cdots(3)(2)(1)}{(n - r)!}$$

$$\tag{2}$$

Note that in equation (2), $(n - r)$ is 1 less than the factor that precedes it, $(n - r + 1)$. Therefore, the numerator in equation (2) is the product of all the positive integers from 1 to n. It is $n!$, and equation (2) can be written

$$_nP_r = \frac{n!}{(n - r)!} \tag{3}$$

EXAMPLE Use equation (3) to evaluate $_5P_3$.

Solution

$$_5P_3 = \frac{5!}{(5 - 3)!} = \frac{5!}{2!} = \frac{5 \cdot 4 \cdot 3 \cdot 2 \cdot 1}{2 \cdot 1} = 5 \cdot 4 \cdot 3 = 60$$

A number such as 7! can be written $7 \cdot 6 \cdot 5!$ or $7 \cdot 6 \cdot 5 \cdot 4!$, etc. In general, we can write $n!$ as $n(n - 1)!$ or $n(n - 1)(n - 2)(n - 3)!$, etc. This technique is useful in evaluating permutations.

EXAMPLE Evaluate $_8P_2$ and $_7P_3$.

Solution

$$_8P_2 = \frac{8!}{(8 - 2)!} = \frac{8!}{6!} = \frac{8 \cdot 7 \cdot 6!}{6!} = 8 \cdot 7 = 56$$

$$_7P_3 = \frac{7!}{(7 - 3)!} = \frac{7!}{4!} = \frac{7 \cdot 6 \cdot 5 \cdot 4!}{4!} = 7 \cdot 6 \cdot 5 = 210$$

Recall that 0! was defined to be 1. We made that definition so that equation (3) can also be used to find $_nP_n$.

EXAMPLE Find the permutations of four objects all used in each arrangement.

Solution Using equation (3),

$$_4P_4 = \frac{4!}{(4 - 4)!} = \frac{4!}{0!} = 4! = 24$$

Thus the definition of zero factorial is made so that the equations for $_nP_n$ and $_nP_r$ are consistent.

A tree diagram may be used to illustrate the permutations of n objects taken r at a time. Each permutation is read along a "branch" of the tree.

EXAMPLE Use a tree diagram to illustrate the permutations of

$$A \quad B \quad C \quad D$$

if each arrangement is to use three different letters, and verify the number of these arrangements by the formula for $_nP_r$.

Solution Figure 3.2.1 contains a listing of 24 permutations resulting from the diagram if each path or branch represents one permutation. To verify that there are actually 24 such arrangements,

$$_nP_r = {}_4P_3 = \frac{4!}{(4-3)!} = 4! = 4\cdot3\cdot2\cdot1 = 24$$

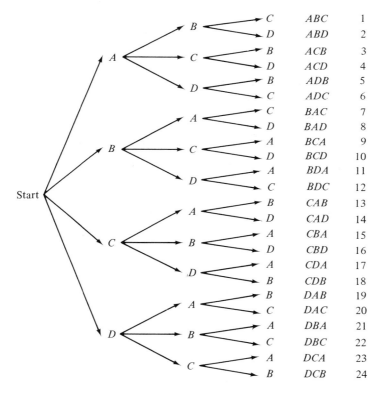

B	C	ABC	1
	D	ABD	2
A C	B	ACB	3
	D	ACD	4
D	B	ADB	5
	C	ADC	6
A	C	BAC	7
	D	BAD	8
B C	A	BCA	9
	D	BCD	10
D	A	BDA	11
	C	BDC	12
A	B	CAB	13
	D	CAD	14
C B	A	CBA	15
	D	CBD	16
D	A	CDA	17
	B	CDB	18
A	B	DAB	19
	C	DAC	20
D B	A	DBA	21
	C	DBC	22
C	A	DCA	23
	B	DCB	24

FIGURE 3.2.1

EXERCISES 3.2

Evaluate Exercises 1–9.

1. $_7P_5$ **2.** $_{12}P_2$ **3.** $_8P_7$

4. $_8P_3$ **5.** $_7P_2$ **6.** $_7P_7$

7. $_4P_4$ **8.** $_4P_5$ **9.** $_6P_{12}$

10. How many seven-digit telephone numbers are possible if:
  a. Any number can be used in any place?
  b. None of the first three digits can be 0?
  c. None of the first three digits can be 0 and the fourth digit cannot be 9?

11. California automobile license plates contain three numerals (from 000 to 999) followed by three letters. How many different plates can be produced?

12. A clothing store's fall line contains two different styles of a two-button jacket, three different styles of a three-button jacket, and four different styles of trousers. Each of the trousers or jackets is available in seven colors. How many distinct outfits can a customer select?

13. How many ways can seven people line up for a photograph?

14. How many ways can three people line up for a photograph if they are selected from a group of seven?

15. How many different five-letter arrangements are possible using the first eight letters of the alphabet with no letter used more than once?

16. Five people are to be seated at a circular table. How many different seating arrangements are possible if different arrangements mean only that different people are next to one another?

17. How many five-digit numbers can be formed from the digits 1, 2, 3, 4, and 5 if the digits can be used only once, and how many four-digit numbers can be so formed that will be less than 4,000?

18. In Exercise 17, how many of the four-digit numbers will be less than 2,500?

19. Prove that at least two people in Sacramento (population 250,000) have the same three initials.

20. Use a tree diagram to show all possible three-digit numbers that can be obtained from the first three counting numbers.

3.3 COMBINATIONS

A corporation executive needs to select a committee of three from a management group of five. He is generally not interested in the committee arrangement—around a table, for instance—but only in the different individuals that make up the committee. That is, he is not interested in forming permutations; he is considering a *combination*.

> **DEFINITION**
>
> A **combination** of r objects selected from a set of n objects is a subset of r objects from the set of n objects.

Note that the definition says nothing about arrangements. The numbers 123, 321, and 312 are three different permutations of three digits from the set of ten, but they are a *single* combination of three digits from the set of ten. Since a combination is a set, the arrangement of its elements is not considered.

EXAMPLE   List the different combinations of three letters from the set $\{A, B, C, D, E\}$.

Solution   By trial and error the combinations are:

$$\begin{array}{ll} \{A, B, C\} & \{A, D, E\} \\ \{A, B, D\} & \{B, C, D\} \\ \{A, B, E\} & \{B, C, E\} \\ \{A, C, D\} & \{B, D, E\} \\ \{A, C, E\} & \{C, D, E\} \end{array}$$

Each of the above subsets is different from all the others, and no different subset of three elements can be listed. Therefore, there are ten combinations of three letters from the set of five.

The above example implies that if an executive wished to form a committee of three from a set of five junior executives, ten different committees are possible.

The notation for the number of different combinations of r objects that can be selected from a set of n objects is

$$_nC_r$$

The symbol $_nC_r$ is sometimes read "the combination of n objects taken r at a time." In the above example it was shown that

$$_5C_3 = 10$$

A formula for $_nC_r$ is easily derived using permutations and the Fundamental Theorem of Counting. Suppose that we have a set of n objects. The number of *permutations* of n objects in arrangements containing r objects is $_nP_r$. Now $_nP_r$ can also be computed using *combinations*. The number of permutations, with r objects in each, can be viewed as a sequence of two events. First form all the *combinations* of r objects from a set of n, then form all of the *permutations* of the r objects in each combination. The number of combinations is $_nC_r$, and each combination contains r objects. For each combination of r objects there are $r!$ permutations possible, using all r of the objects. Therefore, by the Fundamental Theorem of Counting, the total number of *permutations* of r objects from a set of n is $_nC_r \cdot r!$, but it is also given by $_nP_r$. Thus

$$_nC_r \cdot r! = {_nP_r}$$

which can be written

$$_nC_r \cdot r! = \frac{n!}{(n-r)!}$$

Solving for $_nC_r$,

$$_nC_r = \frac{n!}{r!\,(n-r)!} \tag{1}$$

EXAMPLE Evaluate $_5C_3$.

Solution

$$_5C_3 = \frac{5!}{3!\,(5-3)!} = \frac{5!}{3! \cdot 2!} = \frac{5 \cdot 4 \cdot 3!}{3! \cdot 2} = 10$$

EXAMPLE Find the number of combinations that can be made from a set of 7 objects where each of the combinations contains 2 objects.

Solution This example is solved by finding the value of $_7C_2$. From equation (1),

$$_7C_2 = \frac{7!}{2!\,(7-2)!} = \frac{7!}{2! \cdot 5!} = \frac{7 \cdot 6 \cdot 5!}{2 \cdot 5!} = 21$$

There are 21 combinations of 7 objects taken 2 at a time.

Combinations have another standard notation:

$$\binom{n}{r}$$

That is,

$$_nC_r = \binom{n}{r} \quad \text{or} \quad \binom{n}{r} = \frac{n!}{r!\,(n-r)!} = {_nC_r}$$

Recall from an earlier example that

$$_5C_3 = 10$$

then

$$\binom{5}{3} = 10$$

EXAMPLE An automobile manufacturer offers 11 pieces of optional equipment that a customer may or may not order with model X. The average customer orders 6 optionals with model X. How many different model Xs can be manufactured with 6 pieces of optional equipment?

Solution The question is one of deciding how many different subsets containing 6 items can be created from a set of 11. That is, how many sets of 6 different optionals are there?

$$\binom{11}{6} = {}_{11}C_6 = \frac{11!}{6!\,(11-6)!} = \frac{11!}{6!\,5!} = \frac{11\cdot10\cdot9\cdot8\cdot7\cdot6!}{5\cdot4\cdot3\cdot2\cdot1\cdot6!} = 462$$

There are 462 different model Xs possible with 6 of the 11 optional pieces of equipment.

In the card game known as poker, a flush is five cards of the same suit. That is, five cards all diamonds or all hearts or all spades or all clubs would be a flush.

EXAMPLE How many heart flushes are possible from a 52-card poker deck?

Solution There are 13 different cards in the heart suit. Any 5 will form a heart flush. Therefore, we must find the number of 5-card combinations (subsets) possible from a set of 13 cards. We are interested in combinations, not permutations; the cards themselves are important, not their arrangements.

$$\binom{13}{5} = {}_{13}C_5 = \frac{13!}{5!\,(13-5)!} = \frac{13!}{5!\,8!} = \frac{13\cdot12\cdot11\cdot10\cdot9\cdot8!}{5\cdot4\cdot3\cdot2\cdot1\cdot8!} = 1{,}287$$

Evaluate Exercises 1–14.

1. $_5C_2$ **2.** $_7C_4$ **3.** $_7C_6$

4. $_8C_5$ **5.** $_9C_6$ **6.** $_4C_1$

7. $_5C_1$ **8.** $\binom{5}{0}$ **9.** $\binom{8}{0}$

10. $\binom{5}{5}$ **11.** $\binom{6}{6}$ **12.** $\binom{3}{7}$

13. $\binom{6}{2}$ and $\binom{6}{4}$ **14.** $\binom{7}{3}$ and $\binom{7}{4}$

15. Prove that $_nC_r = {}_nC_{n-r}$.

16. How many committees of 4 can be chosen from 100 senators?

17. How many triangles can be formed from five noncollinear points?

18. How many triangles can be formed from six noncollinear points?

19. Four outstanding car salesmen are to be selected from 15 to take part in an in-depth psychological study. How many ways may the 4 be selected?

20. A financier wishes to use the trust departments of 3 different banks. There are 20 banks in his city. How many ways may he choose trust departments?

21. An ecumenical committee is to be made up of 3 Protestants and 3 Catholics. The Protestants will be chosen from a group of 10, while the Catholics will come from a group of 8. In how many ways may the committee be formed?

22. The committee of Exercise 21 is to be enlarged by adding three persons of the Jewish faith to be chosen from a group of 6. Now how many such committees can be made?

23. How many flushes of all types can be created from a deck of 52 poker cards?

24. How many pairs of 3s can be created from a poker deck? How many pairs of 4s?

25. How many different sets of three 2s can be created from a poker deck? How many different sets of three 5s?

26. In poker, a full house is five cards, three of one numerical value and two of another, such as three 5s and two 6s or three kings and two 10s. How many full houses are possible from a poker deck?

27. If you have a penny, a nickel, a dime, a quarter, a half dollar, and a silver dollar, in how many ways can you spend two coins?

28. How many different sums of money can be selected from the set of coins in Exercise 27?

29. A lawn mower sells with six optional items such that customers may buy none or all or only some of these optionals. How many different lawn mowers can be sold?

30. Automobile model Y has five pieces of optional equipment. Any particular model Y can have no optional equipment or one piece of optional equipment or two pieces, up to and including five pieces of optional equipment. How many different model Ys are possible?

31. Relate the answers to Exercises 29 and 30 to 2^n.

32. Is a "combination lock" correctly named in view of the mathematical definition of combination? Discuss.

3.4 ORDERED PARTITIONS

Partitions and combinations are closely related. Suppose that we wish to divide (partition) a history class of 25 students into two work groups of 10 and 15 students each. How many ways can this be done? The problem is to decide the number of ways that a set of 25 objects can be partitioned into subsets of 10 and 15. Let the two subsets be called g_{10} and g_{15}. Starting with g_{10}, there are $\binom{25}{10}$ ways of assigning students to g_{10}, and once g_{10} is filled, the remaining 15 students are g_{15}. Therefore, there are $\binom{25}{10}$ ways that the class can be divided into the two groups. This conclusion uses combinations and the Fundamental Theorem of Counting. More complicated partitioning problems can be solved using the same ideas.

EXAMPLE How many ways can a history class of 25 students be divided into groups of 7, 8, and 10 to work on class projects?

Solution The example asks for the number of ways that a set of 25 elements can be partitioned into three subsets of 7, 8, and 10 elements. Let the subsets be called g_7, g_8, g_{10} and consider three events: first, assigning students to g_7; second, assigning students to g_8; and third, selecting students for g_{10}. Since there are 25 students, they can be assigned to g_7 in $\binom{25}{7}$ ways. After students have been selected for g_7, 18 remain, from which 8 will be selected for g_8. Therefore, there are $\binom{18}{8}$ ways to select members of g_8. After g_8 is chosen, the remaining 10 students are g_{10}. That is, members for g_{10} can be selected in just one way after the other groups have been filled. Therefore, by the Fundamental Theorem of Counting, the number of ways that the class can be partitioned into the three subsets is:

$$\binom{25}{7} \cdot \binom{18}{8} \cdot 1 = \frac{25!}{7!\,18!} \cdot \frac{18!}{8!\,10!}$$

$$= \frac{25!}{7!\,8!\,10!} \tag{1}$$

$$= 21{,}034{,}470{,}600$$

The above is an example of an ordered partition.

DEFINITION

The distribution of n different elements into r disjoint sets is called an **ordered partition** of the n elements.

The above example gives the number of ordered partitions of 25 elements into three sets. The number of ordered partitions for any set of elements can always be found using the Fundamental Theorem of Counting and the formula for combinations. However, it will be convenient to have a formula for the number of ordered partitions.

THEOREM

The number of ordered partitions of n elements into r cells, where the number of elements in each cell is $n_1, n_2, n_3, \ldots, n_r$, is denoted by the symbol $\begin{pmatrix} n \\ n_1, n_2, \ldots, n_r \end{pmatrix}$, and

$$\begin{pmatrix} n \\ n_1, n_2, \ldots, n_r \end{pmatrix} = \frac{n!}{n_1!\, n_2!\, n_3! \ldots n_r!} \tag{2}$$

Formula (2) is certainly true for n elements distributed into *two* cells with n_1 elements and n_2 elements in them. The elements can be assigned to the first cell in $\begin{pmatrix} n \\ n_1 \end{pmatrix}$ ways. Once this is done the remaining elements automatically become the members of the second cell. That is,

$$\begin{pmatrix} n \\ n_1, n_2 \end{pmatrix} = \begin{pmatrix} n \\ n_1 \end{pmatrix}$$

$$= \frac{n!}{n_1!\,(n - n_1)!} \tag{3}$$

but $(n - n_1) = n_2$. Making this substitution in (3), we have

$$\begin{pmatrix} n \\ n_1, n_2 \end{pmatrix} = \frac{n!}{n_1!\, n_2!}$$

We will establish formula (2) for the number of ordered partitions of n objects into three cells whose members number n_1, n_2, and n_3 objects, respectively. There are $\begin{pmatrix} n \\ n_1 \end{pmatrix}$ ways of selecting elements for the first cell. After the first cell is filled, there are $\begin{pmatrix} n - n_1 \\ n_2 \end{pmatrix}$ ways of selecting members for the second cell, and after it is filled, the remaining objects are assigned to the third cell. Thus there is only one way objects may be assigned to the third cell. Therefore,

$$\begin{pmatrix} n \\ n_1, n_2, n_3 \end{pmatrix} = \begin{pmatrix} n \\ n_1 \end{pmatrix} \cdot \begin{pmatrix} n - n_1 \\ n_2 \end{pmatrix} \cdot 1$$

$$= \frac{n!}{n_1!\,(n - n_1)!} \frac{(n - n_1)!}{n_2!\,(n - n_1 - n_2)!} \tag{4}$$

Since $n = n_1 + n_2 + n_3$, $n_3 = n - n_1 - n_2$. Substituting n_3 for $n - n_1 - n_2$ in (4) gives:

$$\binom{n}{n_1, n_2, n_3} = \frac{n!}{n_1!\, n_2!\, n_3!}$$

This proves the theorem for three cells. By mathematical induction it can be proved for any finite number of cells.

EXAMPLE How many ordered partitions of 25 objects into three cells are possible if the cells are to have 7, 8, and 10 objects in them?

Solution This example asks for the value of $\binom{25}{7,\ 8,\ 10}$, and by formula (2),

$$\binom{25}{7, 8, 10} = \frac{25!}{7!\, 8!\, 10!}$$

Note that this agrees with the solution to the first example of this section.

EXAMPLE In how many ways may six executives be assigned to three offices, two to an office?

Solution The problem is to assign six objects (executives) to three cells (offices), two to a cell. This can be done in $\binom{6}{2, 2, 2}$ ways:

$$\binom{6}{2, 2, 2} = \frac{6!}{2!\, 2!\, 2!} = 90$$

There are 90 different office arrangements of the type described.

EXERCISES 3.4

Evaluate Exercises 1–11.

1. $\binom{7}{4,\ 3}$ **2.** $\binom{6}{3,\ 1,\ 2}$ **3.** $\binom{8}{4,\ 2,\ 2}$

4. $\binom{5}{1,\ 2,\ 2}$ **5.** $\binom{4}{4,\ 0}$ **6.** $\binom{8}{8,\ 0,\ 0}$

7. $\binom{200}{199,\ 1}$ **8.** $\binom{507}{505,\ 2,\ 0}$ **9.** $\binom{3}{1,\ 1,\ 1}$

10. $\begin{pmatrix} 7 \\ 4, 3, 2 \end{pmatrix}$ **11.** $\begin{pmatrix} 8 \\ 5, 5, 6 \end{pmatrix}$

12. Prove that if $n = a + b + c + d$, then $\begin{pmatrix} n \\ a, b, c, d \end{pmatrix} = \dfrac{n!}{a!\,b!\,c!\,d!}$

13. What notation of the form $\begin{pmatrix} n \\ n_1, n_2, n_3, \ldots, n_r \end{pmatrix}$ should be used for

$$_5C_4 \cdot {}_4C_3 \cdot {}_3C_2 \cdot {}_2C_1$$

Solve Exercises 14–19 and leave answers in $\dfrac{n!}{n_1!\,n_2!\cdots n_r!}$ *form; you need not find the numerical value.*

14. A union leader has 15 men that he must assign to 3 companies. How many ways can he send 3 to a stamping mill, 5 to a finishing shop, and the rest to an assembly plant?

15. A company has 10 different pieces of equipment that must be moved to the West Coast immediately. The local trucking firm can ship 2 pieces. There is room for 4 on the train, and the rest will have to be sent by air. In how many ways can the equipment be shipped?

16. In how many ways can nine psychologists grouped in groups of three work together?

17. Answer Exercise 16 if two psychologists refuse to work together.

18. Answer Exercise 16 if three psychologists will not work together.

19. At the end of the semester each student in a class of eight is to be assigned a grade of A, B, C, D, or NC. How many ways can this be done if there must be one A, two Bs, two Cs, two Ds, and one NC?

3.5 THE BINOMIAL THEOREM

The binomial theorem offers a technique for expanding $(x + y)^n$, where n is a natural number. Numbers such as $\begin{pmatrix} n \\ r \end{pmatrix}$ occur in the binomial theorem, but this is not the only reason it is presented here. The binomial theorem has important applications in probability theory, the topic of the next chapter.

In order to expand $(a + b)^3$, we can write $(a + b)^3 = (a + b)(a + b)(a + b)$. Now the product $(a + b)(a + b)(a + b)$ could be found by multiplying two factors $(a + b)(a + b)$ and then multiplying that product by $(a + b)$. We choose to accomplish the same result in a different way. Let the number 1 represent the first factor of $(a + b)$ in $(a + b)(a + b)(a + b)$,

2 represent the second factor of $(a + b)$ in $(a + b)(a + b)(a + b)$, and 3 represent the third factor. We can form a product of a's and b's by choosing an a or a b for the first factor of $(a + b)(a + b)(a + b)$, an a or a b for the second factor, and an a or a b for the third. To show exactly which choice is being considered, we form a two-cell partition of $\{1, 2, 3\}$. The left cell of the partition will indicate the factors of $(a + b)(a + b)(a + b)$ from which an a was chosen, and the right-hand cell will indicate the factors of $(a + b)(a + b)$ $(a + b)$ from which a b was chosen. For instance, the partition $(\{1, 2\}, \{3\})$ means that we are considering $a \cdot a \cdot b = a^2b$ where the two a's are from factors 1 and 2 and the b is from factor 3. The partition $(\{1, 3\}, \{2\})$ indicates a product of $a \cdot a \cdot b = a^2b$ where the a's are from factors 1 and 3 and the b is chosen from factor 2. The multiplication of $(a + b)(a + b)(a + b)$ can be performed by adding all the products of a's and b's where an a or a b is chosen from factor 1, then an a or a b is chosen from factor 2, then an a or a b is chosen from factor 3. As mentioned above, the partition $(\{1, 2\}, \{3\})$ indicates a product of a^2b. The question is, in how many ways can a^2b be created? In other words, how many terms of a^2b are added when $(a + b)$ $(a + b)(a + b)$ is multiplied? The answer is that a^2b can be created in $\binom{3}{2,\,1}$ ways since it is the result of having two elements in the left-hand cell of a partition and one element in the right-hand cell. It is an ordered partition of three elements into sets containing two elements and one element. Therefore, the number of times that a^2b will occur when $(a + b)(a + b)(a + b)$ is multiplied is

$$\binom{3}{2,\,1} = \frac{3!}{2! \cdot 1!} = 3$$

Similar arguments can be made for other terms to show that

$$(a + b)^3 = (a + b)(a + b)(a + b) = \binom{3}{3,\,0}a^3 + \binom{3}{2,\,1}a^2b$$

$$+ \binom{3}{1,\,2}ab^2 + \binom{3}{0,\,3}b^3 \qquad (1)$$

$$= a^3 + 3a^2b + 3ab^2 + b^3$$

This agrees with the usual longhand expansion.

Equation (1) is more often written

$$(a + b)^3 = \binom{3}{3}a^3 + \binom{3}{2}a^2b + \binom{3}{1}ab^2 + \binom{3}{0}b^3 \qquad (2)$$

The combination notation for the coefficients in (2) is possible rather than the ordered partition notation in (1) because when just *two* partitions of n objects are being considered, with r objects in the first cell, the number of ordered partitions is $\binom{n}{r}$. Note in (2) that the coefficients $\binom{3}{3}$ and $\binom{3}{0}$ are both equal to 1. This fact leads to the usual method of writing (1) as

$$(a + b)^3 = a^3 + \binom{3}{2}a^2b + \binom{3}{1}ab^2 + b^3$$

Analogous arguments can be given for $(a + b)^n$ which will prove the binomial theorem.

THE BINOMIAL THEOREM

If a and b are real numbers and n is a nonnegative integer, then

$$(a + b)^n = a^n + \binom{n}{n-1}a^{n-1}b + \binom{n}{n-2}a^{n-2}b^2 + \cdots +$$

$$\binom{n}{k}a^kb^{n-k} + \cdots + b^n \quad (3)$$

EXAMPLE   Use the binomial theorem to expand $(x + y)^4$.

Solution

$$(x + y)^4 = x^4 + \binom{4}{3}x^3y + \binom{4}{2}x^2y^2 + \binom{4}{1}xy^3 + y^4$$

$$= x^4 + \frac{4!}{3!\,1!}x^3y + \frac{4!}{2!\,2!}x^2y^2 + \frac{4!}{1!\,3!}xy^3 + y^4$$

$$= x^4 + 4x^3y + 6x^2y^2 + 4xy^3 + y^4$$

The difference of two terms can also be raised to positive integer powers by the binomial theorem.

EXAMPLE   Multiply $(x - 2y)^3$.

Solution   The binomial $(x - 2y)$ can be written $(x + (-2y))$. In the latter form the binomial theorem can be applied:

$$(x + (-2y))^3 = x^3 + \binom{3}{2}x^2(-2y) + \binom{3}{1}x(-2y)^2 + (-2y)^3$$

$$= x^3 - 6x^2y + 12xy^2 - 8y^3$$

Whenever the binomial theorem is applied to the difference of two terms, as in the last example, the signs of the terms in the expansion will alternate, with the second term always being negative.

In (3) notice that for any coefficient $\binom{n}{k}$ the upper number, n, is the power of the binomial, and the lower number, k, is the power of a in the term with coefficient $\binom{n}{k}$. If the terms of (3) are numbered from the left, then the first term is a^n. In the second term b is raised to the first power. In the third term b is raised to the second power. The fourth term would contain b raised to the third power. In general, the exponent of b is 1 less than the term's position in the sequence of terms of the sum. Also notice that in any term the exponents of a and b have a sum of n.

EXAMPLE What is the third term in the expansion of $(3x + 2y)^5$?

Solution The third term will contain a factor of $(2y)^2$ since the exponent must be 1 less than the position of the term in the expansion. Since the sum of the exponents must equal 5, the third term contains a factor of $(3x)^3$. Therefore, the coefficient will be $\binom{5}{3}$, and the third term of the expansion $(3x + 2y)^5$ is

$$\binom{5}{3}(3x)^3 \cdot (2y)^2 = \frac{5!}{3!\,2!} \cdot 27x^3 \cdot 4y^2$$
$$= 10 \cdot 108x^3y^2$$
$$= 1,080x^3y^2$$

EXAMPLE Evaluate the fifth term of the expansion of $(x + y)^7$ when $x = \frac{1}{2}$ and $y = \frac{1}{2}$.

Solution The fifth term will contain a factor of y^4 since the exponent must be 1 less than 5. The sum of the exponents of x and y is 7, so the fifth term will contain a factor of x^3. The coefficient will be of the form $\binom{7}{k}$. where k equals the exponent of x. Therefore, the coefficient is $\binom{7}{3}$, and the fifth term of the expansion of $(x + y)^7$ is

$$\binom{7}{3}x^3y^4 = \frac{7!}{3!\cdot 4!}\, x^3y^4$$
$$= \frac{7 \cdot 6 \cdot 5 \cdot 4!}{3 \cdot 2 \cdot 4!}\, x^3y^4$$
$$= 35x^3y^4$$

If $x = y = \frac{1}{2}$, then

$$35x^3y^4 = 35(\tfrac{1}{2})^3(\tfrac{1}{2})^4$$
$$= 35(\tfrac{1}{2})^7$$
$$= \tfrac{35}{128}$$

The seventeenth century mathematician Blaise Pascal developed a method that yields the first few binomial coefficients without tedious arithmetic. It is called Pascal's triangle.

$(a + b)^0$ 1

$(a + b)^1$ 1   1

$(a + b)^2$ 1   2   1

FIGURE 3.5.1 Pascal's triangle.

Figure 3.5.1 shows the coefficients for $(a + b)^0$, $(a + b)^1$, $(a + b)^2$. The arrows indicate that 2 is computed by adding the 1s. The triangle can be extended to show the coefficients of $(a + b)^n$ for any n. The numbers on the extreme right and left of the triangle will always be 1s. The interior numbers are found by taking the sum of the two numbers above—one slightly to the left and one slightly to the right. Figure 3.5.2 extends the triangle of Figure 3.5.1.

$(a + b)^0$ 1

$(a + b)^1$ 1   1

$(a + b)^2$ 1   2   1

$(a + b)^3$ 1   3   3   1

FIGURE 3.5.2 Pascal's triangle.

The 3s are created by adding $1 + 2$ and then $2 + 1$, as shown by the arrows.

$(a + b)^0$ 1

$(a + b)^1$ 1   1

$(a + b)^2$ 1   2   1

$(a + b)^3$ 1   3   3   1

$(a + b)^4$ 1   4   6   4   1

FIGURE 3.5.3 Pascal's triangle.

Figure 3.5.3 extends the triangle of Figure 3.5.2 to give the coefficients of $(a + b)^4$. The arrows show that the first 4 is the sum of 1 and 3 from the line above. The 6 is the sum of 3 and 3, and the last 4 is the sum of 3 and 1. The pattern of coefficients in (3) shows that in subsequent terms a has decreasing exponents, while b's exponents increase. Noting this pattern and the co-efficients for $(a + b)^4$ in Figure 3.5.3, it is easy to expand $(a + b)^4$:

$$(a + b)^4 = a^4 + 4a^3b + 6a^2b^2 + 4ab^3 + b^4$$

Using this pattern, Pascal's triangle can be extended for any desired value of n. The reader may also verify that each entry in Pascal's triangle corresponds to $\binom{n}{k}$ of the corresponding binomial expansion.

EXERCISES 3.5

1. Extend Pascal's triangle to include coefficients for $(a + b)^8$.

Expand Exercises 2–13 using the binomial theorem.

2. $(x + 1)^3$ **3.** $(x + 4)^4$ **4.** $(x - 1)^5$
5. $(2x + 1)^5$ **6.** $(x - 2)^5$ **7.** $(2x + 3y)^4$
8. $(x - y)^6$ **9.** $(1 - y)^6$ **10.** $(4x + 3y)^4$
11. $(10 + 1)^5$ **12.** $(10 - 1)^5$ **13.** $(100 - 1)^4$
14. Find the third term in the expansion of $(x + y)^5$.
15. Find the sixth term in the expansion of $(x - y)^8$.
16. Find the eighth term in the expansion of $(x - y)^9$.
17. What is the coefficient of the term in the expansion of $(x + y)^9$ that contains factor x^4y^5?
18. Answer Exercise 17 for the term with factor xy^8.
19. Answer Exercise 17 for the term with factor x^8y.
20. In the expansion of $(2x + 3y)^8$, find the coefficient of the term with factor x^3y^5.
21. Answer Exercise 20 for the term with factor x^4y^4.
22. Show in two different ways that $(x + y)^4 = 1$ when $x = y = \frac{1}{2}$.
23. Repeat Exercise 22 for $x = \frac{1}{4}$, $y = \frac{3}{4}$.

3.6 THE MULTINOMIAL THEOREM

The binomial theorem of the last section is easily extended to become the multinomial theorem. A **multinomial** is a first-degree polynomial of two or more terms, and the **multinomial theorem** is a method for raising a multinomial to any positive integral power. To show how the theorem is established we consider

$$(a + b + c)^3$$

Of course $(a + b + c)^3$ could be expanded by multiplying $(a + b + c)$ $(a + b + c)(a + b + c)$. However, as with the binomial theorem, we choose a different approach. Assign the three factors $(a + b + c)(a + b + c)$ $(a + b + c)$ the names $\triangle$, $\square$, and $*$, respectively. The expansion of $(a + b + c)^3$ will contain terms whose factors are powers of a, b, and c. An individual term of the expansion will be chosen by selecting an a, a b, or a c from the first factor of $(a + b + c)(a + b + c)(a + b + c)$, then selecting an a, a b, or a c from the second factor of $(a + b + c)(a + b + c)(a + b + c)$, and finally by selecting an a, a b, or a c from the third factor. Exactly which letters are chosen will be indicated by forming a three-cell partition of $\{\triangle, \square, *\}$. The first cell will show the factors of $(a + b + c)(a + b + c)$ $(a + b + c)$ from which an a is selected, the second cell will name those factors from which a b was selected, and the last cell will indicate those factors from which a c was selected.

For instance, the partition $(\{\triangle, \square\}, \{\ \}, \{*\})$ indicates a^2c since a's are being selected from two factors: $\triangle$ and $\square$. No b's are selected since the second cell is empty, and a single c is used because the third cell contains only $*$. The number of terms in the expansion of $(a + b + c)^3$ with literal factor a^2c is simply the number of three-celled partitions of $\{\triangle, \square, *\}$ that have two elements in the first cell, an empty second cell, and a single element in the third cell. But that number is $\begin{pmatrix} 3 \\ 2, 0, 1 \end{pmatrix} = 3$. Thus the expansion of $(a + b + c)^3$ will contain $3a^2c$. The relationship between numerical coefficient and literal factor is better seen by writing $3a^2c$ as

$$\begin{pmatrix} 3 \\ 2, 0, 1 \end{pmatrix} a^2 b^0 c^1$$

The numbers 2, 0, 1 from $\begin{pmatrix} 3 \\ 2, 0, 1 \end{pmatrix}$ are the exponents of a, b, and c because they result from a partition that indicates the multiplication of two a's, zero b's, and one c. Another term of the expansion of $(a + b + c)^3$ is

$$\begin{pmatrix} 3 \\ 0, 2, 1 \end{pmatrix} a^0 b^2 c^1 = 3b^2c$$

This term can be established in the same manner as the last. The entire expansion for $(a + b + c)^3$ is the sum of all terms of the form

$$\binom{3}{x,\,y,\,z} a^x b^y c^z$$

where $x + y + z = 3$.

THE MULTINOMIAL THEOREM

If $x_1, x_2, \ldots, x_k$ are real numbers and $n, n_1, n_2, \ldots, n_k$ are nonnegative integers, then $(x_1 + x_2 + x_3 + \cdots + x_k)^n$ is expanded by adding all of the terms of the form

$$\binom{n}{n_1,\, n_2,\, \ldots,\, n_k} x_1^{n_1} x_2^{n_2} x_3^{n_3} \cdots x_k^{n_k}$$

where $n_1 + n_2 + n_3 + \cdots + n_k = n$.

EXAMPLE   Expand $(a + b + c)^2$.

Solution

$$(a + b + c)^2 = \binom{2}{2,\,0,\,0} a^2 b^0 c^0 + \binom{2}{0,\,2,\,0} a^0 b^2 c^0 + \binom{2}{0,\,0,\,2} a^0 b^0 c^2$$

$$+ \binom{2}{1,\,1,\,0} a^1 b^1 c^0 + \binom{2}{1,\,0,\,1} a^1 b^0 c^1 + \binom{2}{0,\,1,\,1} a^0 b^1 c^1$$

$$= a^2 + b^2 + c^2 + 2ab + 2ac + 2bc$$

EXAMPLE   Find the coefficient of the term in the expansion of $(a + b + c + d)^7$ that contains the literal factors $a^2 b^2 d^3$.

Solution   The coefficient will result from evaluating $\binom{n}{r,\,s,\,t,\,u}$, where $n = 7$ and r, s, t, and u are the exponents of a, b, c, and d, respectively. The expression $a^2 b^2 d^3$ contains no c. However, it can be written $a^2 b^2 c^0 d^3$, which leads to

$$\binom{7}{2,\,2,\,0,\,3} = \frac{7!}{2!\,2!\,0!\,3!} = \frac{7 \cdot 6 \cdot 5 \cdot 4 \cdot 3!}{2 \cdot 2 \cdot 1 \cdot 3!} = 210$$

EXAMPLE   Find the coefficient of the term with literal factors $w^5 x^3 z$ in the expansion of $(v + w - 2x + 3y - z)^9$.

Solution   The answer is *not* simply $\begin{pmatrix} 9 \\ 0, 5, 3, 0, 1 \end{pmatrix}$ since powers of -2
and 3 and -1, from $-2x$ and $3y$ and $-z$ from $(v + w - 2x + 3y - z)$, can
be factors of the coefficient. The coefficient can be found only by examining
the entire term:

$$\begin{pmatrix} 9 \\ 0, 5, 3, 0, 1 \end{pmatrix} v^0 w^5 (-2x)^3 (3y)^0 (-z)^1 = \frac{9!}{0!\, 5!\, 3!\, 0!\, 1!}\, w^5(-8)x^3(-1)z$$

$$= \frac{9 \cdot 8 \cdot 7 \cdot 6 \cdot 5!}{3 \cdot 2 \cdot 5!}\, 8w^5 x^3 z$$

$$= 4,032 w^5 x^3 z$$

The coefficient is 4,032.

EXERCISES 3.6

Expand Exercises 1–9 using the multinomial theorem.

1. $(a + b + c)^3$ **2.** $(a + b + c + d)^2$
3. $(a + b + c)^4$ **4.** $(x - 2y + z)^3$
5. $(2x - 3y + 2z)^2$ **6.** $(5x + y - 3z)^2$
7. $(2w - x + 2y - z)^2$ **8.** $(a - b)^5$
9. $(a + 2b)^6$

*In Exercises 10–16, find the coefficient of the term with given literal factors that
would occur should the given multinomial be expanded.*

10. xyz^3; $(x + y + z)^5$ **11.** $x^2 y^3 z^5$; $(x + y + z)^{10}$
12. $x^2 y^3 z^5$; $(w + x + y + z)^{10}$ **13.** $a^3 b^2$; $(2a - b - c)^5$
14. $abcd^2$; $(2a + 2b - 3c - 3d)^4$ **15.** x^5; $(x + y + z)^4$
16. $x^3 y^2 z^5$; $(w + x + y + z)^9$

How many terms are in the expansion of each of Exercises 17 and 18?

17. $(x + y + z)^5$ **18.** $(3a - 4b + 7d - 8c)^{10}$

REVIEW EXERCISES

*In Exercises 1 and 2, find **all** partitions of the given set such that no cell is empty.*

1. $\{*, \triangle, \square\}$ **2.** $\{1, 2, 3, 4\}$

3. Find the cross partition of $(\{1, 2, 3\}, \{4, 5\})$ with $(\{1\}, \{2\}, \{3, 4, 5\})$.

Evaluate Exercises 4–15.

4. $_5P_3$ **5.** $_7P_2$ **6.** $_7P_0$

7. $_4C_2$ **8.** $_8C_3$ **9.** $_3C_8$

10. $\begin{pmatrix} 4 \\ 4 \end{pmatrix}$ **11.** $\begin{pmatrix} 4 \\ 6 \end{pmatrix}$ **12.** $\begin{pmatrix} 3 \\ 1, 0, 2 \end{pmatrix}$

13. $\begin{pmatrix} 5 \\ 2, 2, 1 \end{pmatrix}$ **14.** $\begin{pmatrix} 8 \\ 3, 3, 1, 1 \end{pmatrix}$ **15.** $\begin{pmatrix} 11 \\ 4, 2, 3, 1, 1 \end{pmatrix}$

Expand Exercises 16–20.

16. $(x - y)^5$ **17.** $(a + b)^8$ **18.** $(a + b + c)^2$
19. $(a + b + c + d)^2$ **20.** $(2a - b + 3c)^3$

21. What is the coefficient of $x^2y^3z^4$ in the expansion of $(x + y + z)^9$?
22. What is the coefficient of a^2bc in the expansion of $(2a + 3b + 4c)^4$?
23. In a survey of licensed pilots the following information was received:

    29 were licensed for single-engine propeller planes
    41 were licensed for double-engine propeller planes
    25 were licensed for jet aircraft
    15 were licensed for single-engine propeller and double-engine
        propeller planes
    17 were licensed for single-engine propeller and jet aircraft
    12 were licensed for double-engine propeller planes and jets
    10 were licensed for all three types of aircraft
    23 had let their licenses expire

How many pilots were surveyed?
24. In how many different ways can seven people line up for a photograph?
25. A boat manufacturer offers three different hulls in six different colors with a choice of five different engines. None of the hulls may be multi-colored. How many different boats can be produced?
26. In poker how many diamond flushes are possible from a deck of 52 cards?
27. How many different committees of 4 can be selected from a group of 10 people?
28. A desk calculator can be purchased with as many as four pieces of optional equipment. It is possible to order the calculator with some or all

or none of the optional equipment. How many different desk calculators can be purchased?

29. How many ways can 10 students be assigned to 3 rooms if 4 are to be placed in the first room, 2 in the second room, and 4 in the third?

30. Find $(1 + 1 + 2)^3$ in two different ways.

CHAPTER **4**

PROBABILITY

Probability is associated with gambling. Each week the Las Vegas odds are published for various sporting events. In fact, modern probability theory grew from a question about dice that the seventeenth century French gambler the Chevalier de Mere asked of Blaise Pascal. Pascal consulted with Pierre de Fermat, and the letters the two exchanged contained the basic material of today's probability theory.

Today people realize that this theory has many more important applications. Probability is used in insurance, medicine, biology, and business, to name only a few areas. For instance, weather forecasting is based on the probability that certain weather types will occur. Lumber companies hire a weather forecaster to decide when to remove heavy equipment from sites in high elevations in winter. With any forecast they are taking a chance, an expensive gamble.

We all operate within certain probabilities. It is probable that we will safely arrive at work or school each morning, but it is not certain. It is probable that a college education will help a person lead a more successful life, no matter how one defines successful, but it is not certain to do so.

This chapter will show how quantitative measures of probability can be assigned to certain events. These quantitative measures will allow decisions to be made with some degree of confidence. For example, suppose that a certain disease, if introduced into a population, will kill 90 percent of the population, and there is no known treatment for it. A test vaccine is given to twenty individuals who were exposed to the disease, and ten survive. Was the vaccine effective? Could it be that the ten survivors were individuals who would have survived without the vaccine? Questions such as these demand more than simple guesses. A doctor or public health official needs to know the *degree* of confidence he should have in the vaccine—or if he should have any confidence at all. He needs a meaningful quantitative measure of the probability that the vaccine is effective. This chapter will show how these measures can be assigned.

4.1 FINITE SAMPLE SPACES

A formal definition of probability will be given in the next section. That definition will relate a specific result to the number of possible results for the experiment under discussion. In this context an *experiment* will be any action whose consequences can be counted. For instance, the flipping of a coin can and will be called an experiment, since it must result in one of two possible outcomes—either heads or tails.

DEFINITION

A set of nonoverlapping outcomes for an experiment is called a **sample space** for the experiment.

EXAMPLE Give a sample space for the flipping of a coin.

Solution Let the outcome *heads* be denoted by H and the outcome *tails* be denoted T. Then the sample space (assuming a coin cannot land on edge), is

$$\{H, T\}$$

Sample spaces can, of course, be larger than the one in the above example. However, probability theory dealing with infinite sample spaces is quite complicated and is one of the considerations in a formal statistics course. We will restrict our discussion to finite sample spaces.

For any experiment, a sample space is not unique.

EXAMPLE Give two sample spaces for the experiment "rolling a die."

Solution One sample space can be created by listing the number of spots on the faces of the die:

$$S_1 = \{1, 2, 3, 4, 5, 6\}$$

This is a sample space since the numbers 1 through 6 are possible outcomes, and any specific outcome must be one and only one of these. A second sample space is created by noting that the number of spots on any toss of the die is less than or equal to 5 or it is 6. This sample space is

$$S_2 = \{6, \text{not } 6\}$$

Sample spaces S_1 and S_2 have an important difference: the events of S_2 are not "equally probable." Many other sample spaces can be created.

DEFINITIONS

An element of a sample space is called a **sample point.**
An **event** is a subset of a sample space.

EXAMPLE In the sample space $\{1, 2, 3, 4, 5, 6\}$, from rolling a die, how may sample points belong to the event "The number of spots is even"?

Solution The event "The number of spots is even" is the subset $\{2, 4, 6\}$. It contains three sample points.

Tree diagrams can be a useful device for listing the sample spaces for more complicated experiments.

EXAMPLE Give the sample space for the experiment "flipping two coins."

Solution Figure 4.1.1 is the beginning of a tree diagram. It shows a starting point and branches to T and H, the possible outcomes from flipping the first coin.

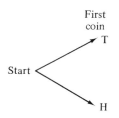

FIGURE 4.1.1 A tree diagram for the tossing of a single coin.

Figure 4.1.2 completes the tree diagram begun in Figure 4.1.1 by showing the possible outcomes from the flip of the second coin; if the first was T, the second could be either T or H; if the first was H, the second could still be either T or H. The sample space is created by reading from *start* to the ends of branches (1), (2), (3), and (4):

$$S = \{(T, T), (T, H), (H, T), (H, H)\}$$

Ordered *n*-tuple notation is standard with sample spaces, but it can be dropped if no ambiguity is likely. The above sample space is often given as

$$S = \{TT, TH, HT, HH\}$$

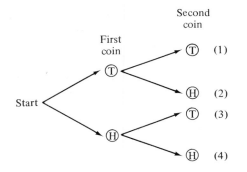

FIGURE 4.1.2   Tree diagram for flipping two coins.

EXAMPLE   Give the sample space for flipping three coins.

Solution   A tree diagram for three coins is easily created by adding two branches to branches (1), (2), (3), and (4) of Figure 4.1.2. The result is Figure 4.1.3. Reading from *start* to the ends of each branch yields the sample space *S*:

$$S = \{TTT, TTH, THT, THH, HTT, HTH, HHT, HHH\}$$

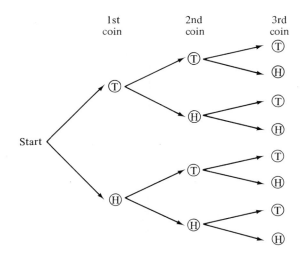

FIGURE 4.1.3   Tree diagram for flipping three coins.

The sample space of the last example above contains eight sample points. The number of points, eight, could have been predicted using the Fundamental Theorem of Counting. There are two possible outcomes for the first flip, two for the second, and two for the third. Then successive flips could have been performed in $2 \cdot 2 \cdot 2 = 8$ ways.

EXERCISES 4.1

1. Make a sample space for an experiment that consists of tossing a coin and then rolling a die.
2. Construct a sample space for tossing four coins.
3. In an urn are four red marbles, five green marbles, and eight blue marbles. Give a sample space for the drawing of a single marble from the urn.
4. Repeat Exercise 3 when two marbles are drawn from the urn.
5. What is the answer to Exercise 3 if three marbles are drawn from the urn? If 20 marbles are drawn and none are replaced?
6. A couple plan to have three children. Give the sample space for this family in terms of the sex and order of birth of the children.
7. What would the sample space in Exercise 6 be if the couple decided to have four children?
8. Use the sum of the spots to make a sample space for rolling a pair of dice.
9. Use ordered pair notation to give a sample space of the rolling of two dice. Include every ordered pair of the form (5, 2). The 5 means that five spots appeared on the first die, and the 2 indicates that two spots turned up on the second die.
10. In the sample space of Exercise 9, how many sample points have ordered pairs that add up to 7? How many add up to 8, to 9, to 11, and to 12?
11. Based on the sample space in Exercise 9, do you feel that a sum of 3 is as likely to be rolled as a sum of 7?
12. Referring to Exercise 9, does a sum of 2 or 3 or 12 spots seem as likely to be rolled as a sum of 7 or 11 spots?

4.2   SIMPLE PROBABILITY

The definition of probability that will shortly be given uses the words *equally likely*. *Equally likely* will remain an undefined, or primitive, term. Intuitively it means that if the outcomes of an experiment are equally likely, then any one outcome can happen as readily as any other. For instance, if a die is made from a perfect cube, then it seems reasonable to assume that any of its faces can come up as readily as any other: 1, 2, 3, 4, 5, or 6 are equally likely. On the other hand, from the exercises of the last section, rolling a sum of 7 with a pair of dice and rolling a sum of 3 with a pair of dice are not equally likely outcomes.

DEFINITION

If an experiment has a finite sample space S with n different equally likely sample points $\{S_1, S_2, S_3, \ldots, S_n\}$ and event E is a subset of S containing points $\{e_1, e_2, e_3, \ldots, e_k\}$, then the **probability** of E, denoted $P(E)$, is

$$P(E) = \frac{n(E)}{n(S)} = \frac{k}{n}$$

The definition states that the probability of E is equal to the number of elements in E, k, divided by the number of elements in the entire sample space, n.

EXAMPLE   Find the probability of tossing a head with a fair coin.

Solution   A fair coin is one such that heads and tails are equally likely. The sample space for the experiment, $\{T, H\}$, contains two elements. The event "tossing a head" has one member, H. Therefore, the probability of tossing a head, $P(H)$, is the number of elements in the event H, which is 1, divided by the number of elements in the sample space, which is 2. Thus

$$P(H) = \tfrac{1}{2}$$

EXAMPLE   A trucking firm has seven employees in its warehouse, three of whom are women. The firm is opening a new plant across town. It must transfer one of its present warehouse employees to the new plant. Assuming that the transfer of any of the seven is equally likely, what is the probability that the transferred person will be male?

Solution   The event "a male is chosen" has four elements. The entire sample space has seven elements. Therefore, the probability that a male is transferred, $P(M)$, is

$$P(M) = \tfrac{4}{7}$$

The probability of any event E is such that $0 \leq P(E) \leq 1$. This is evident from the fact that $P(E) = k/n$, where k is the number of elements in E and n is the number of elements in the sample space. k cannot exceed n

since E is a subset of the sample space. However, k can equal 0, in which case $P(E)$ would equal 0, and if $k = n$, then $P(E) = 1$.

EXAMPLE   What is the probability of rolling a 7 with a standard die?

Solution   Assuming a sample space of equally likely sample points, the event "rolling a 7" has no members, while the sample space for the rolling of a die has six members. Therefore, the probability of rolling a 7 is

$$P(7) = \tfrac{0}{6} = 0$$

A probability of 0 implies that an event cannot happen, as in the above example. A probability of 1 declares that an event is certain to happen whenever the experiment is performed.

EXAMPLE   What is the probability of rolling a number between 0 and 7 with a standard die?

Solution   The event contains all the members of the sample space. Therefore,

$$P(\text{between 0 and 7}) = \tfrac{6}{6} = 1$$

If a die is rolled, it is certain that a number between 0 and 7 will occur.

Of course most probability questions do not deal with an event that cannot happen or with one that is certain. They are more like the next example.

EXAMPLE   The production manager of an automobile assembly plant is going to the motor pool to pick up a car for his personal use. He has 50 cars to choose from. However, 8 of the cars have defective heaters and 5 have defective transmissions. Three have both defects. The rest of the cars are in perfect condition. Assuming that the choice of any car is equally as likely as the choice of any other, what is the probability that he will choose a car with both defects? What is the probability that he will choose a car with at least one defect?

Solution   The sample space for the experiment contains the 50 cars under discussion. The probability that he will choose a car with both defects is $\tfrac{3}{50}$ since there are 3 cars with both defects in a group of 50 cars. The probability that he will choose a car with at least one defect is not as easily computed. Figure 4.2.1 is used to decide the number of defective cars. The H circle represents the cars with defective heaters. The T circle represents the cars with defective transmissions. The intersection of the two sets contains 3 cars. That means that there are 5 cars with defective heaters and good transmissions and there are 2 cars with defective transmissions and good heaters.

Thus $5 + 3 + 2$, or 10, cars have some type of defect. The probability of choosing a defective car, $P(D)$, is

$$P(D) = \tfrac{10}{50} = \tfrac{1}{5}$$

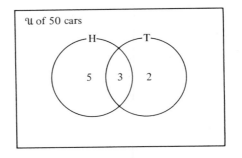

FIGURE 4.2.1

In the last example, $P(D) = \tfrac{1}{5}$ can be interpreted to mean that if the sales manager had the same choice of cars many times, then $\tfrac{1}{5}$ of the time he would choose a car with some defect.

The examples of this section point out, among other things, that computing probabilities requires the construction of sample spaces of equally likely members. A sample space made up of equally likely sample points is called an **equiprobable sample space.**

EXAMPLE   What is the probability of rolling a sum of 7 spots with a pair of fair dice? What is the probability of rolling a sum of 4?

Solution   In order to differentiate between the two dice, suppose that one is red and the other green. The following equiprobable sample space is given in ordered pair notation. The first member of the ordered pair shows the result of rolling the red die. The second shows the result of rolling the green die.

$$
\begin{array}{cccccc}
(1, 1) & (2, 1) & (3, 1) & (4, 1) & (5, 1) & (6, 1) \\
(1, 2) & (2, 2) & (3, 2) & (4, 2) & (5, 2) & (6, 2) \\
(1, 3) & (2, 3) & (3, 3) & (4, 3) & (5, 3) & (6, 3) \\
(1, 4) & (2, 4) & (3, 4) & (4, 4) & (5, 4) & (6, 4) \\
(1, 5) & (2, 5) & (3, 5) & (4, 5) & (5, 5) & (6, 5) \\
(1, 6) & (2, 6) & (3, 6) & (4, 6) & (5, 6) & (6, 6)
\end{array}
$$

There are 36 ordered pairs in the sample space. A sum of 7 spots would result if any one of the 6 pairs (1, 6), (2, 5), (3, 4), (4, 3), (5, 2), or (6, 1) were rolled. Thus there are 6 sample points in the event "a sum of 7 is rolled." Therefore,

$$P(7) = \tfrac{6}{36} = \tfrac{1}{6}$$

A sum of 4 spots results if any one of the pairs (1, 3), (2, 2), or (3, 1) is rolled. Thus

$$P(4) = \tfrac{3}{36} = \tfrac{1}{12}$$

Computing probability can require the ability to count using the combination formula.

EXAMPLE   What is the probability of being dealt 5 cards of the same suit (a flush) from a well-shuffled poker deck?

Solution   It is not practical to list a sample space of all possible poker hands and then pick the number of flushes from it. The total number of poker hands possible is the number of different 5-card hands possible using the 52 cards in a poker deck—that is, the number of combinations, or sets, of 5 cards from a set of 52, or

$$\binom{52}{5} = 2,598,960$$

There are 13 diamonds, 13 hearts, 13 spades, and 13 clubs in a poker deck. Therefore, there are as many diamond flushes as heart flushes, as spade flushes, and as club flushes. If we can compute the number of spade flushes, the total number of flushes will be 4 times this number. The number of flushes in spades is simply the number of different 5-card sets we can create from the 13 spades. But that is $\binom{13}{5}$. Then the total number of flushes is $4\binom{13}{5}$. Since all possible 5-card poker hands number $\binom{52}{5}$, the probability of a flush is:

$$P(F) = \frac{4\binom{13}{5}}{\binom{52}{5}} = \frac{4 \cdot \frac{13!}{8!\,5!}}{\frac{52!}{5!\,47!}}$$

$$= \frac{4 \cdot 13!}{8!} \cdot \frac{47!}{52!}$$

$$= \frac{4 \cdot 13 \cdot 12 \cdot 11 \cdot 10 \cdot 9 \cdot 8!}{8!} \cdot \frac{47!}{52 \cdot 51 \cdot 50 \cdot 49 \cdot 48 \cdot 47!}$$

$$= \frac{4 \cdot 13 \cdot 12 \cdot 11 \cdot 10 \cdot 9}{52 \cdot 51 \cdot 50 \cdot 49 \cdot 48}$$

$$= \frac{33}{16,660}$$

The definition of probability given here, based on numbers of members of an event that is a subset of an equiprobable sample space, is called **a priori probability**. Probability can and at times must be defined *empirically*, from observation. For instance, the probability that a person who is now 65 years old will live to be 70 can be assigned by observing large numbers of people aged 65. If 85% live to be 70, then the event can be assigned a probability of 0.85. A priori and empirical probability definitions are *not* in conflict. In an example from this section it was shown that the probability of rolling a sum of 7 spots with two dice is $\frac{1}{6}$. The same probability could have been assigned empirically by rolling a pair of dice many, many times and counting the number of 7s that occurred. If the dice were fair, that number would approach $\frac{1}{6}$ of the total number of rolls.

EXERCISES 4.2

In Exercises 1–8, a single card is drawn from a well-shuffled poker deck of 52 cards. What is the probability of drawing:

1. A diamond?

2. A king?

3. A diamond or a heart?

4. A 7 or an 8?

5. Not a 2?

6. Not a face card?

7. A 9 or a spade?

8. A 10 or a diamond?

Exercises 9 and 10 are based on the following data:

A real estate company, by checking its records for the past year, finds that the number of single-family dwellings it sold on various days of the week were as follows:

Monday	25
Tuesday	40
Wednesday	70
Thursday	90
Friday	125
Saturday	208
Sunday	100

9. Assuming that sales conditions remain constant, what is the probability that the realty company will sell a single-family dwelling next Monday?

10. What is the probability that a single-family dwelling will be sold next weekend?

In Exercises 11–15, make a sample space for the sex of two children in a family. What is the probability that:

11. There are no boys?   **12.** There are two boys?
13. There is at least one girl?   **14.** There are three girls?
15. There is at least one boy or one girl?

In Exercises 16–18, use the equiprobable sample space for tossing three fair coins to find the probability that:

16. All three are tails   **17.** Two are heads and one is tails
18. At least one comes up heads

For Exercises 19–21, use the sample space given in an example of this section for rolling two dice and give the probability of obtaining:

19. A sum of 12   **20.** A sum of 9
21. A field number (A field number is a sum of 2, 3, 4, 9, 10, 11, or 12.)

22. What justification is there in calling a field bet (see Exercise 21) a "sucker bet"?

23. A store is selling 90 adding machines. Fifteen have defective keys, 4 have defective printers, and 9 have defective stands. One has defective keys and stand but a satisfactory printer. Another has defective keys and printer but a satisfactory stand. Thirteen have only defective keys, 3 have only a defective printer, and 8 have only a defective stand. What is the probability of a buyer selecting an adding machine with all three defects?

24. Five cards are dealt from a well-shuffled poker deck. What is the probability that the five cards form a full house? (A full house is three cards of one denomination and two of another—for example, three aces and two 9s.)

25. Three cards are placed in three different envelopes. One of the cards is black on both sides, another is white on both sides, and the third card is black on one side and white on the other. The envelopes are placed in a box and mixed up. One is selected and opened. The card from it is slid onto a table so that the "down side" cannot be seen. It has a black side showing. What is the probability that the other side is also black? The answer is *not* $\frac{1}{2}$.

4.3 PROBABILITY FORMULAS

More often than not, it is tedious to find probabilities by listing sample spaces because relatively simple experiments can have extensive sample spaces. Computing probabilities can be simplified by using probability formulas.

FORMULA 1

If $\mathcal{U}$ is an equiprobable sample space and $A \subseteq \mathcal{U}$ and the probability of event A occurring is $P(A)$ and the probability of event A *not* occurring is $P(A')$, then $P(A') = 1 - P(A)$.

Suppose that event A has k points and is a subset of a sample space with n sample points. Then $P(A) = k/n$. Now event A' will have $n - k$ points, since $A \cup A'$ is the entire sample space. Therefore,

$$P(A') = \frac{n - k}{n}$$

$$= 1 - \frac{k}{n}$$

$$= 1 - P(A)$$

EXAMPLE What is the probability of *not* rolling a 7, $P(7')$, on a single roll of a pair of dice?

Solution From an earlier example it is known that $P(7) = \frac{1}{6}$. Therefore, by Formula 1,

$$P(7') = 1 - P(7)$$
$$= 1 - \frac{1}{6}$$
$$= \frac{5}{6}$$

EXAMPLE In a certain ethnic minority the probability of an 80-year-old individual living to be 90 is 0.57. What is the probability of an individual who is 80 and a member of the mentioned minority dying sometime in the next 10 years?

Solution The probability of living the next 10 years is $P(10) = 0.57$. In effect, the example asks for the probability of *not* living through the next 10 years.

$$P(10') = 1 - P(10)$$
$$= 1 - 0.57$$
$$= 0.43$$

FORMULA 2

If event A and event B are disjoint subsets of equiprobable sample space S, then the probability that either A or B will happen, $P(A \cup B)$, is

$$P(A \cup B) = P(A) + P(B)$$

Suppose that sample space S has n sample points and event A has k points, while event B has m points. Then since $A \cap B = \varnothing$, $n(A \cup B) = k + m$, and the probability that A or B will occur is

$$P(A \cup B) = \frac{k + m}{n}$$

$$= \frac{k}{n} + \frac{m}{n}$$

$$= P(A) + P(B)$$

EXAMPLE Find the probability of drawing an ace or a deuce on a single draw from a poker deck.

Solution The probability of an ace is $P(A) = \frac{4}{52} = \frac{1}{13}$, and the probability of a deuce is $P(D) = \frac{4}{52} = \frac{1}{13}$. Therefore, by Formula 2,

$$P(A \cup D) = P(A) + P(D)$$

$$= \frac{1}{13} + \frac{1}{13}$$

$$= \frac{2}{13}$$

EXAMPLE If the probability of rolling a sum of 5 spots with a pair of dice is $\frac{4}{36}$ and the probability of rolling a 7 is $\frac{6}{36}$, what is the probability of rolling a 5 or a 7 on a single roll of a pair of dice?

Solution $P(5 \cup 7) = P(5) + P(7)$

$$= \frac{4}{36} + \frac{6}{36}$$

$$= \frac{10}{36}$$

In Formula 2, events A and B are required to be disjoint subsets of S. Disjoint events are called *mutually exclusive* events. This means that if one event occurs, the other is excluded. Both events cannot happen. For instance, the event "drawing an ace" and the event "drawing a king" from a poker deck in a single draw are mutually exclusive. If one occurs, the other is excluded. On the other hand, the event "drawing an ace" and the event "drawing a heart" are not mutually exclusive. It is possible to draw both at once—i.e., the ace of hearts. These two events do not have an empty intersection. Formula 2 is easily modified to compute the probability that event A or event B will happen when $A \cap B \neq \varnothing$.

FORMULA 3

If events A and B are subsets of equiprobable sample space S, then the probability of A or B happening is

$$P(A \cup B) = P(A) + P(B) - P(A \cap B)$$

We will give an example, then the rationale behind Formula 3.

EXAMPLE What is the probability of drawing an ace or a heart from a deck of 52 poker cards in a single draw?

Solution The probability of an ace is $P(A) = \frac{4}{52} = \frac{1}{13}$. The probability of a heart is $P(H) = \frac{13}{52} = \frac{1}{4}$. The events "ace" and "heart" have one point in their intersection: the ace of hearts. Then $P(A \cap H) = \frac{1}{52}$. Therefore, by Formula 3,

$$P(A \cup H) = P(A) + P(H) - P(A \cap H)$$
$$= \frac{1}{13} + \frac{1}{4} - \frac{1}{52}$$
$$= \frac{4}{52} + \frac{13}{52} - \frac{1}{52}$$
$$= \frac{16}{52}$$
$$= \frac{4}{13}$$

To verify in the above example that $P(A \cup H) = \frac{4}{13}$ is true, consider the sample space for $A \cup H$. There are 4 aces in the deck, one of which is the ace of hearts. Besides the ace of hearts, there are 12 other hearts. Then the union of aces and hearts contains $4 + 12$, or 16, members. By the definition of probability, then, $P(A \cup H) = \frac{16}{52} = \frac{4}{13}$, which agrees with the result in the example. If in Formula 3 $A \cap B$ is not empty, $P(A)$ and $P(B)$ "counts" some points twice. Formula 3 uses $P(A \cap B)$ to subtract once the points that were counted twice.

The formulas given so far involve the probability of a single event happening. The last formula of this section has to do with a sequence of events. It deals with finding the probability of achieving outcome A from one experiment and then achieving outcome B from a second experiment. The formula uses the words *independent events*. Independent events are such that if one happens, it does not affect the probability that the other will happen. For instance, if you were to flip a coin and then roll a die, there is no reason to believe that the result of the coin flip affects the result of rolling the die or conversely. They are independent events.

FORMULA 4

If events A and B are independent and subsets of equiprobable sample spaces, then the probability that A will happen and then B will happen, $P(A$ and $B)$, is

$$P(A \text{ and } B) = P(A) \cdot P(B)$$

Formula 4 can be established by assuming that A has j points and is a subset of sample space S_1, with m sample points. Then $P(A) = j/m$. Suppose that B has k points and is a subset of sample space S_2, with n sample points. Then $P(B) = k/n$. A sample space can be created for the sequence of events, one from S_1 and one from S_2, by taking $S_1 \times S_2$. Now $S_1 \times S_2$ will contain $m \cdot n$ sample points. We need to know how many of the sample points of $S_1 \times S_2$ contain one outcome from A and one from B. A sample point from $S_1 \times S_2$ will be an ordered pair. Let (a_i, b_i) represent a sample point with one outcome from A and one from B. By the Fundamental Theorem of Counting, there are $j \cdot k$ such ordered pairs since a_i can happen j ways, $n(A) = j$, and b_i can happen k ways, $n(B) = k$. Therefore, $P(A$ and $B)$ is the number of points in $S_1 \times S_2$ with one member from A and one from B, which is $j \cdot k$, divided by the total number of points in $S_1 \times S_2$, which is $m \cdot n$.

$$P(A \text{ and } B) = \frac{j \cdot k}{m \cdot n}$$

$$= \frac{j}{m} \cdot \frac{k}{n}$$

$$= P(A) \cdot P(B)$$

EXAMPLE Find the probability of flipping a head with a fair coin and then drawing a queen from a well-shuffled deck of poker cards.

Solution The probability of a head is $P(H) = \frac{1}{2}$. The probability of drawing a queen is $P(Q) = \frac{4}{52} = \frac{1}{13}$. They are independent events. Therefore, by Formula 4,

$$P(H \text{ and } Q) = P(H) \cdot P(Q)$$

$$= \frac{1}{2} \cdot \frac{1}{13}$$

$$= \frac{1}{26}$$

EXAMPLE Technician A is installing transistors and capacitors in stereo equipment. The probability that she installs a defective transistor is 6%, and the probability that she installs a defective capacitor is 3%. Find the probability that she installs a defective transistor *and* a defective capacitor.

Solution We are asked to find the probability for the sequence of two events T and C where T and C are independent. By Formula 4,

$$P(T \text{ and } C) = P(T) \cdot P(C)$$

$$= (0.06) \cdot (0.03)$$

$$= 0.0018$$

Thus 18 out of 10,000 stereos would contain both defects.

EXAMPLE Use the data from the preceding example to compute the probability that technician A produces a working stereo and the probability that she produces a defective stereo if defects are the result of installing defective transistors or capacitors only.

Solution If the probability that technician A installs a defective transistor is 6%, then the probability she installs a good transistor is $1 - 0.06$, or 94%. Using the same reasoning, the probability that she installs a good capacitor is 97%. Then the probability that she installs both a good transistor *and* a good capacitor is

$$(0.94)(0.97) = 0.9118, \text{ or } 91.18\%$$

Thus the probability that technician A produces a good stereo is 0.9118. Then the probability that a stereo is not good, $P(g')$, is

$$P(g') = 1 - 0.9118$$
$$= 0.0882$$

The second required probability is 8.82%.

EXERCISES 4.3

1. If the probability of rolling a sum of 3 spots on a single roll of a pair of dice is $\frac{1}{18}$, what is the probability of rolling other than a sum of 3?

2. If the probability that a couple's next child will have blue eyes is $\frac{1}{4}$, what is the probability that the next child will be brown-eyed (not having blue eyes)?

3. The probability that Bill will make a sale tomorrow is 0.08. What is the probability that Bill will not make a sale tomorrow?

4. A coin is flipped twice. What is the probability that at least one will be heads? (*Hint:* At least one head means not both tails.)

In Exercises 5–10, a single card is drawn from a deck of 52 poker cards. What is the probability of drawing:

5. An ace or a queen?   **6.** A diamond or a heart?

7. A 6 or a 7 or an 8?

8. A deuce, a trey, or a one-eyed jack?

9. A spade or a king?   **10.** A jack or a diamond?

A card is drawn from a deck of 52 poker cards and then it is replaced, the cards are shuffled, and a second card is drawn. Give the probabilities in Exercises 11–15.

11. The first card is a diamond and the second is a spade.
12. The first card is a 5 and the second is a jack.
13. The first card is a king and the second is a queen.
14. The first card is a spade and the second is an ace.
15. The first card is the queen of hearts and the second is the ace of diamonds.

16. A student judges that his probability of passing history is 0.85 and his probability of passing economics is 0.70. What is the probability that he will pass both courses? What is the probability that he will pass neither? What is the probability that he will pass one course or the other? What assumptions are required to answer these questions?

17. Answer the questions of Exercise 16 if the probability that he will pass history is 0.90 and the probability that he will pass economics is 0.50.

18. The probability that Joe will make a sale on any day is 12%. What is the probability that he will fail to make a sale for the next three days?

19. The probability that a man will live 30 more years is 30%. The probability that his wife will live 30 more years is 40%. What is the probability that they will both live for 30 years? The probability that at least one will live for 30 years? That neither will live for 30 years?

20. A fair coin is flipped 25 times and comes up heads each time. Which is more likely on the twenty-sixth flip, heads or tails?

21. Explain how $P(A \cup B) = P(A) + P(B) - P(A \cap B)$ can be used to find the probability that A or B will occur if A and B are mutually exclusive.

4.4 THE BIRTHDAY PARADOX

In science and mathematics a number of phenomena occur that run contrary to our intuition. The outstanding example is from Albert Einstein's theory of relativity and is supported by many scientific observations: as an object or person travels at speeds closer to the speed of light, his time slows down as viewed from a stationary observer. In probability theory there are a number of examples that are less spectacular but equally nonintuitive. One is called the birthday paradox. Suppose that fifty people are in a room. Would you be willing to bet your dollar against another dollar that at least two people in the room have the same birthday—the same month and day but not necessarily the same year? Most people would feel that it is not a good bet. The fact is that the probability is 97% that at least two of the fifty do have the same birthday. This percentage is computed by first computing the probability that none of the fifty have the same birthday and then subtracting this probability from 1. For example, if the probability that none of the fifty people have the same birthday is 0.03, then the probability that at least two do have the same birthday is $1 - 0.03 = 0.97$, or 97%.

The probability that at least two people have the same birthday can be computed for any size gathering. We will assume that a year has 365 days and that a person is equally likely to be born on any day of the year. These assumptions are not entirely correct; however, their effect on the probabilities is negligible.

If person A is the only person in a room and person B enters, the probability that B's birthday is different from A's is $\frac{364}{365}$. This is equivalent to saying that the probability that A and B have different birthdays is $\frac{364}{365}$. Now suppose that person C enters the room. The probability that C's birthday is different from both A's and B's is $\frac{363}{365}$. Consider a sequence of events and their probability: the probability that B has a birthday different from A's is $\frac{364}{365}$; the probability that C has a birthday different from both A's and B's is $\frac{363}{365}$. Then the probability that B has a birthday different from A's and C will have a birthday different from both A's and B's is the product $\frac{364}{365} \cdot \frac{363}{365}$. Thus the probability that neither A nor B nor C have the same birthday is $\frac{364}{365} \cdot \frac{363}{365}$, which is to say that the probability that the three will all have different birthdays is $\frac{365}{365} \cdot \frac{363}{365}$.

Now suppose that person D enters the room. The probability that D's birthday is different from A's and B's and C's is $\frac{362}{365}$. Thus the probability that the first three all have different birthdays is $\frac{364}{365} \cdot \frac{363}{365}$, while the probability that the fourth's birthday is different from those of the first three is $\frac{362}{365}$. Then the probability for the sequence of events—the first three all have different birthdays *and* the fourth has a birthday different from the first three—is

$$\tfrac{364}{365} \cdot \tfrac{363}{365} \cdot \tfrac{362}{365}$$

This probability, then, is the probability that four people all have different birthdays. The method used to derive this probability can be used to show that the probability that five people all have different birthdays is

$$\tfrac{364}{365} \cdot \tfrac{363}{365} \cdot \tfrac{362}{365} \cdot \tfrac{361}{365}$$

The probability that r people all have different birthdays is

$$\frac{364}{365} \cdot \frac{363}{365} \cdot \frac{362}{365} \cdot \frac{361}{365} \cdot \cdots \cdot \frac{365 - r + 1}{365}$$

The probability that at least two people in a set of r people *do* have the same birthday is

$$1 - \left(\frac{364}{365} \cdot \frac{363}{365} \cdot \frac{362}{365} \cdot \cdots \cdot \frac{365 - r + 1}{365} \right) \qquad (1)$$

Figure 4.4.1 was created using (1) and a desk calculator. It gives the probabilities that at least two people have the same birthday for groups of different sizes.

Number of People	Probability at Least Two Have Same Birthday
5	0.027
10	0.117
15	0.253
18	0.347
20	0.411
21	0.444
22	0.476
23	0.507
24	0.538
25	0.569
27	0.627
30	0.706
35	0.814
40	0.891
50	0.970
60	0.9951
70	0.99916
80	0.99991
90	0.99999
100	—
125	—
150	—

FIGURE 4.4.1   The birthday paradox.

EXERCISES 4.4

Use Figure 4.4.1 to answer Exercises 1–13.

In Exercises 1–6, what is the probability that at least two people in a room have the same birthday (month and day) if the people in the room number:

1. 10   **2.** 25   **3.** 50
4. 90   **5.** 81   **6.** 102

7. About how many people must be in a room so that the probability that at least two have the same birthday is 50%?
8. Answer Exercise 7 if the probability is to be 70%.
9. Answer Exercise 7 if the probability is to be 90%.

10. What is the probability that at the kickoff in the Super Bowl at least two of the players on the field have the same birthday?

11. What is the probability that *no* two U.S. senators have the same birthday?

12. What is the probability that *no* two members of a 15-member board of regents have the same birthday?

13. What is the probability that *no* two of the first 30 presidents of the United States had the same birthday?

4.5 CONDITIONAL PROBABILITY

The probability that an event will occur may change as more information is learned about it. For example, suppose that a card is slid from a well-shuffled poker deck and placed face down on a table. What is the probability that it is a king? At first glance we might say that the probability is $\frac{4}{52} = \frac{1}{13}$. Suppose it is further known that the card is a face card. Now what is the probability that it is a king? Since there are twelve face cards, four of which are kings, the probability that the card on the table is a king would be $\frac{4}{12}$, or $\frac{1}{3}$. The probability that the card will be a king when it is known that the card is a face card is called a conditional probability. **Conditional probability** is the probability that event A will occur given the additional information that event B has occurred.

> **NOTATION: Conditional Probability**
>
> The probability of event A given that event B has occurred is denoted by
>
> $$P(A \mid B)$$

EXAMPLE Let event A be the event "The card drawn from a poker deck is a 6." Let event B be the event "The card drawn is a 5 or a 6." Find $P(A \mid B)$.

Solution $P(A \mid B)$ asks for the probability of event A given that event B has occurred—that is, the probability that a drawn card will be a 6 when it is known that it is either a 5 or a 6. Since there are four 5s and four 6s:

$$P(A \mid B) = \frac{4}{8} = \frac{1}{2}$$

EXAMPLE The following chart gives the number of responses to the question "Does the European Common Market benefit the United States?" which was asked of 100 college students. The chart lists responses by political party: D for Democrats, R for Republicans, and I for Independents.

	Yes	No	No Opinion
D	40	5	7
R	15	16	5
I	2	2	8

If a student is chosen at random from the 100 polled students, what is the probability that he answered the question in the affirmative given that the student is a Republican, $P(\text{yes} \mid R)$? What is the probability that the student answered no if it is known the student is not a Democrat, $P(\text{no} \mid D')$?

Solution　Republicans number 36, of which 15 answered yes. Therefore,

$$P(\text{yes} \mid R) = \tfrac{15}{36} = \tfrac{5}{12}$$

There are 48 non-Democrats (Republicans and Independents), of which 18 answered no. Therefore,

$$P(\text{no} \mid D') = \tfrac{18}{48} = \tfrac{3}{8}$$

It was stated earlier (Formula 4) that if events A and B are *independent* and subsets of equiprobable sample spaces, then the probability of A and then B occurring is $P(A \text{ and } B) = P(A) \cdot P(B)$.

FORMULA 5

If A and B are *dependent*, then

$$P(A \text{ and } B) = P(A) \cdot P(B \mid A)$$

EXAMPLE　A box contains 80 radio tubes, of which 20 are defective. The defective tubes cannot be detected by sight, and they are uniformly mixed in with the good tubes. What is the probability that an employee will select 2 bad tubes in successive selections from the box if the first tube is not replaced in the box?

Solution　Let A be the event "The first tube selected is defective" and B be the event "The second tube selected is defective." We are asked to compute $P(A \text{ and } B)$. Events A and B are not independent because the outcome of the first selection affects the probability of the second. Therefore, we use Formula 5 for dependent events:

$$P(A \text{ and } B) = P(A) \cdot P(B \mid A)$$

Since there are 20 defective tubes in a total of 80, the probability of selecting a defective tube first is $\tfrac{20}{80}$, or $P(A) = \tfrac{20}{80}$. Now $P(B \mid A) = \tfrac{19}{79}$, since "given that A has occurred" means that a defective tube was selected on the first draw, so there are 19 defective tubes among the remaining 79 tubes.

$$P(A \text{ and } B) = \tfrac{20}{80} \cdot \tfrac{19}{79}$$
$$= \tfrac{1}{4} \cdot \tfrac{19}{79}$$
$$= \tfrac{19}{316}$$

EXERCISES 4.5

1. A card is drawn from a well-shuffled poker deck. What is the probability that it is a king if the card drawn is known to be a face card? What is the probability that it is a king if the card drawn is known to be a card between 4 and 10?

2. In Exercise 1, what is the probability that the card drawn is a 6 if it is known that the card drawn is not a face card? What is the probability that it is a spade if the card drawn cannot be a diamond or a heart or a club?

3. If a pair of fair dice are rolled, what is the probability that their sum is 7 if the first die shows a 2? If the first die shows a 6?

4. In Exercise 3, what is the probability that the sum is 9 if the number shown on the first die is greater than 4? If the number shown on the first die is 1?

Events X, Y, and $X \cap Y$ are outcomes of an event and have the number properties: $n(X) = 60$, $n(Y) = 50$, $n(X \cap Y) = 10$. Use these number properties to compute the probabilities in Exercises 5–16. X and Y are subsets of an equiprobable sample space $\mathscr{U}$, where $n(\mathscr{U}) = 100$.

5. $P(X')$ 6. $P(Y')$ 7. $P(X \text{ and } Y)$

8. $P(X \mid Y)$ 9. $P(Y \mid X)$ 10. $P(X' \mid Y')$

11. $P(Y' \mid X')$ 12. $P(Y \mid X')$ 13. $P(X' \mid Y)$

14. $P(Y' \mid X)$ 15. $P(X \mid Y')$ 16. $1 - P(X \mid Y)$

17. Two cards are drawn from a well-shuffled poker deck. What is the probability that they are both aces if:
    a. The first card is replaced before the second is drawn?
    b. The first card is not replaced?

18. In Exercise 17, answer (a) and (b) for the probability that the first card is a spade and the second card is a jack.

19. A can manufacturing company stores its cans in lots of 1,000. In a particular lot there are 100 defective cans. An inspector chooses 2 cans

at random and inspects them. What is the probability that they are both defective? That they are both not defective?

20. What is the probability that the inspector of Exercise 19 chooses 3 cans that are all defective?

21. A soap company advertises widely on television. It estimates that 40% of the nation has seen its advertisement. It further estimates that if a person sees the advertisement, there is a 10% probability that he will buy the soap. What is the probability that a person selected at random in the country will have seen the advertisement and bought the soap?

22. A cola bottling machine will securely place the caps on 99% of the bottles sent through it when it is correctly adjusted. There is a 10% chance that it is not in adjustment. When it is not in adjustment, 70% of the bottles sent through it are securely capped. What is the probability that a given bottle from the machine will have a secure cap?

4.6   EXPECTED VALUE

Some people study probability for its own sake. For most of us, however, probabilities are a matter of dollars and cents; we deal in probabilities on a monetary basis. It could be that we gamble at Reno or Lake Tahoe or Las Vegas. More often our "bets" are with an insurance company, where the company bets it will keep our premiums without paying us the face value of the policy, and we bet—and hope to lose—that the company will have to pay off before we have made very many premium payments. A department store manager is interested in money when he schedules a sale for the next Wednesday on which the weather bureau predicts a 25% probability of rain. He is taking a chance. If the weather is bad, his overhead for the day may be greater than his sales, and the store will lose money, while if the weather is good, the store will take in much more money than it spends. Based on the weather probabilities, the store manager is "betting" that the store will make money.

Expected value is the phrase used for the relationship between probabilities and the dollars related to certain outcomes. For instance, suppose that you are betting on the occurrence of event A. If event A occurs you win a dollar, and if event A fails to occur you lose a dollar. Further suppose that the probability that A will occur is $\frac{1}{4}$. Then the probability that A fails to occur is $\frac{3}{4}$. Thus if you bet on A, say 100 times, you would expect to win $1 about 25 times and lose $1 about 75 times. Then for 100 typical bets, you would win 25 times and lose 75 times for a net of $-\$50$. That is, you would lose a total of $50 in 100 bets, which is the same as saying you would lose an average of $0.50 per bet. The amount $-\$0.50$ is called the *expected value* of the bet.

DEFINITION

If an equiprobable sample space $\mathscr{U}$ is partitioned into events S_1, $S_2, S_3, \ldots, S_n$ and if the probabilities of these events are $P_1, P_2, P_3, \ldots, P_n$ and if for each of the outcomes one can expect amounts $x_1, x_2, x_3, \ldots, x_n$, then the **expected value**, E, is

$$E = x_1 P_1 + x_2 P_2 + x_3 P_3 + \cdots + x_n P_n$$

EXAMPLE Suppose you are wagering with a friend on the roll of a single die. If you roll 3 spots your friend pays you $6, and if you fail to roll a 3 you pay your friend $1. What is your expected value on this bet?

Solution The probability that you roll a 3 is $\frac{1}{6}$, for which you can expect $+$$6. The probability that you fail to roll a 3 is $\frac{5}{6}$, for which you can expect $-$$1. The $-$ in $-$$1 means that you lose the dollar. Then by the definition for expected value,

$$E = 6 \cdot \tfrac{1}{6} + (-1) \cdot \tfrac{5}{6}$$
$$= 1 - \tfrac{5}{6}$$
$$= \tfrac{1}{6}, \text{ which is approximately } \$0.17$$

An E of approximately $0.17, in the above example, means that if you played the game for a long period of time you would win an average of $0.17 each time you rolled the die. From your point of view that is a favorable bet.

EXAMPLE A Casino in Nevada pays $32 for a $1 bet on a "hard-way 6," and you keep your $1 wager. A hard-way 6 means a pair of dice rolled so that there are 3 spots showing on each die. What is the player's expected value on such a bet?

Solution The probability that the first die rolls 3 is $\frac{1}{6}$. The probability that the second rolls 3 is also $\frac{1}{6}$. Therefore, since the rolls are independent, the probability that the first is 3 and then the second is 3 is $\frac{1}{6} \cdot \frac{1}{6} = \frac{1}{36}$. The probability of failing to roll a hard-way 6 is $1 - \frac{1}{36} = \frac{35}{36}$. Then the probability that the player will win $32 is $\frac{1}{36}$, and the probability that the player will lose $1 is $\frac{35}{36}$. By the definition of expected value,

$$E = (32) \cdot \tfrac{1}{36} + (-1) \cdot \tfrac{35}{36}$$
$$= \tfrac{32}{36} - \tfrac{35}{36}$$
$$= -\tfrac{3}{36}, \text{ approximately } -\$0.08$$

An E of $-$$0.08, in the above example, means that for a very large number of bets on the hard-way 6, the player will lose an average of $0.08 each time he rolls the dice. Why would anyone make such a bet? To beat the probabilities? Of course!

A *fair investment* is one in which the expected value is zero. In a fair investment made a large number of times, you will win about as much as you lose. In effect, there is no "winner" at all.

EXAMPLE   It has recently become popular in California to give small children life insurance policies paid up to age 65. What lump-sum premium should an insurance company charge for a $1,000 life insurance policy to age 65 on a 7-year-old child if the probability that the child will live to be 65 is 90%?

Solution   Let x represent the cost of the premium. Examining the problem from the insurance company's point of view, there is a 90% probability that they will keep the premium of x dollars and a 10% probability that they will have to pay $1,000 on the policy. Then the expected value is

$$E = 0.90x + 0.10(-1,000 + x)$$

For the investment to be fair E must be zero or

$$0.90x - 100 + 0.1x = 0$$

Solving for x,

$$x = \$100$$

A premium of $100 would make the investment fair. It would seem, then, that the insurance company would not be making any money. In fact, a company may charge slightly more than the fair investment premium and make most of its money from investments bought with premium dollars.

EXERCISES 4.6

In Exercises 1–5, you and a friend are wagering on the roll of a single die. Find your expected value if:

1. You win $10 if you roll 1 spot, and you lose $3 if you fail to roll 1.

2. You win $8 if you roll 6, and you lose $1 if you fail to roll 6.

3. You win $5 if you roll 6, and you pay $1 if you fail to roll 6.

4. You win $3 if you roll 5, you win $2 if you roll 3, and you pay $1 if you roll anything else.

5. You win $1 if you roll 1, you win $2 if you roll 2, you win $3 if you roll 3, and you pay $2 if you roll anything else.

6. What is your expected value if you are going to draw a single card from a poker deck and you win $30 if you draw an ace or $20 if you draw a face card, and you must pay $10 if you draw any other card?

7. Answer Exercise 6 if you will be paid $50 if you draw a jack, $25 if you draw a queen, $10 if you draw an ace, $2 if you draw a king, and you must pay $10 if you draw any other card.

8. What is the expected value on a hard-way 8 with a pair of dice (a 4 showing on each die) if you are given $33 if you roll the 8, and you keep your $1, and you lose $1 if you fail to roll two 4s?

9. What is the expected value on a hard-way 10 with a pair of dice (a 5 on each) if you win $35 for rolling the 10 and you lose $1 when failing to make the 10?

10. Nevada roulette wheels have 38 numbers on them. They are numbered 0, 00, 1, 2, . . ., 36. The simplest bet is to place $1 on a number. If the ball lands on that number you win $35 and you keep your $1 bet. If the ball fails to land on the number on which you have bet, you lose your $1. What is the expected value of such a bet?

11. A second type of roulette wager on the wheel described in Exercise 10 is "red or black." There are 18 red numbers, 18 black numbers, and 2 green numbers on the wheel. If you bet $1 on red, you will win $1 and keep yours if the ball lands on red. If it lands on black or green, you lose your $1. What is the expected value of such a bet?

12. In a lottery 1,500 tickets are sold at $2 per ticket. The prize in the lottery is worth $1,000. What is the profit on each ticket sold?

13. What is the expected value in buying one of 1,500 lottery tickets for $2 if the grand prize is worth $1,002 and the second prize is worth $102? There are no other prizes.

14. How should a lottery ticket be priced if it is to be a fair investment, if 1,000 tickets are to be sold, and if the prize is worth $3,000 with the added bonus that the price of the ticket will be refunded to the winner?

15. The probability that a child under the age of 14 years will live to be 25 years old is 98%. Assuming that no other costs are involved, what would be a fair single-payment premium for a $2,000 life insurance policy to age 25 for a child under the age of 14.

16. What would be a fair premium to charge, assuming that no administrative costs or other costs are involved, for a $5,000 life insurance policy to age 65 if there is to be a single-payment premium and the policy is on a man who has a 70% probability of living to age 65?

17. Answer Exercise 16 if the probability that the man will live to age 65 is 50%.

18. Answer Exercise 16 if the probability that the man will live to age 65 is 30%.

19. A construction firm is going to bid on two contracts A and B. The bid on A costs \$1,000 to prepare and will yield \$150,000 profit if accepted. The bid on B will cost \$1,500 to prepare and will yield \$200,000 profit if accepted. The company judges that it has a 40% probability of having its bid on contract A accepted and a 45% probability of having its bid on contract B accepted. What is the expected value of its bid on each contract?

20. A furniture refinishing company is considering bidding on a contract to refinish metal lockers or on a contract to refinish executive desks. It cannot fulfill both contracts. The bid for the locker contract costs \$200 to prepare, it will yield a profit of \$5,000, and the company has a 50% chance of landing it. The desk contract costs \$600 to prepare, it will yield a profit of \$6,000, and there is a 40% chance the bid will be accepted. Based on the expected value of each contract, on which should it bid?

4.7   THE BINOMIAL DISTRIBUTION AND BERNOULLI TRIALS

The experiments of this section will be restricted to a type known as *Bernoulli trials* after the eighteenth century Swiss mathematician Jakob Bernoulli.

DEFINITION

A **Bernoulli trial** is an experiment that has exactly two mutually exclusive outcomes.

For example, a single flip of a coin is a Bernoulli trial since there are two possible outcomes, heads and tails, and they are mutually exclusive. Guessing the sex of an unborn child is a Bernoulli trial since there are two possible outcomes, male or female, and they are mutually exclusive. Bernoulli trials are important in medicine, where a certain drug may or may not cure a specific disease; in business, where a certain item may or may not be correctly fabricated; and in genetics, where a certain gene may or may not be present.

In general, the outcomes of Bernoulli trials are given the names S, for success, and F, for failure. If the probability of S is p, $P(S) = p$, then the probability of F is $1 - p$, $P(F) = 1 - p$. Let $1 - p = q$, then if $P(S) = p$, $P(F) = q$.

THEOREM

The probability of r successes in a sequence of n independent Bernoulli trials, where $P(S) = p$ and $P(F) = 1 - p = q$, is

$$\binom{n}{r} p^r q^{n-r}$$

We establish the theorem by considering a sequence of five Bernoulli trials with three successes and then generalize the results. The problem, then, is to find the probability of three successes in five Bernoulli trials with $P(S) = p$ and $P(F) = q$. Let the ordered five-tuple (S, F, S, S, F) represent one such sequence. The probability of *exactly this* sequence of three successes in five trials is $p \cdot q \cdot p \cdot p \cdot q = p^3 q^2$. Of course three successes could happen like this: (S, S, F, S, F). The probability of the sequence (S, S, F, S, F) is $p \cdot p \cdot q \cdot p \cdot q = p^3 q^2$. Then the probability of (S, F, S, S, F) or (S, S, F, S, F) is $p^3 q^2 + p^3 q^2 = 2p^3 q^2$. This is not the probability we seek, but it shows how that probability can be found.

Any particular sequence of three successes and two failures will have a probability of $p^3 q^2$. Therefore, the number of such sequences times $p^3 q^2$ is the probability of three successes in five trials. To find the number of different sequences of three successes and two failures, examine the following empty five-tuple: (–, –, –, –, –). We seek the number of ways we can place three S's in it and then fill the remaining places with F's. That is the same as asking how many ways can we select three positions for S's out of five places. The answer is the number of combinations of three things selected from a group of five, $\binom{5}{3}$. Then the probability of three successes in five trials is

$$\binom{5}{3} p^3 q^2 = \frac{5!}{3! \, 2!} p^3 q^2$$

$$= \frac{5 \cdot 4 \cdot 3!}{3! \cdot 2} p^3 q^2$$

$$= 10 p^3 q^2$$

The argument for the probability of three successes in five trials can be generalized to r successes in n trials and can establish the theorem.

Note that $\binom{n}{r} p^r q^{n-r}$ is the general term in the binomial theorem, given in the last chapter.

EXAMPLE By noting the family history of a certain couple, it is established that the probability that any child of theirs will be left-handed is $\frac{1}{4}$.

If the couple have four children, what is the probability that two are right-handed and two are left-handed?

Solution If the probability of left-handedness is $\frac{1}{4}$, then the probability of right-handedness is $1 - \frac{1}{4} = \frac{3}{4}$. We will ignore the possibility of ambidexterity. We have Bernoulli trials since right- and left-handedness are exactly two mutually exclusive events. The example, then, asks for the probability of two successes—left-handed children—in four trials—births. By the theorem above the probability is

$$\binom{4}{2}\left(\frac{1}{4}\right)^2\left(\frac{3}{4}\right)^2 = \frac{4!}{2!\,2!}\cdot\frac{1}{16}\cdot\frac{9}{16}$$

$$= \frac{4\cdot3\cdot2!}{2\cdot2!}\cdot\frac{1}{16}\cdot\frac{9}{16}$$

$$= 6\cdot\tfrac{1}{16}\cdot\tfrac{9}{16}$$

$$= 3\cdot\tfrac{1}{8}\cdot\tfrac{9}{16}$$

$$= \tfrac{27}{128}$$

With a short sequence of events, as in the above example, the binomial theorem will quickly give the complete picture. Suppose, then, that in the last example we wished to find the probabilities that all four children are left-handed, three are left-handed, two are left-handed, one is left-handed, or none are left-handed. Let $p = \frac{1}{4}$ and $q = \frac{3}{4}$ and use the binomial theorem to expand $(p + q)^4$:

$$(p + q)^4 = p^4 + 4p^3q + 6p^2q^2 + 4pq^3 + q^4$$

Substitute $p = \frac{1}{4}$ and $q = \frac{3}{4}$:

$$= (\tfrac{1}{4})^4 + 4(\tfrac{1}{4})^3(\tfrac{3}{4}) + 6(\tfrac{1}{4})^2(\tfrac{3}{4})^2 + 4(\tfrac{1}{4})(\tfrac{3}{4})^3 + (\tfrac{3}{4})^4$$

$$= \tfrac{1}{256} + \tfrac{12}{256} + \tfrac{54}{256} + \tfrac{108}{256} + \tfrac{81}{256} \tag{1}$$

The terms of (1) give the probabilities that:

All four are left-handed: $\tfrac{1}{256}$

Three are left-handed: $\tfrac{12}{256} = \tfrac{3}{64}$

Two are left-handed: $\tfrac{54}{256} = \tfrac{27}{128}$

One is left-handed: $\tfrac{108}{256} = \tfrac{27}{64}$

None are left-handed: $\tfrac{81}{256}$

Note that the probabilities of (1) add up to $\tfrac{256}{256} = 1$.

EXAMPLE A fair die is rolled six times. Find the probability of rolling the 1 spot exactly four times. Also find the probability of rolling 1 at least four times in the six rolls.

Solution   The probability of a success, a 1, is $\frac{1}{6}$. The probability of a failure, not 1, is $\frac{5}{6}$. Rolling a 1 or failing to roll a 1 is a Bernoulli trial. The probability of four successes in six trials is

$$\binom{6}{4}\left(\frac{1}{6}\right)^4\left(\frac{5}{6}\right)^2 = \frac{6!}{4!\,2!}\cdot\frac{1}{6^4}\cdot\frac{25}{6^2}$$

$$= \frac{6\cdot5\cdot4!}{4!\cdot2}\cdot\frac{25}{6^6}$$

$$= 15\cdot\frac{25}{6^6}$$

$$= 0.0080 \text{ (approximately)}$$

The probability of exactly four 1s in six rolls is approximately 0.008. The second part of the example asks for the probability of at least four 1s. This means the probability of four 1s or of five 1s or of six 1s. The probability is the sum of the individual probabilities:

$$\binom{6}{4}\left(\frac{1}{6}\right)^4\left(\frac{5}{6}\right)^2 + \binom{6}{5}\left(\frac{1}{6}\right)^5\left(\frac{5}{6}\right) + \binom{6}{6}\left(\frac{1}{6}\right)^6\left(\frac{5}{6}\right)^0$$

$$= \frac{6!}{4!\,2!}\cdot\frac{1}{6^4}\cdot\frac{25}{6^2} + \frac{6!}{5!\,1!}\cdot\frac{5}{6^6} + 1\cdot\frac{1}{6^6}$$

$$= \frac{6\cdot5\cdot4!}{4!\cdot2}\cdot\frac{25}{6^6} + \frac{6\cdot5!}{5!}\cdot\frac{5}{6^6} + \frac{1}{6^6}$$

$$= 15\cdot\frac{25}{6^6} + 6\cdot\frac{5}{6^6} + \frac{1}{6^6}$$

$$= \frac{375}{6^6} + \frac{30}{6^6} + \frac{1}{6^6}$$

$$= \frac{406}{6^6}$$

$$= 0.0087 \text{ (approximately)}$$

The probability of at least four 1s is 0.0087—not very likely.

EXAMPLE   A machine makes ball-point pen refills of which 98% are perfect and 2% are defective. If a quality control inspector selects at random five pens that were created by the machine, what is the probability that two or more are defective?

Solution   The required probability is the probability that exactly two are defective plus the probability that three are defective plus the probability that four are defective plus the probability that all five are defective. However, the arithmetic is simplified by computing the probability that fewer than two are defective and then subtracting this probability from 1. The probability that fewer than two are defective is the sum of the probabilities that one is defective and that none are defective:

$$\binom{5}{1}(0.02)^1(0.98)^4 + \binom{5}{0}(0.02)^0(0.98)^5$$

$$= \frac{5!}{4!\,1!}(0.02)(0.98)^4 + \frac{5!}{0!\,5!}(0.98)^5$$

$$= 5(0.02)(0.98)^4 + (0.98)^5$$

$$= 0.09224 + 0.90392 = 0.99616$$

The probability of fewer than two defective refills out of a sample of five is 0.99616. The probability that two or more refills are defective, which was the question originally posed, is the same as the probability of *not* less than two defective refills, or $1 - 0.99616 = 0.00384$.

The arithmetic of the last two examples is formidable when done by hand. If one is available, it is preferable to use a desk calculator.

EXERCISES 4.7

In Exercises 1–4, the probability that a couple will have a left-handed child is $\frac{1}{4}$. Exclude the possibility of ambidexterity.

1. What is the probability that if the couple have two children, one is left-handed and one is right-handed?
2. What is the probability that if the couple have three children, two are left-handed?
3. What is the probability that in a family of five children only one is left-handed?
4. What is the probability that in a family of five children four are left-handed?

5. A fair coin is tossed twice. What is the probability of not getting heads? Of getting heads once? Of getting heads twice?
6. A fair coin is tossed four times. What is the probability of not getting heads? Of getting heads once? Twice? Three times? Four times? Five times?
7. Assuming that the birth of boys and girls is equally likely, in a family of two children, what is the probability of having a boy and a girl?

8. Under the assumption of Exercise 7, what is the probability in a family of three children of having two boys and a girl?

In Exercises 9 and 10, a card is drawn from a poker deck, then it is replaced and another card is drawn. This process is repeated five times.

9. What is the probability of drawing three aces?
10. What is the probability of drawing four aces?

11. By examining their family histories it is determined that a couple have a $\frac{1}{8}$ probability that any child of theirs will have a slight deformity. If they have three children, what is the probability that at least one has the deformity?
12. In Exercise 11, what is the probability that at least two children have the deformity?
13. It is determined that a couple have a 0.1 probability of giving birth to a genius. If they have five children, what is the probability that the geniuses outnumber the nongeniuses?
14. In Exercise 13, what is the probability that the couple will have at least two geniuses?
15. A package of 100 shotgun shells contains 10 defective shells. If a sample of 5 shells is taken from the package, what is the probability that at least 2 shells are defective?
16. Four light bulbs are defective in a lot of 40. If 4 bulbs are chosen at random, what is the probability that at least 2 are defective?
17. A student is to take a 10-question true-false test. She needs 8 right answers to pass the test. What is the probability that she will pass if she guesses on each question and, because of the studying she has done, her probability of guessing any question correctly is $\frac{3}{4}$? (Set up only.)
18. In a 10-question true-false test, how many questions must a student answer correctly to pass the test if we wish to be 90% certain that he is not just guessing the answers?
19. A couple who are society leaders in a very underpopulated nation wish to have a male child. Assuming that males and females are equally likely, how many children should they plan to have so that the probability of having at least one male is 90%?
20. How many children should the couple of Exercise 19 plan to have if they insist on a 95% probability of having at least one male?

4.8 DECISIONS, DECISIONS

Probabilities are relatively easy to calculate when they are related to rolling dice, flipping coins, or choosing cards. When probabilities are known, decisions are easier to make. For example, a study of probability shows that betting a dollar on a number on a roulette wheel is a losing proposition. The knowledge that the probability of hemophilia is high may not make the decision to have or not to have children an easy one, but it does allow would-be parents and their doctors to prepare. In these examples probabilities can be assigned with some accuracy. It is more often the case, however, that the probabilities are very difficult to assign. When this happens, decisions are difficult, even dangerous, to make.

For instance, suppose that a new vaccine is developed to cure a certain disease. It is administered to ten victims of the disease and seven recover. It is further known that 40% of the victims of the disease recover without using any medication. Is the new vaccine helpful, or could it have been administered to ten people and by chance it happened that seven were going to recover anyway? The question is an important one. If it is likely that seven could have recovered without the vaccine, it would be wasteful in time and money to continue to produce it. It would be equally wasteful in human misery to discard the vaccine if it does improve the percentage of cures. It is not, of course, possible to decide with certainty if the vaccine is effective, nor is it possible to assign to it an exact probability of curing a victim of the disease. It is possible to decide, within certain probabilities, if the vaccine does increase the number of cures. To do this we use the theory of Bernoulli trials and the binomial theorem of the last section.

From the last section it was shown that the probability of r successes in n Bernoulli trials, where $P(S) = p$ and $P(F) = 1 - p = q$, is $\binom{n}{r} p^r q^{n-r}$.

Tables for this probability are available and greatly reduce the arithmetic computations. Table 4.8.1 gives the probability of r successes in 10 Bernoulli trials. The left column gives the value of r from 1 to 10. $P(S)$ is across the top.

EXAMPLE Use Table 4.8.1 to find the probability of 7 successes in 10 Bernoulli trials if the probability of a single success is 0.3.

Solution Read down the left edge and find an r of 7. The required probability is opposite the 7 in the column headed 0.3. The probability of 7 successes in 10 trials when $P(S) = 0.3$ is 0.009.

EXAMPLE Find the probability of at least 8 successes in 10 Bernoulli trials if the probability of a single success is 0.9.

Solution The probability of at least 8 successes is the sum of the probabilities of 8 successes, 9 successes, and 10 successes—that is, from Table 4.8.1,

$$0.194 + 0.387 + 0.349 = 0.930$$

The probability of at least 8 successes in 10 Bernoulli trials with $P(S) = 0.9$ is about 93%.

TABLE 4.8.1 THE PROBABILITY OF r SUCCESSES IN 10 BERNOULLI TRIALS $\binom{10}{r} p^r q^{10-r}$

r \ p	0.1	0.3	0.5	0.7	0.9
0	0.349	0.028	0.001	0.000	0.000
1	0.387	0.121	0.010	0.000	0.000
2	0.194	0.233	0.044	0.001	0.000
3	0.057	0.267	0.117	0.009	0.000
4	0.011	0.200	0.205	0.037	0.000
5	0.001	0.103	0.246	0.103	0.001
6	0.000	0.037	0.205	0.200	0.011
7	0.000	0.009	0.117	0.267	0.057
8	0.000	0.001	0.044	0.233	0.194
9	0.000	0.000	0.010	0.121	0.387
10	0.000	0.000	0.001	0.028	0.349

A table such as Table 4.8.1 can be used to design experiments to aid decision making. For instance, a well-known English lady claims that she can tell by the taste when milk is placed in her cup before tea is added and when the tea has been poured first and then the milk added. She does not claim to be correct every time, but she feels she has a greater than 50% chance of being correct. If she were to taste 10 cups of tea, some with tea poured first and milk added afterwards, others with milk placed in the cup first and then tea added, how many must she correctly identify to show that she has some ability, as she claims? Suppose that we arbitrarily decide that at least 8 correct identifications in 10 trials will convince us that she has some ability in the direction she claims. Referring to Table 4.8.1, if she had no ability, $P(S) = 0.5$, the probability that she could identify at least 8 out of 10 cups is $0.044 + 0.010 + 0.001 = 0.055$. If she could identify when the milk had been added 90% of the time, her probability of being correct 8 of 10 times is $0.194 + 0.387 + 0.349 = 0.930$. Therefore, the 8 of 10 test has only a 5.5% chance of happening if the lady has no ability and a 93% probability of happening if she can identify the appropriate mixture 90% of the time. An excellent test then? Not necessarily. Suppose that the lady has only a 70% ability at choosing. The probability of at least 8 correct choices in 10 is 0.233 + 0.121 + 0.028 = 0.382, or 38.2%. Perhaps a better test would require her to be successful 7 of 10 times. In a 7 of 10 test, by using Table 4.8.1, it can be shown that if $P(S) = 50\%$, she has a 17.2% probability of identifying 7 cups correctly. If $P(S) = 70\%$, she has a 64.9% probability of identifying 7 correctly. And if $P(S) = 90\%$, she has a 98.7% probability of identifying 7 correctly. The 7 of 10 test, then, is such that she would likely fail it if she were only guessing, but she would probably pass it if she had either a 70% ability or a 90% ability, as she claims.

Table 4.8.2 gives the probability of r successes in 15 Bernoulli trials. It is used in exactly the same way that Table 4.8.1 was. For instance, the probability of 10 successes in 15 Bernoulli trials is 0.206 if $P(S) = 0.70$.

TABLE **4.8.2** THE PROBABILITY OF r SUCCESSES IN 15 BERNOULLI TRIALS $\binom{15}{r} p^r q^{15-r}$

r \ p	0.10	0.30	0.50	0.70	0.90
0	0.206	0.005	0.000	0.000	0.000
1	0.343	0.031	0.000	0.000	0.000
2	0.267	0.092	0.003	0.000	0.000
3	0.129	0.170	0.014	0.000	0.000
4	0.043	0.219	0.042	0.001	0.000
5	0.010	0.206	0.092	0.003	0.000
6	0.002	0.147	0.153	0.012	0.000
7	0.000	0.081	0.196	0.035	0.000
8	0.000	0.035	0.196	0.081	0.000
9	0.000	0.012	0.153	0.147	0.002
10	0.000	0.003	0.092	0.206	0.010
11	0.000	0.001	0.042	0.219	0.043
12	0.000	0.000	0.014	0.170	0.129
13	0.000	0.000	0.003	0.092	0.267
14	0.000	0.000	0.000	0.031	0.343
15	0.000	0.000	0.000	0.005	0.206

EXAMPLE A new vaccine is given to 15 victims of disease X. Of the victims of this disease, 30% will recover without receiving any medication. Nine of the 15 recipients of the vaccine recover, and the vaccine is hailed as a major breakthrough. What is the probability that it is of no value?

Solution   The vaccine is of no value if the 9 people who recovered are members of the 30% of the population who will recover without medication. The probability that there are 9 or more successes—people who will recover—in 15 Bernoulli trials when the probability of a single success is 30% is the sum of the probabilities of 9, 10, 11, 12, 13, 14, and 15 from Table 4.8.2 in the 0.30 column:

$$0.012 + 0.003 + 0.001 + 0.000 + 0.000 + 0.000 + 0.000 = 0.016$$

There is only a 1.6% probability that the vaccine is of no value.

EXERCISES 4.8

Use Table 4.8.1 for Exercises 1–5.

1. What is the probability of 4 successes in ten Bernoulli trials if $P(S) = 0.70$?
2. What is the probability of 9 successes in 10 trials if $P(S) = 0.90$?
3. What is the probability of 2 successes in 10 trials if $P(S) = 0.10$?
4. What is the probability of no successes in 10 trials if $P(S) = 0.90$?
5. What is the probability of 11 successes in 10 trials if $P(S) = 0.50$?

Use Table 4.8.2 for Exercises 6–11.

6. What is the probability of 13 successes in 15 Bernoulli trials if $P(S) = 0.50$?
7. What is the probability of 10 successes in 15 trials if $P(S) = 0.10$?
8. What is the probability of 2 successes in 15 trials if $P(S) = 0.70$?
9. What is the probability of 6 successes in 15 trials if $P(S) = 0.30$?
10. What is the probability of 18 successes in 15 trials if $P(S) = 0.30$?

Use either Table 4.8.1 or 4.8.2 for Exercises 11–14.

11. What is the probability of at least 6 successes in 10 trials if $P(S) = 0.70$?
12. Answer Exercise 11 for $P(S) = 0.90$.
13. What is the probability of at least 8 successes in 15 trials if $P(S) = 0.30$?
14. Answer Exercise 13 for $P(S) = 0.90$.

15. Bill believes that he can distinguish Democrats from Republicans by the way they dress. Jim disagrees and they bet a dollar. What is the probability that Bill collects a dollar from Jim and is only guessing if Jim agrees to pay Bill should he guess 9 out of 10 correctly?

16. In Exercise 15, what is the probability that Bill will have to pay Jim a dollar even though he can distinguish Republicans from Democrats 90% of the time?

17. Janet states that she can tell the difference in taste between butter and margarine. Helen disagrees and they bet a case of wine. To test Janet's ability, Helen will offer her 10 pats; 9 are butter and 1 is margarine or 9 are margarine and 1 is butter. Janet must pick the "odd" pat from the 10. This same test will be repeated 10 times. If Janet can pick the odd pat at least 5 times out of 10, she wins the case of wine. Otherwise, she must give Helen a case of wine. If Janet is just guessing, what is the probability that she has to pay Helen?

18. The persons of Exercise 17 repeat the same test over a number of months and discover that Janet can pass the test slightly over 60% of the time. Is she able to tell the difference between butter and margarine?

19. The current treatment of a disease will cure 10% of its victims. It is claimed that a new technique cures 70%. The new technique is to be used on 10 victims of the disease. What number of cures would imply that the new technique is superior to the old?

20. The results of the test in Exercise 19 were inconclusive. It is decided to use the new technique on a larger sample of 15 victims of the disease. Now how many cures would imply that the new treatment is superior to the old, as claimed?

21. Note in Table 4.8.2 that the numbers in the column under 0.10 are identical to the numbers under the column headed 0.90 but in reverse order. Why?

22. The Data Grind Corporation manufactures portable calculators. They have a number of machines that perform the final assembly phase of production. Each machine produces 15 calculators per day, and over the long run, when the machines are in perfect working order, 10% of the calculators are defective. What number of defective calculators from a single machine in one day would imply that the machine is not in perfect working order?

4.9 MARKOV CHAINS

The last two sections dealt with experiments that were Bernoulli trials. Of course, not all experiments are Bernoulli trials. In many cases the events are dependent, and there are more than two possibilities. *Markov chains* are a special technique for computing the probability that a certain event will result at the end of a chain of experiments. A sequence of experiments in which the outcome of one or more of the experiments is not certain is called a **stochastic process**. The sequence of experiments of rolling a die and then flipping a coin is a stochastic process. In fact, all the examples of probabilities presented so far are examples of stochastic processes. However, the phrase is usually reserved for more extensive sequences of experiments. Markov chains or the Markov process is a special technique used with certain finite stochastic processes.

DEFINITION

A sequence of experiments is a **Markov chain** if and only if the outcome of each experiment is one of a finite number of possible outcomes and any particular outcome depends at most on the outcome immediately preceding it.

The outcome of the individual experiments of a Markov chain are sometimes called **states**. For instance, suppose that the mayor of a certain city will either be a Republican or a Democrat. Then there are two possible *states* for any mayoral election: Republican or Democrat. Assuming that no other political parties are introduced into the city, it can be said that after many mayoral elections the city will be in one of the two states. The Markov process is involved with just this question: What is the probability that a given state will occur after a number of trials of the experiment?

The number of states can be extensive. It is convenient to arrange the probabilities that a particular state will change to another state in a **matrix**, a rectangular array of numbers. Matrix theory will be discussed in the next chapter. Much of matrix theory can be used with Markov chains, but here we will simply use a matrix as a device for keeping track of probabilities. For instance, assume that the matrix below represents the probabilities that a certain city will have either a Republican or a Democratic administration.

$$
\begin{array}{c}
\begin{array}{cc}
\quad \textit{Will} & \textit{Will} \\
\textit{Become} & \textit{Become} \\
\quad R & D
\end{array} \\
\begin{array}{c}
\textit{Now } R \\
\textit{Now } D
\end{array}
\begin{bmatrix}
0.4 & 0.6 \\
0.8 & 0.2
\end{bmatrix}
\end{array}
\qquad (1)
$$

The numbers are the probabilities of transitions from one state to another. The 0.4 is the probability that the next administration will be Republican, given that the present one is Republican. The 0.6 is the probability that the next administration will be Democratic given that the present one is Republican. The 0.8 is the probability that the present Democratic administration will be changed to a Republican one after the next election. The 0.2 is the probability that a Democratic administration will remain in office after the next election. The numbers in the matrix are called **transition probabilities**. The matrix is called a **transition matrix**. The pattern of probabilities is constant in transition matrices. The first row gives the probabilities for all possible outcomes given that a state, arbitrarily designated state 1—in this case Republicanism—exists. The second row gives the probabilities for all possible outcomes given that a state, arbitrarily called state 2, exists. The first row of matrix (1) shows the various probabilities of a change in administration given that the present administration is Republican. The second row gives probabilities for a change of administration given that the Democrats are in office. The probabilities in any row must have a sum of 1 since every possible outcome is represented and at least one must occur.

EXAMPLE Using matrix (1), above, and assuming that the city now has a Democratic administration, what is the probability that after two elections the administration will still be Democratic?

Solution Figure 4.9.1 is a tree diagram showing the possible sequence of outcomes for two elections starting with a Democrat in office. We are interested only in the branches that end in D, showing that a Democrat is in office. The numbers on the branches of the figure are assigned from matrix (1). The 0.8 on the branch from D to R is the probability that a Republican will be elected when a Democrat is in office. The 0.6 on the branch from R to D is the probability, from matrix (1), that a Democrat will be elected when the present administration is Republican. The product of 0.8 and 0.6 is the probability that a Republican will be elected after a Democrat *and* then another Democrat will be elected in the following election. The second sequence that ends with a Democrat in office has 0.04 probability of occurring. Thus the probability that a Democrat will be in office is the probability that either the sequence DRD or the sequence DDD occurs, which is the sum of the probabilities:

$$(0.8)(0.6) + (0.2)(0.2) = 0.48 + 0.04 = 0.52$$

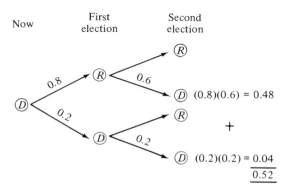

FIGURE 4.9.1 A Democrat is in office at the second election along lines DRD and DDD.

The above example shows the essential features of the Markov process —a transition matrix and a tree diagram to point out the sequence of events that are of interest. The answer is found by multiplying the probabilities along the branches of the tree diagram and then summing these products.

EXAMPLE   Use matrix (1) to find the probability that given there is presently a Republican mayor, there will be a Democrat in office after three elections.

Solution   Figure 4.9.2 is a tree diagram showing the possible sequences of three elections that start with a Republican and end with a Democrat in office. Notice that after the second election the tree diagram was not completely drawn. The branches leading to a Republican on the third election are of no use to us in this example and were omitted. The probability that a Democrat will be in office is 0.456. This was computed by multiplying the probabilities along the branches and then adding the products.

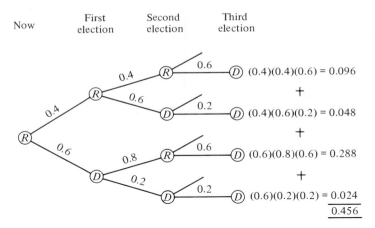

FIGURE 4.9.2   A Democrat is in office after the third election.

Markov chains may use very large matrices. The probabilities are placed in any matrix, regardless of its size, according to a set plan. Below is a 3×3 matrix.

$$\begin{bmatrix} p_{11} & p_{12} & p_{13} \\ p_{21} & p_{22} & p_{23} \\ p_{31} & p_{32} & p_{33} \end{bmatrix}$$

The elements of the matrix represent probabilities. The double subscript shows what row and column of the matrix contain the specific probability.

The element p_{21} is in the second row and first column. That is, the first number of the subscript gives the row and the second the column. Then p_{32} is in the third row and second column. Transition matrices are such that p_{32} is the probability of changing from state 3 to state 2. In general, if p_{ij} is an element of a transition matrix, it gives the probability of changing to state j, given that state i presently exists *and* is located in the ith row and jth column.

EXAMPLE   On any day the weather of San Francisco is one of three types: sunny, rainy, or foggy. Let state 1 be "It is sunny," state 2 be "It is rainy," and state 3 be "It is foggy." Suppose that matrix (2) gives the probabilities for tomorrow's weather type.

	Sunny Tomorrow	Rainy Tomorrow	Foggy Tomorrow	
Sunny Now	0.50	0.10	0.40	
Rainy Now	0.05	0.05	0.90	(2)
Foggy Now	0.25	0.25	0.50	

The first row of matrix (2) gives the probabilities for a sunny, rainy, or foggy day tomorrow, given that today is sunny in San Francisco. The second row contains the probabilities that tomorrow will be sunny, rainy, or foggy if it is rainy today. The third row contains the probabilities for the various weather types occurring tomorrow if it is foggy today.

EXAMPLE   Use matrix (2) to find the probability that it will be foggy two days from now if it is sunny today.

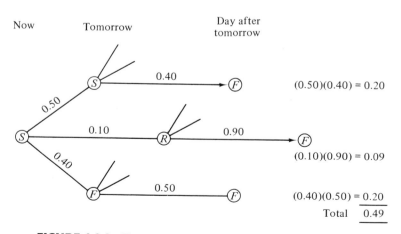

FIGURE 4.9.3   The probability of fog two days after a sunny day.

Solution   Figure 4.9.3 gives the part of the tree diagram necessary to find that the probability that a foggy day will occur two days after a sunny day in San Francisco is 49%.

The election example and the weather example of this section are, of course, overly simplistic. It is not possible to assign probabilities for tomorrow's weather based only on today's weather. Yet by adding more states to the weather the Markov process can be used in forecasting. These other states might be such things as "It is sunny and the barometer is steady, or falling, or rising." Barometric fluctuations in Alaska and Hawaii could also be included because the weather there affects San Francisco's weather. The transition matrix for weather forecasting or for election forecasting would be very large, but its use would involve the basic techniques presented here.

EXERCISES 4.9

In Exercises 1–8, use matrix A

$$A = \begin{bmatrix} \frac{1}{4} & \frac{1}{2} & 0 & \frac{1}{4} \\ \frac{1}{3} & \frac{1}{3} & 0 & \frac{1}{3} \\ \frac{1}{8} & \frac{1}{4} & \frac{1}{4} & \frac{3}{8} \\ 0 & \frac{1}{2} & \frac{1}{2} & 0 \end{bmatrix}$$

and give the probability that:

1. State 1 will occur if state 2 exists.
2. State 3 will occur if the system is in state 1.
3. State 4 will occur if the system is in state 3.
4. State 3 will occur if state 4 exists now.
5. State 2 will occur if state 4 exists.
6. State 4 will occur if the system is in state 1.
7. State 4 will occur immediately after the system is in state 4.
8. State 2 will occur immediately after the system is in state 2.

9. The following transition matrix shows the probabilities that a certain foreign country will be controlled by one of two political parties known as party 1 and 2.

$$\begin{bmatrix} \frac{1}{2} & \frac{1}{2} \\ \frac{1}{4} & \frac{3}{4} \end{bmatrix}$$

What is the probability that party 2 will be in control of the country after three elections if party 1 is in control now?

10. In Exercise 9, what is the probability that party 1 will be in control after three elections?

11. The following matrix gives the probabilities that one of three weather types, known as type I, type II, and type III, will occur on the following day in a certain midwestern town.

$$\begin{bmatrix} \frac{1}{3} & \frac{1}{3} & \frac{1}{3} \\ 0 & 1 & 0 \\ \frac{1}{2} & 0 & \frac{1}{2} \end{bmatrix}$$

What is the probability that weather type II will occur three days after type III occurs?

12. In Exercise 11, what is the probability that type I will occur three days after type III occurs? What is the probability that type III will occur three days after type III occurs?

13. Assume that the probability that a tall man fathers a tall son is 0.70 and that the probability that a short man fathers a short son is 0.80. Disregarding the possibility of medium-height offspring, what is the probability that a tall man will have a tall grandson?

14. In Exercise 13, what is the probability that a short man will have a tall great-grandson?

15. Grape crops are classified each year as vintage, good, or poor. Based upon the previous year's crop only, it is known that the probabilities for a vintage, good, or poor year following a vintage year are 0.5, 0.4, and 0.1, respectively. The probabilities of a vintage, good, or poor year after a good year are, respectively, 0.4, 0.4, 0.2. The probabilities of a vintage or a good year after a poor year are 0.2 and 0.6, respectively. The year 1972 was a vintage year. What is the probability that 1975 will be a vintage year?

16. Using the data of Exercise 15, find the probability that 1975 will be a poor year.

17. In 1970 three computer companies, X, Y, and Z, dominated the field. For practical purposes it can be said that they had all the computer business, with X controlling 70% of the business, Y controlling 20%, and Z controlling 10%. However, every five years customers buy new computer equipment, and there is much switching among customers. The switching has been so extensive that probabilities have been assigned. If a customer is using brand X, there is a 70% probability that he will continue to use it when he buys new equipment, a 15% probability that he will buy brand Y, and a 15% probability that he will buy brand Z. If a customer is using brand Y, there is a 50% probability that he will continue to do so and a 40% probability that he will switch to brand X.

If a customer is using brand Z, there is a 40% probability that he will continue with that brand and a 40% probability that he will switch to brand X. What shares of the computer market will each brand have in 1975 if the probabilities given remain constant?

18. Using the data of Exercise 17, what portion of the computer market does each company have in 1980?

REVIEW EXERCISES

In Exercises 1–4, if a single card is drawn from a well-shuffled poker deck, what is the probability that the drawn card is:

1. A face card?   2. Not a 3?

3. A 9 or a diamond?   4. A spade or a face card?

5. If a coin is flipped three times, what is the probability of getting heads twice? Of getting heads at least twice?

Use the facts that A and B are subsets of an equiprobable sample space $\mathcal{U}$, $n(\mathcal{U})$ = 100, $n(A) = 50$, $n(B) = 70$, $n(A \cap B) = 20$ to compute Exercises 6–11.

6. $P(A)$ 7. $P(A \cup B)$ 8. $P(A$ and $B)$
9. $P(A \mid B)$ 10. $P(B \mid A)$ 11. $P(A' \mid B')$

12. A student judges that his probability of passing mathematics is 70% and his probability of passing history is 80%. What is his probability of passing both? Of passing at least one course?

13. Use Figure 4.4.1 to find the probability that at least two members of the United States Senate have the same birthday (month and day, not year).

14. In a lot of 500 light bulbs 20 are defective. An inspector checks 2 light bulbs at random from the lot. What is the probability that both are defective?

15. You and a friend are wagering on the roll of a single die. You will pay your friend $1 if you fail to roll a 5 and he will pay you $6 if you roll a 5. What is your expected value?

16. In Nevada, for a $1 bet you can collect $34 and keep your $1 if you roll two spots with a pair of dice on a single roll. If you fail to roll the two spots you lose your $1. What is your expected value?

17. The probability that a child under the age of 10 will live to age 23 is 99%. Assuming that no other costs are involved, what would be a fair single premium for a $1,000 life insurance policy to age 23 for the child?

18. The probability that any child of a certain couple will be a genius is 20%. What is the probability that they have two geniuses in a family of five children?

19. Use Table 4.9.2 to find the probability of at least 7 successes in 15 Bernoulli trials if $P(S) = 0.70$.

20. The current treatment of disease X cures 70% of its victims. It is claimed that a new treatment will cure 90%. The new technique is to be used on 15 victims of the disease. How many must be cured to show that the new technique is superior to the old?

21. The transition matrix below gives the probabilities that a certain stock will go up or down. It never stays the same. State 1 is *up* and state 2 is *down*.

$$\begin{bmatrix} 0.50 & 0.50 \\ 0.40 & 0.60 \end{bmatrix}$$

What is the probability that the stock will be up three days after it was down?

22. Data Grind stock will go up or down or stay the same. Based only on its state the day before, it is found that if Data Grind is up, then the probability that it will go up or down the next day is 60% and 10%, respectively. If the stock is down, the probability that it will be up or down the next day is 50% and 50%, respectively. If Data Grind stock did not change, the probability that it will be up the next day is 100%. Find the probability that Data Grind stock will be up four days after the price of the stock did not change.

CHAPTER **5**

LINEAR ALGEBRA

This chapter will consider one of those branches of mathematics that has found wide applications in business and economics and in the biological and physical sciences. Its formal mathematical name is linear algebra, and it is formally defined as a class of mathematical structures following a prescribed and well-defined pattern. It is beyond the scope of our discussion to consider this formal structure. Instead, we will examine some theory and applications of vectors and matrices. Matrix theory and its special case, vector theory, are the stereotypes from which the abstract mathematical structure of linear algebra was developed.

The growth of the application of vector and matrix theory to economic problems has occurred for a number of reasons. First, as we will see, the vector and matrix forms provide an ideal means for compactly and symbolically describing many situations. Second, the procedures used on matrix forms are analogous to those which would have to be applied to the economic concepts they model for analysis. The third and perhaps most far-reaching factor has been the development of the digital computer. The relationship between the economic and mathematical structures has been known for a relatively long time. However, the calculations one encounters can be long and involved when the matrix models are used. In fact, they are too long and too involved to be practical with only hand calculation. Modern data processing systems change the whole concept of a "long and involved" calculation. Thus the once impractical matrix methods have become the simplest and fastest problem-solving methods available in many cases.

In light of these considerations, the reader is warned that some of the procedures that follow are somewhat lengthy. We hope to examine how these methods work and how they may be applied. In a real-world, practical sense, some form of automatic data processing equipment is needed for their application.

5.1   VECTORS AND VECTOR ARITHMETIC

Many readers will have already encountered the vector concept, usually in an elementary physics course. There a vector is usually viewed as something that requires a magnitude and direction of action for description. Vector quantities are viewed as directed line segments or arrows, and their arithmetic is often defined in geometric terms. Vectors as applied to economic theory are usually more complex than their physical counterparts; thus they are usually too complex for this informal geometric approach. This is because their descriptions require more than three dimensions, and our geometric intuition cannot extend to such cases.

DEFINITION

An **n-dimensional row vector** is an ordered set of n real numbers arranged in a row.

NOTATION

$$[a_1, a_2, \ldots, a_n]$$

will be used to denote an n-dimensional row vector. The individual real numbers, $a_1, a_2, a_3, \ldots, a_n$, are called the **components** or **elements** of the vector, and a_1 is read "a sub one," a_2 as "a sub two," and so forth.

EXAMPLES

$[3, 6, -2]$ denotes a three-dimensional row vector with the first component 3, the second component 6, and the third component -2. $[1, 0, 1, -1, 4, -\frac{1}{2}, 6]$ denotes a seven-dimensional row vector.

DEFINITION

Two n-dimensional row vectors $[a_1, a_2, \ldots, a_n]$ and $[b_1, b_2, \ldots, b_n]$ are equal if and only if $a_1 = b_1, a_2 = b_2, a_3 = b_3, \ldots, a_n = b_n$. Alternatively, $a_i = b_i$ for all $i = 1, 2, 3, \ldots, n$.

EXAMPLES

$[1, 2, -4] = [1, 2, -4]$ because $1 = 1$, $2 = 2$, and $-4 = -4$.

$[2, 7, 3, -5] = [x, y, z, w]$ if and only if $2 = x$, $7 = y$, $3 = z$, and $-5 = w$.

$[-4, 0, -3.1] \neq [-4, 1, -3.1]$ because the second components do not match.

$[1, 0, 1, 1] \neq [1, 0, 1, 1, 1]$ because one is a four-dimensional vector and the other is a five-dimensional vector.

$[-4, 1, 3] \neq [1, -4, 3]$ because the components are not equal in order.

Sometimes we consider column vectors.

DEFINITION

An **n-dimensional column vector** is an ordered set of n real numbers arranged in a column.

NOTATION

$$\begin{bmatrix} a_1 \\ a_2 \\ \vdots \\ a_n \end{bmatrix}$$

will denote an n-dimensional column vector. The real numbers $a_1, a_2, \ldots, a_n$ are the **components** or **elements** of the column vector.

DEFINITION

Two n-dimensional column vectors

$$\begin{bmatrix} a_1 \\ a_2 \\ \vdots \\ a_n \end{bmatrix} \quad \text{and} \quad \begin{bmatrix} b_1 \\ b_2 \\ \vdots \\ b_n \end{bmatrix}$$

are equal if and only if $a_1 = b_1$, $a_2 = b_2, \ldots, a_n = b_n$. Alternatively, $a_i = b_i$ for all $i = 1, 2, \ldots, n$.

In other words, two column vectors are equal if and only if they have the same dimension and their components taken in order are equal. Likewise, two row vectors are equal if and only if they have the same dimension and their components taken in order are equal.

EXAMPLE
$$\begin{bmatrix} -1 \\ 2 \\ \frac{1}{2} \end{bmatrix} = \begin{bmatrix} -1 \\ 2 \\ x \end{bmatrix}$$

if and only if $x = \frac{1}{2}$.

EXAMPLE $\begin{bmatrix} 2 \\ 7 \end{bmatrix} \neq [2 \quad 7]$ because one is a column vector and the other a row vector.

The individual real numbers are called **scalars** to distinguish them from the vectors. While *we will consider only real numbers as scalars in this text*, it is possible to construct a system of vectors using complex numbers as the scalars. As a general convention we will use capital letters as the names of vectors and lowercase letters as their scalar components.

EXAMPLE If $A = [a_1, a_2, a_3, a_4]$ and $B = [1, 6, -1, 4]$, under what conditions does $A = B$?

Solution $A = B$ if and only if $a_1 = 1$, $a_2 = 6$, $a_3 = -1$, and $a_4 = 4$.

EXAMPLE If $C = [c_1, c_2, c_3]$ and

$$D = \begin{bmatrix} d_1 \\ d_2 \\ d_3 \end{bmatrix}$$

under what conditions does $C = D$?

Solution $C \neq D$ because C is a row vector and D is a column vector.

We are now able to define arithmetic operations on vectors.

DEFINITION

Let $A = [a_1, a_2, \ldots, a_n]$ and $B = [b_1, b_2, \ldots, b_n]$. Then

$$A + B = [a_1 + b_1, a_2 + b_2, a_3 + b_3, \ldots, a_n + b_n]$$

In other words, the sum of two n-dimensional row vectors is the n-dimensional row vector found by pairwise adding the components of the original vectors.

EXAMPLE If $A = [1, 3, -5, 2]$ and $B = [0, 4, -4, -2]$, find $A + B$.

Solution $A + B = [1 + 0, 3 + 4, -5 + (-4), 2 + (-2)]$
$$= [1, 7, -9, 0]$$

EXAMPLE   If $C = [1, 3, 0, 7]$ and $D = [2, 1, 6]$, find $C + D$.

Solution $C + D$ is undefined because the dimensions of C and D do not match.

DEFINITION

Let

$$A = \begin{bmatrix} a_1 \\ a_2 \\ \vdots \\ a_n \end{bmatrix} \quad \text{and} \quad B = \begin{bmatrix} b_1 \\ b_2 \\ \vdots \\ b_n \end{bmatrix}$$

Then

$$A + B = \begin{bmatrix} a_1 + b_1 \\ a_2 + b_2 \\ \vdots \\ a_n + b_n \end{bmatrix}$$

EXAMPLE   If

$$A = \begin{bmatrix} 1 \\ 3 \\ -5 \\ 2 \end{bmatrix} \quad \text{and} \quad B = \begin{bmatrix} 0 \\ 4 \\ -4 \\ -2 \end{bmatrix}$$

find $A + B$.

Solution

$$A + B = \begin{bmatrix} 1 + 0 \\ 3 + 4 \\ -5 + (-4) \\ 2 + (-2) \end{bmatrix} = \begin{bmatrix} 1 \\ 7 \\ -9 \\ 0 \end{bmatrix}$$

EXAMPLE   If

$$C = \begin{bmatrix} 1 \\ 3 \\ 0 \\ 7 \end{bmatrix} \quad \text{and} \quad D = \begin{bmatrix} 2 \\ 1 \\ 6 \end{bmatrix}$$

find $C + D$.

Solution $C + D$ is not defined because the dimension of C is 4 and that of D is 3.

EXAMPLE If

$$A = \begin{bmatrix} 1 \\ 3 \\ -5 \\ 2 \end{bmatrix}$$

and $B = [0, 4, -4, -2]$, find $A + B$.

Solution $A + B$ is not defined. Even though both A and B are vectors of dimension 4, A is a column vector and B is a row vector; thus their sum is not defined.

Vector addition obeys the same rules as the addition of real numbers because its components are real numbers. That is, if A, B, and C are all n-dimensional vectors (all row or all column),

$$A + B = B + A \qquad \text{(Commutative law)}$$

and

$$A + (B + C) = (A + B) + C \qquad \text{(Associative law)}$$

EXAMPLE Let $A = [1, 2]$, $B = [3, 4]$, and $C = [5, 6]$. Then

$$A + B = [1 + 3, 2 + 4] = [4, 6]$$
$$B + A = [3 + 1, 4 + 2] = [4, 6] = A + B$$

and

$$
\begin{aligned}
A + (B + C) &= [1, 2] + ([3, 4] + [5, 6]) \\
&= [1, 2] + ([3 + 5, 4 + 6]) \\
&= [1, 2] + [8, 10] \\
&= [1 + 8, 2 + 10] \\
&= [9, 12]
\end{aligned}
$$

and

$$
\begin{aligned}
(A + B) + C &= ([1, 2] + [3, 4]) + [5, 6] \\
&= ([1 + 3, 2 + 4]) + [5, 6] \\
&= [4, 6] + [5, 6] \\
&= [4 + 5, 6 + 6] \\
&= [9, 12] = A + (B + C)
\end{aligned}
$$

The definition for the difference of two vectors follows that for vector sum.

DEFINITION

If $A = [a_1, a_2, \ldots, a_n]$ and $B = [b_1, b_2, \ldots, b_n]$ are two n-dimensional row vectors,

$$A - B = [a_1 - b_1, a_2 - b_2, \ldots, a_n - b_n]$$

DEFINITION

If

$$A = \begin{bmatrix} a_1 \\ \vdots \\ a_n \end{bmatrix} \quad \text{and} \quad B = \begin{bmatrix} b_1 \\ \vdots \\ b_n \end{bmatrix}$$

are two n-dimensional column vectors, then

$$A - B = \begin{bmatrix} a_1 - b_1 \\ a_2 - b_2 \\ \vdots \\ a_n - b_n \end{bmatrix}$$

EXAMPLE   If

$$A = \begin{bmatrix} 1 \\ 3 \\ -4 \\ 2 \end{bmatrix}, \quad B = \begin{bmatrix} 0 \\ 3 \\ 2 \\ -3 \end{bmatrix}$$

$C = [1, 2, 4]$, and $D = [1, 2, 4, 7]$, find $A - B$, $A - C$, $C - D$.

Solution

$$A - B = \begin{bmatrix} 1 \\ 3 \\ -4 \\ 2 \end{bmatrix} - \begin{bmatrix} 0 \\ 3 \\ 2 \\ -3 \end{bmatrix} = \begin{bmatrix} 1 - 0 \\ 3 - 3 \\ -4 - 2 \\ 2 - (-3) \end{bmatrix} = \begin{bmatrix} 1 \\ 0 \\ -6 \\ 5 \end{bmatrix}$$

$A - C$ and $C - D$ are undefined.

DEFINITION

Any n-dimensional vector with all zero components will be denoted $\bar{0}$. Its dimensionality will be chosen to match the context of the discussion.

EXAMPLE   If $A = [1, 3, 4, -5]$, find $A + \bar{0}$.

Solution Since A is a four-dimensional vector,

$$\bar{0} = [0, 0, 0, 0]$$

and

$$\begin{aligned}
A + \bar{0} &= [1, 3, 4, -5] + [0, 0, 0, 0] \\
&= [1 + 0, 3 + 0, 4 + 0, -5 + 0] \\
&= [1, 3, 4, -5] \\
&= A
\end{aligned}$$

EXAMPLE If

$$B = \begin{bmatrix} 1 \\ -2 \\ 7 \end{bmatrix}$$

find a three-dimensional column vector

$$X = \begin{bmatrix} x_1 \\ x_2 \\ x_3 \end{bmatrix}$$

such that $B + X = \bar{0}$.

Solution From the context,

$$\bar{0} = \begin{bmatrix} 0 \\ 0 \\ 0 \end{bmatrix}$$

Thus $B + X = \bar{0}$ is

$$\begin{bmatrix} 1 \\ -2 \\ 7 \end{bmatrix} + \begin{bmatrix} x_1 \\ x_2 \\ x_3 \end{bmatrix} = \begin{bmatrix} 0 \\ 0 \\ 0 \end{bmatrix}$$

$$\begin{bmatrix} 1 + x_1 \\ -2 + x_2 \\ 7 + x_3 \end{bmatrix} = \begin{bmatrix} 0 \\ 0 \\ 0 \end{bmatrix}$$

Applying the definition of vector equality, this is true if and only if

$$\begin{aligned}
1 + x_1 &= 0 \\
-2 + x_2 &= 0 \\
7 + x_3 &= 0
\end{aligned}$$

or

$$\begin{aligned}
x_1 &= -1 \\
x_2 &= 2 \\
x_3 &= -7
\end{aligned}$$

or

$$X = \begin{bmatrix} -1 \\ 2 \\ -7 \end{bmatrix}$$

The vectors symbolized by $\bar{0}$ are identities for vector addition. That is, for any vector A,

$$A + \bar{0} = \bar{0} + A = A$$

What kinds of problems require vector descriptions? The following examples are illustrations.

EXAMPLE A company has five major divisions:

α division
β division
γ division
δ division
ϵ division

It keeps track of the number of man-hours required to manufacture a given product by assigning a 5-dimensional row vector with the first component the required number of hours from its α division, the second component the number of hours required from the β division, etc. Thus if

$$A = [3, 6, 0, 4, 0]$$

is the vector describing the manufacture of some item, this indicates that the manufacture of this item requires

3 man-hours from α division
6 man-hours from β division
0 man-hours from γ division
4 man-hours from δ division
0 man-hours from ϵ division

EXAMPLE Suppose that the federal government identifies the cost of 6 key commodities with the use of a six-dimensional column vector. Each component corresponds to the cost of a fixed amount of each item. If in May the cost vector is

$$\begin{bmatrix} 1.10 \\ 2.39 \\ 4.00 \\ 2.99 \\ 4.63 \\ 10.17 \end{bmatrix}$$

and in June it is

$$\begin{bmatrix} 1.16 \\ 2.41 \\ 4.00 \\ 2.93 \\ 4.70 \\ 10.10 \end{bmatrix}$$

compute the change in the cost of these items in vector form.

Solution The change would be given by the vector difference:

$$\begin{bmatrix} 1.16 \\ 2.41 \\ 4.00 \\ 2.93 \\ 4.70 \\ 10.10 \end{bmatrix} - \begin{bmatrix} 1.10 \\ 2.39 \\ 4.00 \\ 2.99 \\ 4.63 \\ 10.17 \end{bmatrix} = \begin{bmatrix} 0.06 \\ 0.02 \\ 0 \\ -0.06 \\ 0.07 \\ -0.07 \end{bmatrix}$$

EXERCISES 5.1

Determine the values of the literal expressions in Exercises 1–6 so that the indicated vector equality holds.

1. $[1, 4, x] = [1, 4, -3]$

2.
$$\begin{bmatrix} 1 \\ 2 \\ 2x \\ y \end{bmatrix} = \begin{bmatrix} 1 \\ 2 \\ y \\ 2x \end{bmatrix}$$

3.
$$[-1, 0, -1, 0] = \begin{bmatrix} -1 \\ 0 \\ x \\ 0 \end{bmatrix}$$

4. $[(x + 1)^2, y, 3] = [x^2 + 2x + 1, 3, y]$

5. $[(x + 1)(x - 1), 3] = [x^2 - 1, 3]$

6.
$$\begin{bmatrix} 1 \\ 4 \\ -2 \end{bmatrix} = \begin{bmatrix} 1 \\ x \\ -2 \\ y \end{bmatrix}$$

In Exercises 7–12, find the indicated sums and differences.

7. $[1, -4, 2, 3] + [-4, -3, 2, 1]$

8. $\begin{bmatrix} -2 \\ -1 \\ -7 \\ 7 \end{bmatrix} - \begin{bmatrix} 6 \\ 0 \\ 2 \\ -3 \end{bmatrix}$

9. $\begin{bmatrix} -1 \\ 4 \\ 4 \end{bmatrix} + \left(\begin{bmatrix} 6 \\ 1 \\ 6 \end{bmatrix} - \begin{bmatrix} 9 \\ 9 \\ -5 \end{bmatrix} \right)$

10. $[\frac{1}{2}, -\frac{1}{3}, 3, -5] - [\frac{1}{6}, \frac{1}{3}, \frac{1}{2}, -\frac{4}{5}]$

11. $[4, 2, 9, -5] - [-4, -2, -9, 5, 1]$

12. $([2, 3] + [-4, 3]) - ([-4, 2] + [0, -3])$

In Exercises 13 and 14, if $A = [1, -4, 6]$ and B is a three-dimensional row vector, find B such that:

13. $A + B = \bar{0}$ **14.** $A - B = \bar{0}$

If A, B, and C are two-dimensional row vectors with $A = [1, 2]$ and $B = [-2, 1]$, determine what conditions must exist on C if the statements in Exercises 15–18 are true.

15. $A + (B - C) = (A + B) - C$ **16.** $A - (B + C) = (A - B) + C$
17. $A - (B - C) = (A - B) - C$ **18.** $A + (B + C) = (A + B) + C$

Exercises 19–21 are based on the following:

    A company has four sales divisions identified as Eastern, Southern, Midwestern, and Western. Set up a vector notation indicating the sales of each division in thousands of dollars per quarter (three-month period).

19. Write the following information in vector notation:

Division	1st	2nd	3rd	4th
		Quarter		
Eastern	1,700	2,000	2,250	1,975
Southern	1,500	1,800	1,600	1,915
Midwestern	2,000	2,230	2,275	2,540
Western	2,275	2,500	2,600	2,575

20. Using the vector notation developed in Exercise 19, describe in vector notation the annual sales of the company.

21. Using the vector notation developed in Exercise 19, describe in vector notation the net change in sales between two successive quarters.

5.2 VECTORS AND PRODUCTS

We will define two types of multiplication involving vectors.

DEFINITION

Let $A = [a_1, a_2, \ldots, a_n]$ be an n-dimensional row vector and c be any scalar (real number). Then the product cA is defined as

$$cA = [ca_1, ca_2, ca_3, ca_4, \ldots, ca_n]$$

This type of multiplication is called **scalar multiplication of a vector,** and it always results in a vector.

EXAMPLE   Let $B = [1, 5, 7, -4]$. Find $3B$.

Solution $3B = 3[1, 5, 7, -4] = [3 \cdot 1, 3 \cdot 5, 3 \cdot 7, 3 \cdot (-4)]$

$$= [3, 15, 21, -12]$$

EXAMPLE   If $X = [x_1, x_2]$, find X such that $[1, -5] + 2X = \bar{0}$.

Solution   The vector equation becomes

$$[1, -5] + 2[x_1, x_2] = [0, 0]$$

$$[1, -5] + [2x_1, 2x_2] = [0, 0]$$

$$[1 + 2x_1, -5 + 2x_2] = [0, 0]$$

Applying vector equality,

$$1 + 2x_1 = 0 \quad \text{and} \quad -5 + 2x_2 = 0$$

$$2x_1 = -1 \qquad\qquad 2x_2 = 5$$

$$x_1 = -\tfrac{1}{2} \qquad\qquad x_2 = \tfrac{5}{2}$$

Thus $X = [-\tfrac{1}{2}, \tfrac{5}{2}]$.

Scalar multiplication of column vectors follows the same pattern.

DEFINITION

Let

$$A = \begin{bmatrix} a_1 \\ a_2 \\ \vdots \\ a_n \end{bmatrix}$$

be an n-dimensional column vector and c be any scalar (real number). Then

$$cA = c \begin{bmatrix} a_1 \\ a_2 \\ \vdots \\ a_n \end{bmatrix} = \begin{bmatrix} ca_1 \\ ca_2 \\ \vdots \\ ca_n \end{bmatrix}$$

EXAMPLE A company with five divisions uses vectors to denote the number of man-hours required for the manufacture of certain items. Specifically,

$$A = \text{time vector for item } a = \begin{bmatrix} 1 \\ 0 \\ 3 \\ 0 \\ 5 \end{bmatrix}$$

$$B = \text{time vector for item } b = \begin{bmatrix} 1 \\ 0 \\ 2 \\ 1 \\ 3 \end{bmatrix}$$

$$C = \text{time vector for item } c = \begin{bmatrix} 0 \\ 2 \\ 0 \\ 2 \\ 1 \end{bmatrix}$$

Write in vector form the man-hour vector describing the manufacture of 60 of item a, 40 of item b, and 52 of item c.

Solution In vector form the solution is

$$60A + 40B + 52C = 60 \begin{bmatrix} 1 \\ 0 \\ 3 \\ 0 \\ 5 \end{bmatrix} + 40 \begin{bmatrix} 1 \\ 0 \\ 2 \\ 1 \\ 3 \end{bmatrix} + 52 \begin{bmatrix} 0 \\ 2 \\ 0 \\ 2 \\ 1 \end{bmatrix}$$

$$= \begin{bmatrix} 60 \\ 0 \\ 180 \\ 0 \\ 300 \end{bmatrix} + \begin{bmatrix} 40 \\ 0 \\ 80 \\ 40 \\ 120 \end{bmatrix} + \begin{bmatrix} 0 \\ 104 \\ 0 \\ 104 \\ 52 \end{bmatrix}$$

$$= \begin{bmatrix} 60 + 40 + 0 \\ 0 + 0 + 104 \\ 180 + 80 + 0 \\ 0 + 40 + 104 \\ 300 + 120 + 52 \end{bmatrix} = \begin{bmatrix} 100 \\ 104 \\ 260 \\ 144 \\ 472 \end{bmatrix}$$

The second kind of vector multiplication is defined only between two vectors of the same dimension, but one a row vector and one a column vector. The result is a scalar.

DEFINITION

Let $A = [a_1, a_2, \ldots, a_n]$ be an n-dimensional row vector and

$$B = \begin{bmatrix} b_1 \\ \vdots \\ b_n \end{bmatrix}$$

be an n-dimensional column vector. The **dot** or **scalar** or **inner** product of A and B, $A \cdot B$, is given by

$$A \cdot B = a_1 b_1 + a_2 b_2 + a_3 b_3 + \cdots + a_n b_n$$

EXAMPLE Let $A = [1, 2, -4, 3]$ and

$$B = \begin{bmatrix} -1 \\ 3 \\ 8 \\ 6 \end{bmatrix}$$

Find $A \cdot B$.

Solution

$$A \cdot B = [1, 2, -4, 3] \cdot \begin{bmatrix} -1 \\ 3 \\ 8 \\ 6 \end{bmatrix} = 1(-1) + 2(3) + (-4)(8) + 3(6)$$

$$= -1 + 6 - 32 + 18$$
$$= -9$$

EXAMPLE   Let $C = [1, -1]$ and $D = \begin{bmatrix} -2 \\ 2 \end{bmatrix}$. Find $C \cdot D$ and $D \cdot C$.

Solution

$$C \cdot D = [1, -1] \cdot \begin{bmatrix} -2 \\ 2 \end{bmatrix} = 1(-2) + (-1)(2) = -4$$

$$D \cdot C = \begin{bmatrix} -2 \\ 2 \end{bmatrix} \cdot [1, -1], \text{ which is not defined.}$$

EXAMPLE   For A and D as in the two previous examples, find $A \cdot D$.

Solution $A \cdot D = [1, 2, -4, 3] \cdot \begin{bmatrix} -2 \\ 2 \end{bmatrix}$, which is not defined because the two vectors do not match.

EXAMPLE   Find and compare $3(A \cdot B)$ and $(3A) \cdot B$ if $A = [1, -4]$ and $B = \begin{bmatrix} \frac{1}{2} \\ 6 \end{bmatrix}$.

Solution

$$A \cdot B = [1, -4] \cdot \begin{bmatrix} \frac{1}{2} \\ 6 \end{bmatrix}$$
$$= 1(\tfrac{1}{2}) + (-4)(6)$$
$$= \tfrac{1}{2} - 24 = -\tfrac{47}{2}$$
$$3(A \cdot B) = 3(-\tfrac{47}{2}) = -\tfrac{141}{2}$$
$$3A = 3[1, -4] = [3, -12]$$
$$(3A) \cdot B = [3, -12] \cdot \begin{bmatrix} \frac{1}{2} \\ 6 \end{bmatrix}$$
$$= 3(\tfrac{1}{2}) + (-12)6$$
$$= \tfrac{3}{2} - 72 = -\tfrac{141}{2}$$

Notice that the dot product of two vectors is always a scalar (real number).

EXAMPLE   In the five-division company described in the example on page 178, the following cost vector gives the cost per man-hour per division (entries in $):

$$\underline{D} = [12, 7.50, 15, 31, 20]$$

Find the cost of manufacturing each of the three items a, b, and c.

Solution   The cost for each item can be found by "dotting" the cost vector into vectors A, B, and C. That is,

Cost to make item $a = D \cdot A$

$$= [12, 7.50, 15, 31, 20] \cdot \begin{bmatrix} 1 \\ 0 \\ 3 \\ 0 \\ 5 \end{bmatrix}$$

$$= (12 \times 1) + (7.50 \times 0) + (15 \times 3)$$
$$+ (31 \times 0) + (20 \times 5)$$

$$= 12 + 45 + 100$$

$$= \$157$$

Cost to make item $b = D \cdot B$

$$= [12, 7.50, 15, 31, 20] \cdot \begin{bmatrix} 1 \\ 0 \\ 2 \\ 1 \\ 3 \end{bmatrix}$$

$$= (12 \times 1) + (7.50 \times 0) + (15 \times 2)$$
$$+ (31 \times 1) + (20 \times 3)$$

$$= 12 + 30 + 31 + 60 = \$133$$

Cost to make item $c = D \cdot C$

$$= [12, (7.50, 15, 31, 20] \cdot \begin{bmatrix} 0 \\ 2 \\ 0 \\ 2 \\ 1 \end{bmatrix}$$

$$= (12 \times 0) + (7.50 \times 2) + (15 \times 0)$$
$$+ (31 \times 2) + (20 \times 1)$$

$$= 15 + 62 + 20 = \$97$$

One can form terms which can use both the products of vectors and their sums. This is done through the distributive laws.

> **THEOREM: Distributive Law for the Product of Scalars and Vectors**
>
> Let A and B be n-dimensional vectors of the same type (i.e., both row or both column vectors). If c is any scalar,
>
> $$c(A + B) = cA + cB$$

> **THEOREM: Distributive Law for the Product of Two Vectors**
>
> Let A and B be n-dimensional column vectors and C be an n-dimensional row vector. Then
>
> $$C \cdot (A + B) = C \cdot A + C \cdot B$$

We will provide a proof of this second distributive law for vectors.

Proof: Let

$$A = \begin{bmatrix} a_1 \\ \vdots \\ a_n \end{bmatrix}, \quad B = \begin{bmatrix} b_1 \\ \vdots \\ b_n \end{bmatrix}$$

and $C = [c_1, c_2, \ldots, c_n]$. Then

$$A + B = \begin{bmatrix} a_1 + b_1 \\ a_2 + b_2 \\ \vdots \\ a_n + b_n \end{bmatrix}$$

and

$$C \cdot (A + B) = [c_1, \ldots, c_n] \cdot \begin{bmatrix} a_1 + b_1 \\ \vdots \\ a_n + b_n \end{bmatrix}$$

$$= c_1(a_1 + b_1) + c_2(a_2 + b_2) + \cdots + c_n(a_n + b_n)$$

$$= c_1 a_1 + c_1 b_1 + c_2 a_2 + c_2 b_2 + \cdots + c_n a_n + c_n b_n$$

$$C \cdot A = [c_1 \cdots c_n] \cdot \begin{bmatrix} a_1 \\ \vdots \\ a_n \end{bmatrix} = c_1 a_1 + c_2 a_2 + \cdots + c_n a_n$$

$$C \cdot B = [c_1 \cdots c_n] \cdot \begin{bmatrix} b_1 \\ \vdots \\ b_n \end{bmatrix} = c_1 b_1 + c_2 b_2 + \cdots + c_n b_n$$

$$C \cdot A + C \cdot B = c_1 a_1 + c_2 a_2 + \cdots + c_n a_n + c_1 b_1 + \cdots + c_n b_n$$

Rearranging terms,

$$= c_1 a_1 + c_1 b_1 + c_2 a_2 + c_2 b_2 + \cdots + c_n a_n + c_n b_n$$
$$= C \cdot (A + B)$$

The proof of the first distributive law is left as an exercise.

EXERCISES 5.2

Find the indicated products and sums in Exercises 1–6.

1. $3[7, 2, 9, -5] - 2[-1, 4, 3, 6]$

2. $\frac{1}{2}[4, 3, -7, 5] + 3[-1, 0, -1, 0]$

3.
$$6 \begin{bmatrix} 1 \\ 4 \\ 3 \end{bmatrix} - 2 \begin{bmatrix} 3 \\ -5 \\ 0 \end{bmatrix} + 3 \begin{bmatrix} 2 \\ 2 \\ 2 \end{bmatrix}$$

4. $0.5 \begin{bmatrix} 2 \\ 1 \end{bmatrix} - 0.3 \begin{bmatrix} 0.3 \\ 0.1 \end{bmatrix} + 2 \begin{bmatrix} 1 \\ 2 \end{bmatrix}$

5.
$$2[4, 3, 2] + \begin{bmatrix} -2 \\ 1 \\ 3 \end{bmatrix}$$

6. $4[2, 3] - 2[1, 5] - 3 \begin{bmatrix} 2 \\ 1 \end{bmatrix}$

Let $X = [x_1, x_2, x_3]$. Find X in Exercises 7–10.

7. $2X - [4, -1, 3] = \bar{0}$

8. $4[0, 1, 0] = 2X$

9. $3X + [4, 4, 4] = 2X - [3, 3, 3]$

10. $0[3, 4, 6] = 0X$

Exercises 11–14 are based on the following data:
   A company has three divisions. The company uses a time/man-hour vector (load vector) to describe work loads for various projects. The divisions are identified as α division, β division, and γ division. Load vectors are three-dimensional column vectors with the first component the load for α division, the second component the load for β division, and the third component the load for γ division.

Exercises 11 and 12 are based on the following data:

    For one of the products it makes, the company assigns a load vector of

$$\begin{bmatrix} 1.5 \\ 6.3 \\ 2.1 \end{bmatrix}$$

11. What is the total number of man-hours required to make five of these items?

12. The company has 14 hours in α division, 72 hours in β division, and 18 hours in γ division. Consider the item whose manufacture is described in Exercise 11. What is the maximum number of this item that can be manufactured with the available time?

Exercises 13 and 14 are based on the following data:

    Considering three items a, b, and c, the company assigns the following work load vectors:

$$item\ a:\ \begin{bmatrix} 0.7 \\ 4.3 \\ 1.0 \end{bmatrix}, \quad item\ b:\ \begin{bmatrix} 0 \\ 3.7 \\ 2.1 \end{bmatrix}, \quad item\ c:\ \begin{bmatrix} 1.1 \\ 0.7 \\ 3.6 \end{bmatrix}$$

13. Find a vector describing the number of man-hours required in each division to make 30 of item a, 40 of item b, and 16 of item c.

14. If N of item a and M of item b are to be made, the company finds a total man-hour load vector of

$$\begin{bmatrix} 3.5 \\ 47.4 \\ 19.7 \end{bmatrix}$$

Find N and M. *Hint:*

$$N \begin{bmatrix} 0.7 \\ 4.3 \\ 1.0 \end{bmatrix} + M \begin{bmatrix} 0 \\ 3.7 \\ 2.1 \end{bmatrix} = \begin{bmatrix} 3.5 \\ 47.4 \\ 19.7 \end{bmatrix}$$

Find the indicated products and sums in Exercises 15–20.

15. $[-2, 1, 3] \cdot \begin{bmatrix} 4 \\ 6 \\ 2 \end{bmatrix}$

16. $7 + [-5, 2] \cdot \begin{bmatrix} -3 \\ 3 \end{bmatrix}$

17. $6 \left([0, 1, 0] \cdot \begin{bmatrix} 1 \\ 0 \\ 1 \end{bmatrix} \right)$

18. $3\left([2, 1] \cdot \begin{bmatrix} 1 \\ 2 \end{bmatrix}\right) + 4\left([-2, 1] \cdot \begin{bmatrix} 3 \\ 2 \end{bmatrix}\right)$

19. $[6, 7, -2] \cdot \begin{bmatrix} 0 \\ 2 \\ 1 \end{bmatrix} + [3, 1] \cdot \begin{bmatrix} 1 \\ 3 \end{bmatrix} - 6$

20. $[1, 2, 3, 4, \ldots, n] \cdot \begin{bmatrix} 1 \\ 2 \\ \vdots \\ n \end{bmatrix}$ if $n = 15$

Exercises 21–24 are based on the man-hour load vectors described in Exercises 13 and 14. The company has a cost/man-hour vector of $[10, 8.50, 21]$ *(entries in \$).*

21. What is the total cost of making one unit of item *a*?

22. What is the total cost of making one unit of item *b*?

23. What is the total cost of making one unit of item *c*?

24. Which is more expensive to manufacture: 10 units of item *a* or 5 units each of *b* and *c*?

Each of Exercises 25–29 represents a theorem about vectors and their products. Lowercase letters stand for scalars (real numbers) and capital letters represent vectors. In each case present an argument in case the vectors have dimension 3.

25. $(a + b)A = aA + bA$ 　　　　　**26.** $(ab)A = a(bA)$

27. $a(A \cdot B) = (aA) \cdot B$ 　　　　　**28.** $a(A + B) = aA + aB$

29. $(A + B) \cdot C = A \cdot C + B \cdot C$

30. Does $(A \cdot B)C$ make sense? (*Hint:* Let A and B be appropriate two-dimensional vectors and "try" it.)

5.3 MATRICES

A column or row vector is a special case of a more general mathematical concept, the matrix.

DEFINITION

A rectangular array of scalars (real numbers)

$$\begin{bmatrix} a_{11} & a_{12} & \cdots & a_{1m} \\ a_{21} & & & \\ \vdots & & & \\ a_{n1} & \cdots\cdots\cdots & & a_{nm} \end{bmatrix}$$

is called a **matrix**. The horizontal lines of numbers are called the **rows** of the matrix, the vertical lines the **columns**. A matrix having n rows and m columns is called an $n \times m$ matrix (read "n by m").

It is understood that the individual elements or components of the matrix are identified with a double subscript. That is, a_{11} (read "a sub 1, 1") is the number in the first row and first column. a_{12} (read "a sub 1, 2") is the element in the first row and second column. In general, a_{ij} (read "a sub i, j") refers to the number in the ith row and jth column.

EXAMPLE   In the 3×2 matrix below, identify each element by row and column.

$$\begin{bmatrix} 3 & 7 \\ 2 & -4 \\ 0 & 3 \end{bmatrix}$$

Solution

3 is the element in row 1, column 1 $= a_{11}$.
7 is the element in row 1, column 2 $= a_{12}$.
2 is the element in row 2, column 1 $= a_{21}$.
-4 is the element in row 2, column 2 $= a_{22}$.
0 is the element in row 3, column 1 $= a_{31}$.
3 is the element in row 3, column 2 $= a_{32}$.

EXAMPLE $\begin{bmatrix} 1 & 6 & -4 & 0 \\ 2 & 8 & 3 & 7 \end{bmatrix}$ is an example of a 2×4 matrix.

EXAMPLE   If $A = [a_1, a_2, \ldots, a_n]$ is an n-dimensional row vector, is it also a matrix?

Solution   Yes. A can be thought of as an example of a $1 \times n$ matrix. Similarly,

$$B = \begin{bmatrix} b_1 \\ \vdots \\ b_n \end{bmatrix}$$

an n-dimensional column vector, is also an example of an $n \times 1$ matrix.

From this point on, the commas normally used to separate the components or elements of a row vector will be omitted.

Row vectors are also referred to as **row matrices**, and column vectors are referred to as **column matrices**. Other special matrices are also given special names.

DEFINITION

A matrix with the same number of rows as columns is called a **square matrix**.

EXAMPLE　If

$$A = \begin{bmatrix} a_{11} & \cdots & a_{1m} \\ \vdots & & \\ a_{n1} & \cdots & a_{nm} \end{bmatrix}$$

what conditions on n and m would make it a square matrix?

Solution　A will be a square matrix if and only if $n = m$.

In the additional definitions and theorems which will be developed about matrices, we want to make sure that in any cases where these definitions overlap with those applied previously to vectors, no changes have been made. That is, we want to set up matrix theory in such a way that vectors behave the same way if they are viewed as vectors or as $1 \times n$ or $n \times 1$ matrices. The following definition of matrix equality will illustrate this point. As with vectors, we are using the convention of designating capital letters as the names of matrices and lowercase letters for matrix elements or components (scalars).

DEFINITION

Let A and B be two $n \times m$ matrices with

$$A = \begin{bmatrix} a_{11} & \cdots & a_{1m} \\ \vdots & & \\ a_{n1} & \cdots & a_{nm} \end{bmatrix} \quad \text{and} \quad B = \begin{bmatrix} b_{11} & \cdots & b_{1m} \\ \vdots & & \\ b_{n1} & \cdots & b_{nm} \end{bmatrix}$$

Then $A = B$ if and only if $a_{ij} = b_{ij}$ for all $i = 1$ to $i = n$ and all $j = 1$ to $j = m$.

The definition states that two matrices are equal if and only if they are the same size—i.e., both have the same number of rows and columns—and each element or component of one equals the corresponding element or component of the other.

EXAMPLE   Under what conditions does $\begin{bmatrix} 3 & 2 & -5 \\ x & y & z \end{bmatrix} = \begin{bmatrix} a & b & c \\ -4 & 0 & 3 \end{bmatrix}$?

Solution   The two matrices are equal if and only if $a = 3$, $b = 2$, $c = -5$, $x = -4$, $y = 0$, and $z = 3$.

The reader should note that two vectors are equal when viewed as matrices or when thought of as vectors.

DEFINITION

A matrix with all zero elements is called **a zero** or **null matrix** and will be denoted as $\bar{0}$. The number of rows or columns involved will be taken from the context of the discussion.

EXAMPLE   Find a 3×5 zero matrix.

Solution

$$\bar{0} = \begin{bmatrix} 0 & 0 & 0 & 0 & 0 \\ 0 & 0 & 0 & 0 & 0 \\ 0 & 0 & 0 & 0 & 0 \end{bmatrix}$$

DEFINITION: Matrix Sum

Let A and B be two $n \times m$ matrices such that

$$A = \begin{bmatrix} a_{11} & \cdots & a_{1m} \\ \vdots & & \\ a_{n1} & \cdots & a_{nm} \end{bmatrix} \quad \text{and} \quad B = \begin{bmatrix} b_{11} & \cdots & b_{1m} \\ \vdots & & \\ b_{n1} & \cdots & b_{nm} \end{bmatrix}$$

Then

$$A + B = \begin{bmatrix} a_{11} + b_{11} & a_{12} + b_{12} & \cdots & a_{1m} + b_{1m} \\ a_{21} + b_{21} & \cdots & & a_{2m} + b_{2m} \\ \vdots & & & \\ a_{n1} + b_{n1} & \cdots & & a_{nm} + b_{nm} \end{bmatrix}$$

EXAMPLE   If

$$A = \begin{bmatrix} 3 & 4 & 6 \\ -4 & 0 & 7 \\ 8 & 9 & 3 \end{bmatrix} \quad \text{and} \quad B = \begin{bmatrix} 0 & 6 & -1 \\ -1 & 3 & -1 \\ 4 & -2 & 4 \end{bmatrix}$$

find $A + B$.

Solution

$$A + B = \begin{bmatrix} 3 + & 0 & 4 + & 6 & 6 + (-1) \\ -4 + (-1) & 0 + & 3 & 7 + (-1) \\ 8 + & 4 & 9 + (-2) & 3 + & 4 \end{bmatrix}$$

$$= \begin{bmatrix} 3 & 10 & 5 \\ -5 & 3 & 6 \\ 12 & 7 & 7 \end{bmatrix}$$

DEFINITION

Let A and B be two $n \times m$ matrices:

$$A = \begin{bmatrix} a_{11} & \cdots & a_{1m} \\ \vdots & & \\ a_{n1} & \cdots & a_{nm} \end{bmatrix} \quad \text{and} \quad B = \begin{bmatrix} b_{11} & \cdots & b_{1m} \\ \vdots & & \\ b_{n1} & \cdots & b_{nm} \end{bmatrix}$$

Then

$$A - B = \begin{bmatrix} a_{11} - b_{11} & \cdots & a_{1m} - b_{1m} \\ \vdots & & \\ a_{n1} - b_{n1} & \cdots & a_{nm} - b_{nm} \end{bmatrix}$$

EXAMPLE If $P = \begin{bmatrix} 3 & 6 & 7 \\ 2 & 4 & -5 \end{bmatrix}$ and $Q = \begin{bmatrix} -1 & 6 & -2 \\ 4 & 3 & -5 \end{bmatrix}$, find $P - Q$.

Solution

$$P - Q = \begin{bmatrix} 4 & 0 & 9 \\ -2 & 1 & 0 \end{bmatrix}$$

EXAMPLE If $R = \begin{bmatrix} 3 & 2 \\ 7 & 1 \end{bmatrix}$ and $S = \begin{bmatrix} 1 & 2 & 3 \\ 4 & 6 & 8 \end{bmatrix}$, find $R - S$ and $R + S$.

Solution Since R and S are not the same size, neither their difference nor their sum is defined.

EXAMPLE Let $B = \begin{bmatrix} 3 & 6 \\ -4 & 2 \end{bmatrix}$ and $X = \begin{bmatrix} x_{11} & x_{12} \\ x_{21} & x_{22} \end{bmatrix}$. Find X if $X + B = \bar{0}$.

Solution The matrix equation is

$$\begin{bmatrix} x_{11} & x_{12} \\ x_{21} & x_{22} \end{bmatrix} + \begin{bmatrix} 3 & 6 \\ -4 & 2 \end{bmatrix} = \begin{bmatrix} 0 & 0 \\ 0 & 0 \end{bmatrix}$$

or

$$\begin{bmatrix} x_{11} + & 3 & x_{12} + 6 \\ x_{21} + (-4) & x_{22} + 2 \end{bmatrix} = \begin{bmatrix} 0 & 0 \\ 0 & 0 \end{bmatrix}$$

which is true if and only if

$$
\begin{aligned}
x_{11} + 3 &= 0 \quad \text{or} \quad x_{11} = -3 \\
x_{12} + 6 &= 0 \quad \text{or} \quad x_{12} = -6 \\
x_{21} - 4 &= 0 \quad \text{or} \quad x_{21} = 4 \\
x_{22} + 2 &= 0 \quad \text{or} \quad x_{22} = -2
\end{aligned}
$$

Thus

$$X = \begin{bmatrix} -3 & -6 \\ 4 & -2 \end{bmatrix}$$

EXERCISES 5.3

1. Let $A = \begin{bmatrix} 1 & 4 & -3 & 2 \\ 6 & 8 & 3 & 1 \\ 7 & 5 & 9 & -6 \end{bmatrix}$

   Find: a. The element in the second row and third column
   b. The element in the third row and second column

2. Let $B = \begin{bmatrix} b_{11} & b_{12} & b_{13} & b_{14} \\ b_{21} & b_{22} & b_{23} & b_{24} \end{bmatrix}$

   If $b_{ij} = i + j$ for all $i = 1$ or 2, $j = 1$ to 4, find the following specific elements of B:
   a. b_{13}
   b. b_{24}
   c. b_{42}

In Exercises 3–11, find all x, y, and z such that the indicated matrix equality holds.

3. $\begin{bmatrix} x & 3 \\ 4 & z \end{bmatrix} = \begin{bmatrix} 2 & y \\ 4 & -7 \end{bmatrix}$

4. $\begin{bmatrix} x + y & 3 \\ x - y & 2 \end{bmatrix} = \begin{bmatrix} 2 & z \\ 0 & 2 \end{bmatrix}$

5. $\begin{bmatrix} x^2 - 3x & 4 & y^2 \\ 2 & 0 & 7 \end{bmatrix} = \begin{bmatrix} 10 & 4 & 4 \\ 2 & 0 & 7 \end{bmatrix}$

6. $\begin{bmatrix} x^2 - 2x & x \\ 0 & z \end{bmatrix} = \begin{bmatrix} -1 & 1 \\ y & 0 \end{bmatrix}$

7. $\begin{bmatrix} x & y \\ 3 & z \end{bmatrix} = \begin{bmatrix} y & x \\ z & 3 \end{bmatrix}$

8. $\begin{bmatrix} x & y & z \\ 3 & 2 & -5 \\ 6 & -2 & 0 \end{bmatrix} = \begin{bmatrix} x & 3 & 6 \\ y & 2 & -2 \\ z & -5 & 0 \end{bmatrix}$

9. $\begin{bmatrix} x & z & -4 & 2 \\ 8 & y & 3 & 7 \end{bmatrix} = \begin{bmatrix} z & x & -2 & 2 \\ 8 & y & 3 & 7 \end{bmatrix}$

10. $\begin{bmatrix} 1 \\ 5 \\ -4 \\ 2 \end{bmatrix} = \begin{bmatrix} x - y \\ 2x + y \\ -2x \\ 2x - 2y \end{bmatrix}$

11. $[x^2 - 2x \quad y^2 - 6y \quad z^2 - 8z] = [-1 \quad -9 \quad -16]$

Find the indicated sums and differences in Exercises 12–15.

12. $\begin{bmatrix} -2 & 1 & 6 & 8 \\ 4 & 3 & 7 & -5 \end{bmatrix} + \begin{bmatrix} -4 & -3 & 5 & 6 \\ 6 & 7 & 1 & 2 \end{bmatrix}$

13. $\begin{bmatrix} 2 & 1 & 7 \\ 5 & -1 & -5 \\ -3 & 6 & 0 \end{bmatrix} + \begin{bmatrix} -4 & 1 & 3 \\ 7 & -1 & -4 \\ 4 & -2 & -1 \end{bmatrix}$

14. $\begin{bmatrix} 2 & -5 \\ 3 & 7 \\ -4 & 6 \\ 0 & 2 \end{bmatrix} - \begin{bmatrix} 8 & 1 \\ 3 & -2 \\ -2 & -2 \\ 0 & 2 \end{bmatrix}$

15. $\begin{bmatrix} 1 & 2 & 3 & 4 \\ 5 & 6 & 7 & 8 \end{bmatrix} - \begin{bmatrix} 8 & 7 & 6 & 5 \\ 4 & 3 & 2 & 1 \end{bmatrix}$

In Exercises 16–19, assume that A, B, and C are n × m matrices. If the statement is true, demonstrate it for 2 × 3 matrices.

16. In general, does $A + (B + C) = (A + B) + C$?
17. In general, does $A - (B - C) = (A - B) - C$?
18. In general, does $A - (B - C) = (A - B) + C$?
19. In general, does $A + B = B + A$?

5.4   MATRIX PRODUCTS AND THE IDENTITY MATRIX

As with vector products, we will consider two kinds of multiplication involving matrices: a scalar times a matrix and a matrix times a matrix.

DEFINITION

Let

$$A = \begin{bmatrix} a_{11} & \cdots & a_{1m} \\ \vdots & & \\ a_{n1} & \cdots & a_{nm} \end{bmatrix}$$

be an $n \times m$ matrix and c be any scalar (real number). Then

$$cA = \begin{bmatrix} ca_{11} & ca_{12} & \cdots & ca_{1m} \\ ca_{21} & & & \\ \vdots & & & \\ ca_{n1} & \cdots\cdots\cdots & & ca_{nm} \end{bmatrix}$$

EXAMPLE   If $A = \begin{bmatrix} 3 & -4 & 2 \\ 6 & -3 & 1 \end{bmatrix}$, find $7A$ and $-3A$.

Solution

$$7A = \begin{bmatrix} 7\times 3 & 7\times(-4) & 7\times 2 \\ 7\times 6 & 7\times(-3) & 7\times 1 \end{bmatrix}$$

$$= \begin{bmatrix} 21 & -28 & 14 \\ 42 & -21 & 7 \end{bmatrix}$$

$$-3A = \begin{bmatrix} -9 & 12 & -6 \\ -18 & 9 & -3 \end{bmatrix}$$

EXAMPLE   If $B = \begin{bmatrix} -2 & 1 \\ -4 & 6 \end{bmatrix}$ and $C = \begin{bmatrix} 3 & -4 \\ 8 & 2 \end{bmatrix}$, find $4B + 3C$.

Solution

$$4B = \begin{bmatrix} -8 & 4 \\ -16 & 24 \end{bmatrix}, \quad 3C = \begin{bmatrix} 9 & -12 \\ 24 & 6 \end{bmatrix}$$

Therefore,

$$4B + 3C = \begin{bmatrix} 1 & -8 \\ 8 & 30 \end{bmatrix}$$

DEFINITION

Let

$$A = \begin{bmatrix} a_{11} & \cdots & a_{1m} \\ \vdots & & \\ a_{n1} & \cdots & a_{nm} \end{bmatrix}$$

be an $n \times m$ matrix and

$$B = \begin{bmatrix} b_{11} & \cdots & b_{1p} \\ \vdots & & \\ b_{m1} & \cdots & b_{mp} \end{bmatrix}$$

an $m \times p$ matrix. Then $C = AB$ is an $n \times p$ matrix

$$C = \begin{bmatrix} c_{11} & \cdots & c_{1p} \\ \vdots & & \\ c_{n1} & \cdots & c_{np} \end{bmatrix}$$

such that $c_{ij} = a_{i1}b_{1j} + a_{i2}b_{2j} + a_{i3}b_{3j} + \cdots + a_{im}b_{mj}$ for all $i = 1$ to $i = m$ and all $j = 1$ to $j = p$.

Notice that matrix multiplication is defined between two matrices *only* if the number of columns of the first matrix equals the number of rows of the second. The product matrix has the same number of rows as the first matrix and the same number of columns as the second. Also note that c_{ij} is the dot product of the ith row and jth column of A and B, respectively, in $C = AB$.

EXAMPLE If

$$A = \begin{bmatrix} -2 & 1 & 4 \\ 6 & 3 & 5 \end{bmatrix} \quad \text{and} \quad B = \begin{bmatrix} 3 & 2 & 1 \\ -4 & 0 & 3 \\ 6 & 7 & 5 \end{bmatrix}$$

find AB.

Solution Since A is a 2×3 matrix and B is a 3×3 matrix, multiplication is defined. The product will be a 2×3 matrix.

$$
\begin{aligned}
AB &= \begin{bmatrix} -2 & 1 & 4 \\ 6 & 3 & 5 \end{bmatrix} \begin{bmatrix} 3 & 2 & 1 \\ -4 & 0 & 3 \\ 6 & 7 & 5 \end{bmatrix} \\[2mm]
&= \begin{bmatrix} (-2)(3)+1(-4)+4(6) & (-2)(2)+1(0)+4(7) & (-2)1+1(3)+4(5) \\ 6(3)+3(-4)+5(6) & 6(2)+3(0)+5(7) & 6(1)+3(3)+5(5) \end{bmatrix} \\[2mm]
&= \begin{bmatrix} -6-4+24 & -4+0+28 & -2+3+20 \\ 18-12+30 & 12+0+35 & 6+9+25 \end{bmatrix} \\[2mm]
&= \begin{vmatrix} 14 & 24 & 21 \\ 36 & 47 & 40 \end{vmatrix}
\end{aligned}
$$

EXAMPLE Let

$$P = [2 \quad 1 \quad -3] \quad \text{and} \quad Q = \begin{bmatrix} 3 \\ 7 \\ 5 \end{bmatrix}$$

Find PQ and QP if possible.

Solution

$$PQ = [2 \quad 1 \quad -3] \begin{bmatrix} 3 \\ 7 \\ 5 \end{bmatrix}$$

is defined because P is a 1×3 matrix and Q a 3×1. The product of a 1×3 and 3×1 matrix should be a 1×1 matrix.

$$[2 \quad 1 \quad -3] \begin{bmatrix} 3 \\ 7 \\ 5 \end{bmatrix} = [(2)(3) + 1(7) + (-3)5]$$

$$= [6 + 7 - 15] = [-2]$$

$$QP = \begin{bmatrix} 3 \\ 7 \\ 5 \end{bmatrix} [2 \quad 1 \quad -3]$$

and should be a 3×3 matrix.

$$\begin{bmatrix} 3 \\ 7 \\ 5 \end{bmatrix} [2 \quad 1 \quad -3] = \begin{bmatrix} 3 \times 2 & 3 \times 1 & 3 \times (-3) \\ 7 \times 2 & 7 \times 1 & 7 \times (-3) \\ 5 \times 2 & 5 \times 1 & 5 \times (-3) \end{bmatrix}$$

$$= \begin{bmatrix} 6 & 3 & -9 \\ 14 & 7 & -21 \\ 10 & 5 & -15 \end{bmatrix}$$

If the reader has been attentive, she or he will have noticed that PQ in the previous example was the product of two vectors, within the definition of the vector inner product. The result was a 1×1 matrix. So that we do not have to work with two possible vector products, we will agree that a 1×1 matrix will be considered a scalar. Thus $PQ = [-2] = -2$.

EXAMPLE   Let $A = \begin{bmatrix} 1 & 4 \\ 6 & 3 \end{bmatrix}$ and $B = \begin{bmatrix} -2 & 5 \\ 1 & 6 \end{bmatrix}$. Find AB and BA.

Solution

$$AB = \begin{bmatrix} 1 & 4 \\ 6 & 3 \end{bmatrix} \begin{bmatrix} -2 & 5 \\ 1 & 6 \end{bmatrix} = \begin{bmatrix} 1(-2) + 4(1) & 1(5) + 4(6) \\ 6(-2) + 3(1) & 6(5) + 3(6) \end{bmatrix}$$

$$= \begin{bmatrix} 2 & 29 \\ -9 & 48 \end{bmatrix}$$

$$BA = \begin{bmatrix} -2 & 5 \\ 1 & 6 \end{bmatrix} \begin{bmatrix} 1 & 4 \\ 6 & 3 \end{bmatrix} = \begin{bmatrix} -2(1) + 5(6) & -2(4) + 5(3) \\ 1(1) + 6(6) & 1(4) + 6(3) \end{bmatrix}$$

$$= \begin{bmatrix} 28 & 7 \\ 37 & 22 \end{bmatrix}$$

Notice that even though AB and BA both exist in this example, they are not equal. In general, if A and B are square matrices of the same size, $AB \neq BA$ even though both exist. Naturally there are special cases when $AB = BA$. The remaining definitions and concepts of this section will consider square matrices and some of their special properties.

DEFINITIONS

In an $n \times n$ square matrix

$$A = \begin{bmatrix} a_{11} & a_{12} & \cdots & a_{1n} \\ \vdots & & & \\ a_{n1} & \cdots \cdots \cdots & & a_{nn} \end{bmatrix}$$

the diagonal of elements running from a_{11} in the upper left corner to a_{nn} in the lower right corner is called the **principal diagonal** of the matrix.

A square matrix with 1s along its principal diagonal and 0 everywhere else will be called the **identity matrix**. The size of such a matrix will be taken from the context, and all such matrices will be denoted I.

EXAMPLE   Write out the specific form of I if from the context you know it is a 6×6 matrix.

Solution

$$I = \begin{bmatrix} 1 & 0 & 0 & 0 & 0 & 0 \\ 0 & 1 & 0 & 0 & 0 & 0 \\ 0 & 0 & 1 & 0 & 0 & 0 \\ 0 & 0 & 0 & 1 & 0 & 0 \\ 0 & 0 & 0 & 0 & 1 & 0 \\ 0 & 0 & 0 & 0 & 0 & 1 \end{bmatrix}$$

The reason for calling I the identity matrix becomes clear with the following theorem.

THEOREM

If A is any $n \times n$ matrix, then

$$AI = IA = A$$

Proof: First let us test this for a 3×3 matrix and use the pattern to argue the theorem for other cases. Let

$$A = \begin{bmatrix} a_{11} & a_{12} & a_{13} \\ a_{21} & a_{22} & a_{23} \\ a_{31} & a_{32} & a_{33} \end{bmatrix}$$

Then

$$IA = \begin{bmatrix} 1 & 0 & 0 \\ 0 & 1 & 0 \\ 0 & 0 & 1 \end{bmatrix} \begin{bmatrix} a_{11} & a_{12} & a_{13} \\ a_{21} & a_{22} & a_{23} \\ a_{31} & a_{32} & a_{33} \end{bmatrix}$$

$$= \begin{bmatrix} 1 \cdot a_{11} + 0 \cdot a_{21} + 0 \cdot a_{31} & 1 \cdot a_{12} + 0 \cdot a_{22} + 0 \cdot a_{32} & 1 \cdot a_{13} + 0 \cdot a_{23} + 0 \cdot a_{33} \\ 0 \cdot a_{11} + 1 \cdot a_{21} + 0 \cdot a_{31} & 0 \cdot a_{12} + 1 \cdot a_{22} + 0 \cdot a_{32} & 0 \cdot a_{13} + 1 \cdot a_{23} + 0 \cdot a_{33} \\ 0 \cdot a_{11} + 0 \cdot a_{21} + 1 \cdot a_{31} & 0 \cdot a_{12} + 0 \cdot a_{22} + 1 \cdot a_{32} & 0 \cdot a_{13} + 0 \cdot a_{23} + 1 \cdot a_{33} \end{bmatrix}$$

$$= \begin{bmatrix} a_{11} & a_{12} & a_{13} \\ a_{21} & a_{22} & a_{23} \\ a_{31} & a_{32} & a_{33} \end{bmatrix}$$

The reader can verify that $AI = A$.
In general, if

$$A = \begin{bmatrix} a_{11} & \cdots & a_{1n} \\ \vdots & & \\ a_{n1} & \cdots & a_{nn} \end{bmatrix}$$

we can write I as

$$I = \begin{bmatrix} \delta_{11} & \delta_{12} & \cdots & \delta_{1n} \\ \vdots & & & \\ \delta_{n1} & \cdots \cdots \cdots & & \delta_{nn} \end{bmatrix}$$

where

$$\delta_{ij} = 0 \quad \text{if} \quad i \neq j$$
$$\delta_{ij} = 1 \quad \text{if} \quad i = j$$

If

$$IA = B = \begin{bmatrix} b_{11} & \cdots & b_{1n} \\ \vdots & & \\ b_{n1} & \cdots & b_{nn} \end{bmatrix}$$

then $b_{ij} = \delta_{i1}a_{1j} + \delta_{i2}a_{2j} + \cdots + \delta_{in}a_{nj}$.

The only nonzero value for δ_{i1}, δ_{i2}, etc. occurs when the second subscript of δ_{ik} equals i. Thus

$$b_{ij} = 0 \cdot a_{1j} + 0 \cdot a_{2j} + \cdots + 1 \cdot a_{ij} + \cdots + 0 \cdot a_{nj}$$
$$= a_{ij}$$

and $b_{ij} = a_{ij}$ for all possible i and j or

$$B = A$$

and

$$IA = A$$

The proof that $AI = A$ is similar.

We can also define powers of a square matrix.

DEFINITION

If A is an $n \times n$ square matrix,

$$A^2 = AA$$
$$A^3 = AA^2$$

In general, $A^k = AA^{k-1}$

for all positive integers k.

EXAMPLE   If $A = \begin{bmatrix} 1 & 3 \\ -2 & 0 \end{bmatrix}$, find A^2, A^3, and A^4.

Solution

$$A^2 = AA = \begin{bmatrix} 1 & 3 \\ -2 & 0 \end{bmatrix}\begin{bmatrix} 1 & 3 \\ -2 & 0 \end{bmatrix} = \begin{bmatrix} -5 & 3 \\ -2 & -6 \end{bmatrix}$$

$$A^3 = AA^2 = \begin{bmatrix} 1 & 3 \\ -2 & 0 \end{bmatrix}\begin{bmatrix} -5 & 3 \\ -2 & -6 \end{bmatrix} = \begin{bmatrix} -11 & -15 \\ 10 & -6 \end{bmatrix}$$

$$A^4 = AA^3 = \begin{bmatrix} 1 & 3 \\ -2 & 0 \end{bmatrix}\begin{bmatrix} -11 & -15 \\ 10 & -6 \end{bmatrix} = \begin{bmatrix} 19 & -33 \\ 22 & 30 \end{bmatrix}$$

EXAMPLE   For a real number a, if $a^2 = 0$, then $a = 0$. If A is a 2×2 matrix and $A^2 = \bar{0}$, does $A = \bar{0}$?

Solution   Consider $A = \begin{bmatrix} 0 & 1 \\ 0 & 0 \end{bmatrix}$. Clearly $A \neq \bar{0}$, but

$$A^2 = \begin{bmatrix} 0 & 1 \\ 0 & 0 \end{bmatrix}\begin{bmatrix} 0 & 1 \\ 0 & 0 \end{bmatrix} = \begin{bmatrix} 0 & 0 \\ 0 & 0 \end{bmatrix} = \bar{0}$$

Thus if $A^2 = \bar{0}$, A may or may not be equal to $\bar{0}$.

EXERCISES 5.4

Find the indicated products and sums in Exercises 1–10.

1. $3\begin{bmatrix} 2 & 5 \\ -4 & 6 \end{bmatrix} - 2\begin{bmatrix} 8 & 3 \\ -2 & -2 \end{bmatrix}$

2. $4\begin{bmatrix} -2 & 1 \\ 0 & 1 \end{bmatrix} + 3\begin{bmatrix} 1 & -2 \\ 1 & 0 \end{bmatrix}$

3. $\begin{bmatrix} 3 & 2 & -4 \\ 6 & 3 & 7 \end{bmatrix} \begin{bmatrix} 1 & 2 \\ -2 & 1 \\ 0 & 3 \end{bmatrix}$

4. $\begin{bmatrix} 1 & 4 & 6 & 7 \end{bmatrix} \begin{bmatrix} -2 & -7 \\ 0 & 5 \\ 1 & 3 \\ 3 & 6 \end{bmatrix}$

5. $\begin{bmatrix} 1 \\ 2 \\ 3 \end{bmatrix} \begin{bmatrix} 1 & 2 \end{bmatrix}$

6. $\begin{bmatrix} 1 & 2 \end{bmatrix} \begin{bmatrix} 1 & 3 \\ -4 & 2 \end{bmatrix}$

7. $\begin{bmatrix} 1 & 2 & 7 \end{bmatrix} \begin{bmatrix} 7 \\ 1 \\ 2 \end{bmatrix}$

8. $\begin{bmatrix} 7 \\ 1 \\ 2 \end{bmatrix} \begin{bmatrix} 1 & 2 & 7 \end{bmatrix}$

9. $\begin{bmatrix} 1 \\ 4 \\ 3 \\ 2 \end{bmatrix} \begin{bmatrix} 1 & -2 & -3 & 7 \end{bmatrix}$

10. $\begin{bmatrix} 1 & 0 & 1 & 0 \\ 0 & 1 & 0 & 1 \\ 1 & 0 & 1 & 0 \end{bmatrix} \begin{bmatrix} 1 & 0 & 1 & 0 & 1 \\ 0 & 1 & 0 & 1 & 0 \\ 1 & 0 & 1 & 0 & 1 \\ 0 & 1 & 0 & 1 & 0 \end{bmatrix}$

In Exercises 11–22, consider A, B, and C three matrices.

11. Under what conditions on the size of A, B, and C is $A(B + C)$ defined?
12. Assuming that $A(B + C)$ is defined, does $A(B + C) = AB + AC$?
13. Under what conditions do both AB and BA exist?
14. Under what conditions on size do $A(BC)$ and $(AB)C$ both exist?
15. If they both exist, does $A(BC) = (AB)C$?
16. Does $(A + B)^2 = A^2 + 2AB + B^2$?
17. Does $(A - B)^2 = (A - B)(A - B)$?
18. Based on the results of Exercises 16 and 17, can one conclude that usual algebraic procedures such as factoring continue to work for matrix arithmetic? Explain.

19. Find any 2×2 matrices C and D such that $C \neq D$, $C \cdot D = \bar{0}$, and $C \neq \bar{0}$ and $D \neq \bar{0}$.

20. Find any 3×3 matrix $B \neq \bar{0}$ such that $B^2 = \bar{0}$.

21. Prove that for an $n \times n$ matrix A, $AI = A$.

22. Find two 3×3 matrices A and B such that $AB = BA$, with $A \neq \bar{0}$ and $B \neq \bar{0}$.

5.5 MATRIX INVERSES

In this section we will consider a very special property of certain square matrices, the existence of a matrix "inverse." We will also consider an algorithm, or mechanical procedure, for finding these matrix inverses. In the next section we will examine an application of matrix inverses.

> **DEFINITION**
>
> Let A be an $n \times n$ square matrix. If there exists an $n \times n$ matrix A^{-1} such that
>
> $$AA^{-1} = A^{-1}A = I$$
>
> the $n \times n$ identity matrix, then A^{-1} is called the inverse matrix for matrix A.

Note that if A^{-1} is the inverse of matrix A, then A is the inverse of matrix A^{-1}. Since not all $n \times n$ matrices have inverses, we have the following definition.

> **DEFINITION**
>
> If A is an $n \times n$ matrix such that its inverse A^{-1} exists, then A is said to be a **nonsingular matrix**. If A^{-1} fails to exist, A is said to be **singular**.

A simple example will illustrate the fact that not all square matrices have inverses.

EXAMPLE Show that the 2×2 matrix $\begin{bmatrix} 2 & 1 \\ 4 & 2 \end{bmatrix}$ does not have an inverse.

Solution Assume that $A = \begin{bmatrix} 2 & 1 \\ 4 & 2 \end{bmatrix}$ and A^{-1} exists.

Let
$$A^{-1} = \begin{bmatrix} a & b \\ c & d \end{bmatrix}$$

Then
$$AA^{-1} = I$$

or
$$\begin{bmatrix} 2 & 1 \\ 4 & 2 \end{bmatrix} \begin{bmatrix} a & b \\ c & d \end{bmatrix} = \begin{bmatrix} 1 & 0 \\ 0 & 1 \end{bmatrix}$$

or carrying out the multiplication of A and A^{-1},

$$\begin{bmatrix} 2a + c & 2b + d \\ 4a + 2c & 4b + 2d \end{bmatrix} = \begin{bmatrix} 1 & 0 \\ 0 & 1 \end{bmatrix}$$

Applying the definition of matrix equality,

$$2a + c = 1 \qquad 2b + d = 0$$
$$4a + 2c = 0 \qquad 4b + 2d = 1$$

Consider the first two equations:

$$2a + c = 1$$
$$4a + 2c = 0$$

Dividing the second equation by 2, we have

$$2a + c = 1$$
$$2a + c = 0$$

Clearly this is impossible. The conclusion is that A^{-1} cannot exist.

The following example shows that inverses do exist.

EXAMPLE   Find the inverse of the 2×2 matrix $A = \begin{bmatrix} 1 & 1 \\ 1 & -1 \end{bmatrix}$.

Solution   Let

$$A^{-1} = \begin{bmatrix} a & b \\ c & d \end{bmatrix}$$

then
$$AA^{-1} = I$$

or
$$\begin{bmatrix} 1 & 1 \\ 1 & -1 \end{bmatrix} \begin{bmatrix} a & b \\ c & d \end{bmatrix} = \begin{bmatrix} 1 & 0 \\ 0 & 1 \end{bmatrix}$$

Carrying out the matrix multiplication on the left,

$$\begin{bmatrix} a + c & b + d \\ a - c & b - d \end{bmatrix} = \begin{bmatrix} 1 & 0 \\ 0 & 1 \end{bmatrix}$$

Applying matrix equality,

$$a + c = 1$$
$$a - c = 0$$
$$b + d = 0$$
$$b - d = 1$$

The first pair of equations can be solved for a and c using the methods of ordinary algebra:

$$a + c = 1$$
$$a - c = 0$$

Eliminating c by adding the left sides and the right sides, we have

$$2a = 1$$
$$a = \tfrac{1}{2}$$

If $a = \tfrac{1}{2}$, then

$$\tfrac{1}{2} + c = 1$$
$$c = \tfrac{1}{2}$$

Performing the same method on the second pair of equations,

$$b + d = 0$$
$$b - d = 1$$

yields

$$2b = 1$$
$$b = \tfrac{1}{2}$$
$$\tfrac{1}{2} + d = 0$$
$$d = -\tfrac{1}{2}$$

Then

$$A^{-1} = \begin{bmatrix} a & b \\ c & d \end{bmatrix} = \begin{bmatrix} \tfrac{1}{2} & \tfrac{1}{2} \\ \tfrac{1}{2} & -\tfrac{1}{2} \end{bmatrix}$$

Checking,

$$AA^{-1} = \begin{bmatrix} 1 & 1 \\ 1 & -1 \end{bmatrix}\begin{bmatrix} \tfrac{1}{2} & \tfrac{1}{2} \\ \tfrac{1}{2} & -\tfrac{1}{2} \end{bmatrix}$$

$$= \begin{bmatrix} 1(\tfrac{1}{2}) + 1(\tfrac{1}{2}) & 1(\tfrac{1}{2}) + 1(-\tfrac{1}{2}) \\ 1(\tfrac{1}{2}) + (-1)(\tfrac{1}{2}) & 1(\tfrac{1}{2}) + (-1)(-\tfrac{1}{2}) \end{bmatrix}$$

$$= \begin{bmatrix} 1 & 0 \\ 0 & 1 \end{bmatrix}$$

$$A^{-1}A = \begin{bmatrix} \tfrac{1}{2} & \tfrac{1}{2} \\ \tfrac{1}{2} & -\tfrac{1}{2} \end{bmatrix}\begin{bmatrix} 1 & 1 \\ 1 & -1 \end{bmatrix} = \begin{bmatrix} \tfrac{1}{2}(1) + \tfrac{1}{2}(1) & \tfrac{1}{2}(1) + \tfrac{1}{2}(-1) \\ \tfrac{1}{2}(1) + (-\tfrac{1}{2})1 & \tfrac{1}{2}(1) + (-\tfrac{1}{2})(-1) \end{bmatrix}$$

$$= \begin{bmatrix} 1 & 0 \\ 0 & 1 \end{bmatrix}$$

It is reasonably clear that the method used on the two previous examples will find the inverse of a 2 × 2 matrix or establish that the matrix fails to have an inverse. It is also clear that this method is not practical for finding the inverse of a 3 × 3 or larger square matrix. In the case of a 3 × 3 matrix, the reader can verify that this procedure would lead to three sets of three equations involving three unknowns, whose solution is no small task.

To get around this problem the remainder of this section will consider an algorithm, or mechanical procedure, which can be used to find the inverse of a nonsingular matrix.

THEOREM

If A is a nonsingular $n \times n$ matrix, then A^{-1} is unique.

Proof: Assume that there exist two distinct $n \times n$ matrices A^{-1} and $(A^{-1})'$ such that $A^{-1} \neq (A^{-1})'$ and

$$AA^{-1} = A(A^{-1})' = I = A^{-1}A = (A^{-1})'A$$

Then
$$A^{-1} = A^{-1}I = A^{-1}[A(A^{-1})']$$
$$= (A^{-1}A)(A^{-1})'$$
$$= I(A^{-1})'$$
$$= (A^{-1})'$$

Thus A^{-1} is unique.

DEFINITION

The following operations on matrices are called **row operations:**

1. Interchanging any two rows of the matrix
2. Multiplying the elements of any row by a nonzero scalar
3. Adding any multiple of one row, term by term, to the elements in any other row

EXAMPLE   Consider the matrix

$$C = \begin{bmatrix} 1 & 4 & 2 & 3 \\ 6 & 1 & -5 & 4 \\ 7 & 5 & 3 & -7 \end{bmatrix}$$

Then using row operation (1), we might interchange the first and second rows. The result is

$$\begin{bmatrix} 6 & 1 & -5 & 4 \\ 1 & 4 & 2 & 3 \\ 7 & 5 & 3 & -7 \end{bmatrix}$$

One might use row operation (2) and multiply row 3 of C by -4. The result is

$$\begin{bmatrix} 1 & 4 & 2 & 3 \\ 6 & 1 & -5 & 4 \\ -28 & -20 & -12 & 28 \end{bmatrix}$$

An example of row operation (3) might be adding -6 times the elements of row 1 of C to row 2. The result is

$$\begin{bmatrix} 1 & 4 & 2 & 3 \\ 6 + (-6) & 1 + (-24) & -5 + (-12) & 4 + (-18) \\ 7 & 5 & 3 & -7 \end{bmatrix}$$

$$\begin{bmatrix} 1 & 4 & 2 & 3 \\ 0 & -23 & -17 & -14 \\ 7 & 5 & 3 & -7 \end{bmatrix}$$

Row operations form the basis for our matrix inversion algorithm.

Matrix Inversion Algorithm

Let A be any $n \times n$ matrix such that

$$A = \begin{bmatrix} a_{11} & a_{12} & \cdots & a_{1n} \\ \vdots & & & \\ a_{n1} & \cdots\cdots & & a_{nn} \end{bmatrix}$$

and I be the corresponding $n \times n$ identity matrix. If by a finite sequence of row operations we can transform the $n \times 2n$ "augmented" matrix

$$\begin{bmatrix} a_{11} & a_{12} & \cdots & a_{1n} & 1 & 0 & \cdots & 0 \\ a_{21} & a_{22} & \cdots & a_{2n} & 0 & 1 & \cdots & 0 \\ \vdots & & & & \vdots & & & \\ a_{n1} & a_{n2} & \cdots & a_{nn} & 0 & \cdots\cdots & & 1 \end{bmatrix}$$

into the form

$$\begin{bmatrix} 1 & 0 & \cdots & 0 & a'_{11} & a'_{12} & \cdots & a'_{1n} \\ 0 & 1 & \cdots & 0 & a'_{21} & \cdots\cdots & & a'_{2n} \\ \vdots & & & & \vdots & & & \\ 0 & \cdots\cdots & & 1 & a'_{n1} & \cdots\cdots & & a'_{nn} \end{bmatrix}$$

then the matrix

$$A^{-1} = \begin{bmatrix} a'_{11} & a'_{12} & \cdots & a'_{1n} \\ \vdots & & & \\ a'_{n1} & \cdots\cdots & & a'_{nn} \end{bmatrix}$$

will be the inverse of A. If at any time during the process the left-hand "half" has a row of all zeros, the process fails, and A is singular, i.e., A fails to have an inverse.

EXAMPLE   Use the algorithm to find the inverse of $\begin{bmatrix} 1 & 1 \\ 1 & -1 \end{bmatrix}$.

Solution   The augmented matrix is

$$\begin{bmatrix} 1 & 1 & | & 1 & 0 \\ 1 & -1 & | & 0 & 1 \end{bmatrix}$$

If -1 times the elements of the first row are added to the second row, we have

$$\begin{bmatrix} 1 & 1 & | & 1 & 0 \\ 1 + (-1) & -1 + (-1) & | & 0 + (-1) & 1 + 0 \end{bmatrix} = \begin{bmatrix} 1 & 1 & | & 1 & 0 \\ 0 & -2 & | & -1 & 1 \end{bmatrix}$$

This is a type (3) row operation.

The second row can be multiplied by $-\frac{1}{2}$. This is a type (2) row operation, and the result is

$$\begin{bmatrix} 1 & 1 & | & 1 & 0 \\ 0 & 1 & | & \frac{1}{2} & -\frac{1}{2} \end{bmatrix}$$

Then performing another type (3) row operation, we add (-1) times the elements in the second row to the first row. The result is

$$\begin{bmatrix} 1 + 0(-1) & 1 + (-1) & | & 1 + (-\frac{1}{2}) & 0 + (\frac{1}{2}) \\ 0 & 1 & | & \frac{1}{2} & -\frac{1}{2} \end{bmatrix} = \begin{bmatrix} 1 & 0 & | & \frac{1}{2} & \frac{1}{2} \\ 0 & 1 & | & \frac{1}{2} & -\frac{1}{2} \end{bmatrix}$$

According to the algorithm, the matrix

$$\begin{bmatrix} \frac{1}{2} & \frac{1}{2} \\ \frac{1}{2} & -\frac{1}{2} \end{bmatrix}$$

is the inverse of

$$\begin{bmatrix} 1 & 1 \\ 1 & -1 \end{bmatrix}$$

This agrees with our previous result.

EXAMPLE   Use the inversion algorithm to find the inverse of $\begin{bmatrix} 2 & 1 \\ 4 & 2 \end{bmatrix}$.

Solution

$$\begin{bmatrix} 2 & 1 & | & 1 & 0 \\ 4 & 2 & | & 0 & 1 \end{bmatrix}$$

is the augmented matrix. Dividing the first row by 2, we have

$$\begin{bmatrix} 1 & \frac{1}{2} & | & \frac{1}{2} & 0 \\ 4 & 2 & | & 0 & 1 \end{bmatrix}$$

Multiplying the first row by -4 and adding it to the second row, we have

$$\begin{bmatrix} 1 & \frac{1}{2} & | & \frac{1}{2} & 0 \\ 4-4 & 2-2 & | & 0-2 & 1-0 \end{bmatrix} = \begin{bmatrix} 1 & \frac{1}{2} & | & \frac{1}{2} & 0 \\ 0 & 0 & | & -2 & 1 \end{bmatrix}$$

Since the second row consists of all zeros on the left, the matrix

$$\begin{bmatrix} 2 & 1 \\ 4 & 2 \end{bmatrix}$$

does not have an inverse. This agrees with the previous example.

EXAMPLE   Find (if it exists) the inverse of

$$\begin{bmatrix} 1 & 1 & 1 \\ 2 & -1 & 1 \\ 3 & 1 & 5 \end{bmatrix}$$

Solution   The augmented matrix is

$$\left[\begin{array}{ccc|ccc} 1 & 1 & 1 & 1 & 0 & 0 \\ 2 & -1 & 1 & 0 & 1 & 0 \\ 3 & 1 & 5 & 0 & 0 & 1 \end{array}\right]$$

Adding -2 times the elements in the first row to the elements of row 2, we have

$$\left[\begin{array}{ccc|ccc} 1 & 1 & 1 & 1 & 0 & 0 \\ 0 & -3 & -1 & -2 & 1 & 0 \\ 3 & 1 & 5 & 0 & 0 & 1 \end{array}\right]$$

Adding -3 times the elements in the first row to those in the third row, we have

$$\left[\begin{array}{ccc|ccc} 1 & 1 & 1 & 1 & 0 & 0 \\ 0 & -3 & -1 & -2 & 1 & 0 \\ 0 & -2 & 2 & -3 & 0 & 1 \end{array}\right]$$

Multiplying the second row by $-\frac{1}{3}$, the matrix becomes

$$\left[\begin{array}{ccc|ccc} 1 & 1 & 1 & 1 & 0 & 0 \\ 0 & 1 & \frac{1}{3} & \frac{2}{3} & -\frac{1}{3} & 0 \\ 0 & -2 & 2 & -3 & 0 & 1 \end{array}\right]$$

Multiplying the elements of row 2 by -1 and adding them to row 1, we have

$$\left[\begin{array}{ccc|ccc} 1 & 0 & \frac{2}{3} & \frac{1}{3} & \frac{1}{3} & 0 \\ 0 & 1 & \frac{1}{3} & \frac{2}{3} & -\frac{1}{3} & 0 \\ 0 & -2 & 2 & -3 & 0 & 1 \end{array}\right]$$

Multiplying the elements of the second row by 2 and adding them to row 3, we have

$$\left[\begin{array}{ccc|ccc} 1 & 0 & \frac{2}{3} & \frac{1}{3} & \frac{1}{3} & 0 \\ 0 & 1 & \frac{1}{3} & \frac{2}{3} & -\frac{1}{3} & 0 \\ 0 & 0 & \frac{8}{3} & -\frac{5}{3} & -\frac{2}{3} & 1 \end{array}\right]$$

Multiplying row 3 by $\frac{3}{8}$, we have

$$\begin{bmatrix} 1 & 0 & \frac{2}{3} & \frac{1}{3} & \frac{1}{3} & 0 \\ 0 & 1 & \frac{1}{3} & \frac{2}{3} & -\frac{1}{3} & 0 \\ 0 & 0 & 1 & -\frac{5}{8} & -\frac{1}{4} & \frac{3}{8} \end{bmatrix}$$

If $-\frac{2}{3}$ times row 3 is added to row 1, we have

$$\begin{bmatrix} 1 & 0 & 0 & \frac{3}{4} & \frac{1}{2} & -\frac{1}{4} \\ 0 & 1 & \frac{1}{3} & \frac{2}{3} & -\frac{1}{3} & 0 \\ 0 & 0 & 1 & -\frac{5}{8} & -\frac{1}{4} & \frac{3}{8} \end{bmatrix}$$

Finally adding $-\frac{1}{3}$ times row 3 to row 2,

$$\begin{bmatrix} 1 & 0 & 0 & \frac{3}{4} & \frac{1}{2} & -\frac{1}{4} \\ 0 & 1 & 0 & \frac{7}{8} & -\frac{1}{4} & -\frac{1}{8} \\ 0 & 0 & 1 & -\frac{5}{8} & -\frac{1}{4} & \frac{3}{8} \end{bmatrix}$$

Therefore, the inverse of

$$\begin{bmatrix} 1 & 1 & 1 \\ 2 & -1 & 1 \\ 3 & 1 & 5 \end{bmatrix} \quad \text{is} \quad \begin{bmatrix} \frac{3}{4} & \frac{1}{2} & -\frac{1}{4} \\ \frac{7}{8} & -\frac{1}{4} & -\frac{1}{8} \\ -\frac{5}{8} & -\frac{1}{4} & \frac{3}{8} \end{bmatrix}$$

Checking,

$$\begin{bmatrix} 1 & 1 & 1 \\ 2 & -1 & 1 \\ 3 & 1 & 5 \end{bmatrix} \begin{bmatrix} \frac{3}{4} & \frac{1}{2} & -\frac{1}{4} \\ \frac{7}{8} & -\frac{1}{4} & -\frac{1}{8} \\ -\frac{5}{8} & -\frac{1}{4} & \frac{3}{8} \end{bmatrix}$$

$$= \begin{bmatrix} 1(\frac{3}{4})+1(\frac{7}{8})+1(-\frac{5}{8}) & 1(\frac{1}{2})+1(-\frac{1}{4})+1(-\frac{1}{4}) & 1(-\frac{1}{4})+1(-\frac{1}{8})+1(\frac{3}{8}) \\ 2(\frac{3}{4})+(-1)(\frac{7}{8})+1(-\frac{5}{8}) & 2(\frac{1}{2})+(-1)(-\frac{1}{4})+1(-\frac{1}{4}) & 2(-\frac{1}{4})+(-1)(-\frac{1}{8})+1(\frac{3}{8}) \\ 3(\frac{3}{4})+1(\frac{7}{8})+5(-\frac{5}{8}) & 3(\frac{1}{2})+1(-\frac{1}{4})+5(-\frac{1}{4}) & 3(-\frac{1}{4})+1(-\frac{1}{8})+5(\frac{3}{8}) \end{bmatrix}$$

$$= \begin{bmatrix} 1 & 0 & 0 \\ 0 & 1 & 0 \\ 0 & 0 & 1 \end{bmatrix}$$

It is left as an exercise to show that

$$\begin{bmatrix} \frac{3}{4} & \frac{1}{2} & -\frac{1}{4} \\ \frac{7}{8} & -\frac{1}{4} & -\frac{1}{8} \\ -\frac{5}{8} & -\frac{1}{4} & \frac{3}{8} \end{bmatrix} \begin{bmatrix} 1 & 1 & 1 \\ 2 & -1 & 1 \\ 3 & 1 & 5 \end{bmatrix} = \begin{bmatrix} 1 & 0 & 0 \\ 0 & 1 & 0 \\ 0 & 0 & 1 \end{bmatrix}$$

EXAMPLE   Find the inverse of

$$\begin{bmatrix} 2 & 1 & -1 \\ 4 & 2 & 3 \\ 6 & 3 & 2 \end{bmatrix}$$

Solution Starting with the augmented matrix and applying row operations,

$$\begin{bmatrix} 2 & 1 & -1 & | & 1 & 0 & 0 \\ 4 & 2 & 3 & | & 0 & 1 & 0 \\ 6 & 3 & 2 & | & 0 & 0 & 1 \end{bmatrix}$$

$$\begin{bmatrix} 1 & \frac{1}{2} & -\frac{1}{2} & | & \frac{1}{2} & 0 & 0 \\ 4 & 2 & 3 & | & 0 & 1 & 0 \\ 6 & 3 & 2 & | & 0 & 0 & 1 \end{bmatrix}$$

$$\begin{bmatrix} 1 & \frac{1}{2} & -\frac{1}{2} & | & \frac{1}{2} & 0 & 0 \\ 0 & 0 & 5 & | & -2 & 1 & 0 \\ 6 & 3 & 2 & | & 0 & 0 & 1 \end{bmatrix}$$

$$\begin{bmatrix} 1 & \frac{1}{2} & -\frac{1}{2} & | & \frac{1}{2} & 0 & 0 \\ 0 & 0 & 5 & | & -2 & 1 & 0 \\ 0 & 0 & 5 & | & -3 & 0 & 1 \end{bmatrix}$$

If -1 times row 2 is added to row 3, the result is

$$\begin{bmatrix} 1 & \frac{1}{2} & -\frac{1}{2} & | & \frac{1}{2} & 0 & 0 \\ 0 & 0 & 5 & | & -2 & 1 & 0 \\ 0 & 0 & 0 & | & -1 & -1 & 1 \end{bmatrix}$$

The zeros in row 3 indicate that no inverse exists.

Normally the steps are compressed by "zeroing" one column at each writing of the augmented matrix. This is illustrated in the following inversion of a 4 × 4 matrix.

EXAMPLE Find the inverse of

$$\begin{bmatrix} 1 & -1 & -1 & 0 \\ 2 & 0 & 1 & 1 \\ 1 & -1 & 0 & 1 \\ 1 & 0 & -1 & 1 \end{bmatrix}$$

Solution The augmented matrix is

$$\begin{bmatrix} 1 & -1 & -1 & 0 & | & 1 & 0 & 0 & 0 \\ 2 & 0 & 1 & 1 & | & 0 & 1 & 0 & 0 \\ 1 & -1 & 0 & 1 & | & 0 & 0 & 1 & 0 \\ 1 & 0 & -1 & 1 & | & 0 & 0 & 0 & 1 \end{bmatrix}$$

Using multiples of the first row to "zero" the remaining elements of the first column, we have

$$\begin{bmatrix} 1 & -1 & -1 & 0 & | & 1 & 0 & 0 & 0 \\ 0 & 2 & 3 & 1 & | & -2 & 1 & 0 & 0 \\ 0 & 0 & 1 & 1 & | & -1 & 0 & 1 & 0 \\ 0 & 1 & 0 & 1 & | & -1 & 0 & 0 & 1 \end{bmatrix}$$

Dividing row 2 by 2,

$$\begin{bmatrix} 1 & -1 & -1 & 0 & | & 1 & 0 & 0 & 0 \\ 0 & 1 & \frac{3}{2} & \frac{1}{2} & | & -1 & \frac{1}{2} & 0 & 0 \\ 0 & 0 & 1 & 1 & | & -1 & 0 & 1 & 0 \\ 0 & 1 & 0 & 1 & | & -1 & 0 & 0 & 1 \end{bmatrix}$$

Using row 2 to eliminate or zero the remaining elements of the second column, we get

$$\begin{bmatrix} 1 & 0 & \frac{1}{2} & \frac{1}{2} & | & 0 & \frac{1}{2} & 0 & 0 \\ 0 & 1 & \frac{3}{2} & \frac{1}{2} & | & -1 & \frac{1}{2} & 0 & 0 \\ 0 & 0 & 1 & 1 & | & -1 & 0 & 1 & 0 \\ 0 & 0 & -\frac{3}{2} & \frac{1}{2} & | & 0 & -\frac{1}{2} & 0 & 1 \end{bmatrix}$$

Using row 3 to eliminate the other elements of column 3,

$$\begin{bmatrix} 1 & 0 & 0 & 0 & | & \frac{1}{2} & \frac{1}{2} & -\frac{1}{2} & 0 \\ 0 & 1 & 0 & -1 & | & \frac{1}{2} & \frac{1}{2} & -\frac{3}{2} & 0 \\ 0 & 0 & 1 & 1 & | & -1 & 0 & 1 & 0 \\ 0 & 0 & 0 & 2 & | & -\frac{3}{2} & -\frac{1}{2} & \frac{3}{2} & 1 \end{bmatrix}$$

Dividing row 4 by 2,

$$\begin{bmatrix} 1 & 0 & 0 & 0 & | & \frac{1}{2} & \frac{1}{2} & -\frac{1}{2} & 0 \\ 0 & 1 & 0 & -1 & | & \frac{1}{2} & \frac{1}{2} & -\frac{3}{2} & 0 \\ 0 & 0 & 1 & 1 & | & -1 & 0 & 1 & 0 \\ 0 & 0 & 0 & 1 & | & -\frac{3}{4} & -\frac{1}{4} & \frac{3}{4} & \frac{1}{2} \end{bmatrix}$$

Using row 4 to eliminate the remaining elements of column 4,

$$\begin{bmatrix} 1 & 0 & 0 & 0 & | & \frac{1}{2} & \frac{1}{2} & -\frac{1}{2} & 0 \\ 0 & 1 & 0 & 0 & | & -\frac{1}{4} & \frac{1}{4} & -\frac{3}{4} & \frac{1}{2} \\ 0 & 0 & 1 & 0 & | & -\frac{1}{4} & \frac{1}{4} & \frac{1}{4} & -\frac{1}{2} \\ 0 & 0 & 0 & 1 & | & -\frac{3}{4} & -\frac{1}{4} & \frac{3}{4} & \frac{1}{2} \end{bmatrix}$$

It is left to the reader to verify that

$$\begin{bmatrix} \frac{1}{2} & \frac{1}{2} & -\frac{1}{2} & 0 \\ -\frac{1}{4} & \frac{1}{4} & -\frac{3}{4} & \frac{1}{2} \\ -\frac{1}{4} & \frac{1}{4} & \frac{1}{4} & -\frac{1}{2} \\ -\frac{3}{4} & -\frac{1}{4} & \frac{3}{4} & \frac{1}{2} \end{bmatrix}$$

is the inverse of

$$\begin{bmatrix} 1 & -1 & -1 & 0 \\ 2 & 0 & 1 & 1 \\ 1 & -1 & 0 & 1 \\ 1 & 0 & -1 & 1 \end{bmatrix}$$

EXERCISES 5.5

1. Verify that

$$\begin{bmatrix} \frac{3}{4} & \frac{1}{2} & -\frac{1}{4} \\ \frac{7}{8} & -\frac{1}{4} & -\frac{1}{8} \\ -\frac{5}{8} & -\frac{1}{4} & \frac{3}{8} \end{bmatrix} \begin{bmatrix} 1 & 1 & 1 \\ 2 & -1 & 1 \\ 3 & 1 & 5 \end{bmatrix} = \begin{bmatrix} 1 & 0 & 0 \\ 0 & 1 & 0 \\ 0 & 0 & 1 \end{bmatrix}$$

2. Verify that if

$$A = \begin{bmatrix} 1 & -1 & -1 & 0 \\ 2 & 0 & 1 & 1 \\ 1 & -1 & 0 & 1 \\ 1 & 0 & -1 & 1 \end{bmatrix} \quad \text{and} \quad A^{-1} = \begin{bmatrix} \frac{1}{2} & \frac{1}{2} & -\frac{1}{2} & 0 \\ -\frac{1}{4} & \frac{1}{4} & -\frac{3}{4} & \frac{1}{2} \\ -\frac{1}{4} & \frac{1}{4} & \frac{1}{4} & -\frac{1}{2} \\ -\frac{3}{4} & -\frac{1}{4} & \frac{3}{4} & \frac{1}{2} \end{bmatrix}$$

$$AA^{-1} = A^{-1}A = I.$$

Find the inverses, if they exist, of each of the matrices in Exercises 3–6.

3. $\begin{bmatrix} 2 & 3 \\ -1 & 4 \end{bmatrix}$

4. $\begin{bmatrix} 5 & -5 \\ 1 & -1 \end{bmatrix}$

5. $\begin{bmatrix} 3 & 1 & -1 \\ 1 & 4 & 1 \\ 3 & -1 & 1 \end{bmatrix}$

6. $\begin{bmatrix} 1 & -1 & 1 \\ -1 & 3 & 4 \\ 1 & 3 & 2 \end{bmatrix}$

Find the inverses of each of the nonsingular matrices in Exercises 7–18.

7. $\begin{bmatrix} 2 & 0 & 1 \\ 0 & 2 & 1 \\ 2 & 1 & -2 \end{bmatrix}$

8. $\begin{bmatrix} 3 & -2 & 1 \\ 2 & 3 & -1 \\ 2 & -2 & 1 \end{bmatrix}$

9. $\begin{bmatrix} 2 & -1 & 0 \\ 1 & 0 & -1 \\ 0 & -1 & 1 \end{bmatrix}$

10. $\begin{bmatrix} 1 & -1 & -1 \\ 2 & 4 & -3 \\ 5 & -1 & -2 \end{bmatrix}$

11. $\begin{bmatrix} 1 & 0 & 2 \\ 1 & 0 & -2 \\ 2 & -2 & 4 \end{bmatrix}$

12. $\begin{bmatrix} 1 & -1 & 2 \\ 1 & 2 & 3 \\ 1 & -3 & 3 \end{bmatrix}$

13. $\begin{bmatrix} 1 & 1 & 1 \\ 1 & 2 & 2 \\ -1 & 4 & 3 \end{bmatrix}$

14. $\begin{bmatrix} 3 & 2 & -1 \\ -1 & 1 & 2 \\ 2 & 1 & 1 \end{bmatrix}$

15. $\begin{bmatrix} 1 & 3 & 4 \\ -1 & 2 & 3 \\ 2 & 2 & 5 \end{bmatrix}$

16. $\begin{bmatrix} 1 & 1 & 1 & 1 \\ 2 & -1 & 3 & 1 \\ 1 & 1 & 2 & 2 \\ 3 & 2 & 1 & -1 \end{bmatrix}$

17. $\begin{bmatrix} 1 & 2 & -1 & 1 \\ -1 & -3 & 4 & 2 \\ 2 & 1 & 1 & -3 \\ 3 & -1 & -2 & 3 \end{bmatrix}$
18. $\begin{bmatrix} 1 & 1 & 0 & 0 & 0 \\ 0 & 1 & 1 & 0 & 0 \\ 0 & 0 & 1 & 1 & 0 \\ 0 & 0 & 0 & 1 & 1 \\ 0 & 0 & 0 & 0 & 1 \end{bmatrix}$

19. Suppose that A is a nonsingular matrix and B and C are matrices such that $AB = AC$. Does $B = C$? Explain.

20. Let A and B be nonsingular 3×3 matrices. Show that $(AB)^{-1} = B^{-1}A^{-1}$.

5.6 MATRIX SOLUTIONS TO SYSTEMS OF EQUATIONS *

One of the most powerful applications of matrix inversion is in the solution of systems of linear equations.

Consider the problem of finding the solution to a system of two linear equations in two unknowns

$$a_{11}x_1 + a_{12}x_2 = b_1$$
$$a_{21}x_1 + a_{22}x_2 = b_2$$

where a_{11}, a_{12}, a_{21}, a_{22}, b_1, and b_2 are constants and x_1 and x_2 are the unknowns. This system of equations can be written in matrix and vector form. Let

$$A = \begin{bmatrix} a_{11} & a_{12} \\ a_{21} & a_{22} \end{bmatrix}, \quad X = \begin{bmatrix} x_1 \\ x_2 \end{bmatrix}, \quad \text{and} \quad B = \begin{bmatrix} b_1 \\ b_2 \end{bmatrix}$$

Then the system of equations can be written

$$AX = B$$

since

$$AX = \begin{bmatrix} a_{11} & a_{12} \\ a_{21} & a_{22} \end{bmatrix}\begin{bmatrix} x_1 \\ x_2 \end{bmatrix} = \begin{bmatrix} a_{11}x_1 + a_{12}x_2 \\ a_{21}x_1 + a_{22}x_2 \end{bmatrix}$$

Thus $AX = B$ becomes

$$\begin{bmatrix} a_{11}x_1 + a_{12}x_2 \\ a_{21}x_1 + a_{22}x_2 \end{bmatrix} = \begin{bmatrix} b_1 \\ b_2 \end{bmatrix}$$

Applying the definition of matrix equality,

$$AX = B$$

---

* See Chapter A for nonmatrix solutions to systems of linear equations.

if and only if

$$a_{11}x_1 + a_{12}x_2 = b_1$$
$$a_{21}x_1 + a_{22}x_2 = b_2$$

Suppose then that A is a nonsingular matrix, i.e., that A^{-1} exists. Then multiplying both sides of the matrix equation by A^{-1}, we have

$$A^{-1}(AX) = A^{-1}B$$
$$(A^{-1}A)X = A^{-1}B$$
$$IX = A^{-1}B$$
$$X = A^{-1}B$$

EXAMPLE Use matrix inversion to solve the following system of equations:

$$x_1 + 3x_2 = 4$$
$$2x_1 - x_2 = -1$$

Solution In matrix form the system can be written

$$\begin{bmatrix} 1 & 3 \\ 2 & -1 \end{bmatrix} \begin{bmatrix} x_1 \\ x_2 \end{bmatrix} = \begin{bmatrix} 4 \\ -1 \end{bmatrix}$$

Letting

$$A = \begin{bmatrix} 1 & 3 \\ 2 & -1 \end{bmatrix}$$

we can use the matrix inversion algorithm to find A^{-1}:

$$\left[\begin{array}{cc|cc} 1 & 3 & 1 & 0 \\ 2 & -1 & 0 & 1 \end{array} \right]$$

$$\left[\begin{array}{cc|cc} 1 & 3 & 1 & 0 \\ 0 & -7 & -2 & 1 \end{array} \right]$$

$$\left[\begin{array}{cc|cc} 1 & 3 & 1 & 0 \\ 0 & 1 & \frac{2}{7} & -\frac{1}{7} \end{array} \right]$$

$$\left[\begin{array}{cc|cc} 1 & 0 & \frac{1}{7} & \frac{3}{7} \\ 0 & 1 & \frac{2}{7} & -\frac{1}{7} \end{array} \right]$$

Thus

$$A^{-1} = \begin{bmatrix} \frac{1}{7} & \frac{3}{7} \\ \frac{2}{7} & -\frac{1}{7} \end{bmatrix}$$

Multiplying both sides of the matrix equation by A^{-1},

$$\begin{bmatrix} \frac{1}{7} & \frac{3}{7} \\ \frac{2}{7} & -\frac{1}{7} \end{bmatrix} \begin{bmatrix} 1 & 3 \\ 2 & -1 \end{bmatrix} \begin{bmatrix} x_1 \\ x_2 \end{bmatrix} = \begin{bmatrix} \frac{1}{7} & \frac{3}{7} \\ \frac{2}{7} & -\frac{1}{7} \end{bmatrix} \begin{bmatrix} 4 \\ -1 \end{bmatrix}$$

$$\begin{bmatrix} \frac{1}{7}+\frac{6}{7} & \frac{3}{7}-\frac{3}{7} \\ \frac{2}{7}-\frac{2}{7} & \frac{6}{7}+\frac{1}{7} \end{bmatrix} \begin{bmatrix} x_1 \\ x_2 \end{bmatrix} = \begin{bmatrix} \frac{4}{7}-\frac{3}{7} \\ \frac{8}{7}+\frac{1}{7} \end{bmatrix}$$

$$\begin{bmatrix} 1 & 0 \\ 0 & 1 \end{bmatrix} \begin{bmatrix} x_1 \\ x_2 \end{bmatrix} = \begin{bmatrix} \frac{1}{7} \\ \frac{9}{7} \end{bmatrix}$$

$$\begin{bmatrix} x_1 + 0x_2 \\ 0x_1 + x_2 \end{bmatrix} = \begin{bmatrix} \frac{1}{7} \\ \frac{9}{7} \end{bmatrix}$$

$$\begin{bmatrix} x_1 \\ x_2 \end{bmatrix} = \begin{bmatrix} \frac{1}{7} \\ \frac{9}{7} \end{bmatrix}$$

Thus $x_1 = \frac{1}{7}$ and $x_2 = \frac{9}{7}$ is the solution to the system. This can be checked by substitution:

$$\begin{cases} \frac{1}{7} + 3(\frac{9}{7}) = 4 \\ 2(\frac{1}{7}) - \frac{9}{7} = -1 \end{cases}$$

or

$$\begin{cases} \dfrac{1 + 27}{7} = 4 \\ \dfrac{2 - 9}{7} = -1 \end{cases}$$

$$\begin{cases} \frac{28}{7} = 4 \\ -\frac{7}{7} = -1 \end{cases}$$

$$\begin{cases} 4 = 4 \\ -1 = -1 \end{cases}$$

The procedure works equally well with systems that have three, four, or more unknowns.

EXAMPLE Solve this system of equations:

$$x_1 + x_2 + x_3 = 5$$

$$2x_1 - x_2 + x_3 = -2$$

$$3x_1 + x_2 + 5x_3 = 6$$

Solution In matrix form this system can be written

$$\begin{bmatrix} 1 & 1 & 1 \\ 2 & -1 & 1 \\ 3 & 1 & 5 \end{bmatrix} \begin{bmatrix} x_1 \\ x_2 \\ x_3 \end{bmatrix} = \begin{bmatrix} 5 \\ -2 \\ 6 \end{bmatrix}$$

The matrix of coefficients can be inverted, i.e., its inverse can be found using the algorithm. In fact, this was done in the previous section, and the inverse was found to be

$$\begin{bmatrix} \frac{3}{4} & \frac{1}{2} & -\frac{1}{4} \\ \frac{7}{8} & -\frac{1}{4} & -\frac{1}{8} \\ -\frac{5}{8} & -\frac{1}{4} & \frac{3}{8} \end{bmatrix}$$

Thus

$$\begin{bmatrix} \frac{3}{4} & \frac{1}{2} & -\frac{1}{4} \\ \frac{7}{8} & -\frac{1}{4} & -\frac{1}{8} \\ -\frac{5}{8} & -\frac{1}{4} & \frac{3}{8} \end{bmatrix} \begin{bmatrix} 1 & 1 & 1 \\ 2 & -1 & 1 \\ 3 & 1 & 5 \end{bmatrix} \begin{bmatrix} x_1 \\ x_2 \\ x_3 \end{bmatrix} = \begin{bmatrix} \frac{3}{4} & \frac{1}{2} & -\frac{1}{4} \\ \frac{7}{8} & -\frac{1}{4} & -\frac{1}{8} \\ -\frac{5}{8} & -\frac{1}{4} & \frac{3}{8} \end{bmatrix} \begin{bmatrix} 5 \\ -2 \\ 6 \end{bmatrix}$$

$$\begin{bmatrix} 1 & 0 & 0 \\ 0 & 1 & 0 \\ 0 & 0 & 1 \end{bmatrix} \begin{bmatrix} x_1 \\ x_2 \\ x_3 \end{bmatrix} = \begin{bmatrix} \frac{15}{4} - 1 - \frac{6}{4} \\ \frac{35}{8} + \frac{2}{4} - \frac{6}{8} \\ -\frac{25}{8} + \frac{2}{4} + \frac{18}{8} \end{bmatrix}$$

$$\begin{bmatrix} x_1 \\ x_2 \\ x_3 \end{bmatrix} = \begin{bmatrix} \frac{5}{4} \\ \frac{33}{8} \\ -\frac{3}{8} \end{bmatrix}$$

or

$$x_1 = \frac{5}{4}$$
$$x_2 = \frac{33}{8}$$
$$x_3 = -\frac{3}{8}$$

EXERCISES 5.6

Use matrix inversion to solve the systems of equations in Exercises 1–20.

1. $2x_1 - 4x_2 = 3$
$\quad x_1 + x_2 = -1$

2. $\frac{1}{2}x_1 + x_2 = 3$
$\quad 3x_1 - 4x_2 = 7$

3. $3x_1 - x_2 = -1$
$\quad x_1 - \frac{1}{3}x_2 = -1$

4. $x_2 - x_1 = 3$
$\quad 2x_2 - 4x_1 = 3$

5. $x_1 + x_2 + x_3 = 1$
$\quad -x_1 + 2x_2 + x_3 = 0$
$\quad x_1 + 3x_2 + 5x_3 = 1$

6. $x_1 - x_2 + x_3 = 2$
$\quad -x_1 + 3x_2 + 4x_3 = 3$
$\quad x_1 + 3x_2 + x_3 = -1$

7. $2x_2 + x_3 = -4$
$\quad 2x_1 + x_3 = 2$
$\quad x_1 + 2x_2 - 2x_3 = 1$

8. $-2x_1 + x_2 - 3x_3 = 2$
$\quad 3x_1 - x_2 + 2x_3 = 0$
$\quad -2x_1 + x_2 + 2x_3 = 1$

9. $-x_1 + 2x_2 = 0$
  $x_1 - x_3 = 0$
  $-x_1 + x_3 = 0$

10. $x_1 + x_2 - x_3 = 0$
  $4x_1 + 2x_2 - 3x_3 = 1$
  $-x_1 + 5x_2 - 2x_3 = 2$

11. $x_1 + 2x_2 = -1$
  $x_1 - 2x_2 = -1$
  $2x_1 + 4x_2 - 2x_3 = -1$

12. $x_1 + 2x_2 - x_3 = 4$
  $x_1 - 3x_2 + 2x_3 = 2$
  $x_1 + 3x_2 - 3x_3 = 2$

13. $x_1 + 2x_2 + 2x_3 = -2$
  $x_1 + x_2 + x_3 = 4$
  $-x_1 + 4x_2 + 3x_3 = 1$

14. $-x_1 + x_2 + 3x_3 = 0$
  $3x_1 + 2x_2 - x_3 = 1$
  $2x_1 + x_2 + x_3 = 1$

15. $4x_1 + x_2 + 3x_3 = 5$
  $3x_1 - x_2 + 2x_3 = 4$
  $5x_1 + 2x_2 + 2x_3 = 1$

16. $x_1 - 2x_2 + 3x_3 + 2x_4 = 1$
  $x_1 + 2x_2 - 3x_3 + x_4 = 2$
  $-x_1 - x_2 - x_3 - x_4 = 3$
  $2x_1 - 4x_2 - x_3 - 3x_4 = 4$

17. $2x_1 + x_2 - x_3 + x_4 = 1$
  $-3x_1 - x_2 + x_3 + 2x_4 = 0$
  $x_1 + 2x_2 + x_3 - 3x_4 = 1$
  $-x_1 + 3x_2 - 2x_3 + 3x_4 = 2$

18. $x_1 + x_2 + x_3 + x_4 = -3$
  $2x_1 + 3x_2 - x_3 + x_4 = -3$
  $x_1 + 2x_2 + x_3 + 2x_4 = -2$
  $3x_1 + x_2 + 2x_3 - x_4 = -1$

19. $kx_1 + lx_2 = 0$
  $mx_1 + nx_2 = 0$ ($k, l, m,$ and n are constants)

20. $kx_1 + lx_2 = 1$
  $lx_1 + kx_2 = 1$

5.7   APPLICATIONS

Matrix theory and inversion have practical applications. Consider the following examples.

EXAMPLE   The Good Health Vitamin Company (GHV) has 3 divisions—processing, testing, and packaging. The number of hours required of each division to make 1 lb each of vitamins A, B, and E is given in the following table:

| | | Vitamin | |
Division	A	B	E
Processing	1	2	1
Testing	2	1	1
Packaging	3	1	3

Processing has 1,100 hours available, testing has 800 hours, and packaging has 1,300 hours. How much of each type of vitamin should GHV make?

Solution Let

$$x_1 = \text{the number of lb of vitamin A to be made}$$
$$x_2 = \text{the number of lb of vitamin B}$$
$$x_3 = \text{the number of lb of vitamin E}$$

Then for the processing division,

$$x_1 + 2x_2 + x_3 = 1{,}100$$

For the testing division,

$$2x_1 + x_2 + x_3 = 800$$

For the packaging division,

$$3x_1 + x_2 + 3x_3 = 1{,}300$$

In matrix form,

$$\begin{bmatrix} 1 & 2 & 1 \\ 2 & 1 & 1 \\ 3 & 1 & 3 \end{bmatrix} \begin{bmatrix} x_1 \\ x_2 \\ x_3 \end{bmatrix} = \begin{bmatrix} 1{,}100 \\ 800 \\ 1{,}300 \end{bmatrix}$$

Inverting the matrix of coefficients,

$$\left[\begin{array}{ccc|ccc} 1 & 2 & 1 & 1 & 0 & 0 \\ 2 & 1 & 1 & 0 & 1 & 0 \\ 3 & 1 & 3 & 0 & 0 & 1 \end{array}\right]$$

$$\left[\begin{array}{ccc|ccc} 1 & 2 & 1 & 1 & 0 & 0 \\ 0 & -3 & -1 & -2 & 1 & 0 \\ 0 & -5 & 0 & -3 & 0 & 1 \end{array}\right]$$

$$\left[\begin{array}{ccc|ccc} 1 & 2 & 1 & 1 & 0 & 0 \\ 0 & 1 & \frac{1}{3} & \frac{2}{3} & -\frac{1}{3} & 0 \\ 0 & -5 & 0 & -3 & 0 & 1 \end{array}\right]$$

$$\left[\begin{array}{ccc|ccc} 1 & 0 & \frac{1}{3} & -\frac{1}{3} & \frac{2}{3} & 0 \\ 0 & 1 & \frac{1}{3} & \frac{2}{3} & -\frac{1}{3} & 0 \\ 0 & 0 & \frac{5}{3} & \frac{1}{3} & -\frac{5}{3} & 1 \end{array}\right]$$

$$\left[\begin{array}{ccc|ccc} 1 & 0 & \frac{1}{3} & -\frac{1}{3} & \frac{2}{3} & 0 \\ 0 & 1 & \frac{1}{3} & \frac{2}{3} & -\frac{1}{3} & 0 \\ 0 & 0 & 1 & \frac{1}{5} & -1 & \frac{3}{5} \end{array}\right]$$

$$\left[\begin{array}{ccc|ccc} 1 & 0 & 0 & -\frac{2}{5} & 1 & -\frac{1}{5} \\ 0 & 1 & 0 & \frac{3}{5} & 0 & -\frac{1}{5} \\ 0 & 0 & 1 & \frac{1}{5} & -1 & \frac{3}{5} \end{array}\right]$$

Therefore, the solution is given by

$$
\begin{bmatrix} x_1 \\ x_2 \\ x_3 \end{bmatrix} = \begin{bmatrix} -\frac{2}{5} & 1 & -\frac{1}{5} \\ \frac{3}{5} & 0 & -\frac{1}{5} \\ \frac{1}{5} & -1 & \frac{3}{5} \end{bmatrix} \begin{bmatrix} 1{,}100 \\ 800 \\ 1{,}300 \end{bmatrix}
$$

$$
= \begin{bmatrix} -\frac{2}{5}(1{,}100) + 800 - \frac{1}{5}(1{,}300) \\ \frac{3}{5}(1{,}100) + 0(800) - \frac{1}{5}(1{,}300) \\ \frac{1}{5}(1{,}100) - 1(800) + \frac{3}{5}(1{,}300) \end{bmatrix}
$$

$$
= \begin{bmatrix} 100 \\ 400 \\ 200 \end{bmatrix}
$$

They should make 100 lb of vitamin A, 400 lb of vitamin B, and 200 lb of vitamin E.

EXAMPLE Two companies are involved in a business venture with three distinct outcomes. The probability of each of the three outcomes is unknown, but company A will receive $300 if the first outcome occurs, $600 if the second outcome takes place, and $1,000 if the third outcome takes place. This company knows that on the average it expects to make $725 (expected value). The corresponding values of the second company, company B, are $400, $400, and $900, with an expected value of $650. What are the probabilities?

Solution The three outcomes form a sample space for the venture. Thus if x_1 is the probability of the first outcome, x_2 the probability of the second outcome, and x_3 the probability of the third outcome,

$$x_1 + x_2 + x_3 = 1$$

Since company A has an expected return of $725,

$$300x_1 + 600x_2 + 1{,}000x_3 = 725$$

For company B,

$$400x_1 + 400x_2 + 900x_3 = 650$$

In matrix form,

$$
\begin{bmatrix} 1 & 1 & 1 \\ 300 & 600 & 1{,}000 \\ 400 & 400 & 900 \end{bmatrix} \begin{bmatrix} x_1 \\ x_2 \\ x_3 \end{bmatrix} = \begin{bmatrix} 1 \\ 725 \\ 650 \end{bmatrix}
$$

The inverse of the matrix of coefficients is

$$
\begin{bmatrix} \frac{14}{15} & -\frac{1}{300} & \frac{1}{375} \\ \frac{13}{15} & \frac{1}{300} & -\frac{7}{1500} \\ -\frac{4}{5} & 0 & \frac{1}{500} \end{bmatrix}
$$

Thus

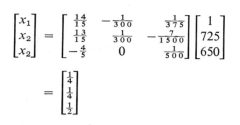

$$\begin{bmatrix} x_1 \\ x_2 \\ x_2 \end{bmatrix} = \begin{bmatrix} \frac{14}{15} & -\frac{1}{300} & \frac{1}{375} \\ \frac{13}{15} & \frac{1}{300} & -\frac{7}{1500} \\ -\frac{4}{5} & 0 & \frac{1}{500} \end{bmatrix} \begin{bmatrix} 1 \\ 725 \\ 650 \end{bmatrix}$$

$$= \begin{bmatrix} \frac{1}{4} \\ \frac{1}{4} \\ \frac{1}{2} \end{bmatrix}$$

Therefore, the probability of the first outcome is $\frac{1}{4}$, that of the second $\frac{1}{4}$, and that of the last outcome $\frac{1}{2}$.

EXAMPLE A shipping firm wishes to load a shipping container with three types of cargo. The following table lists various parameters of the cargo, per unit of cargo:

	Type		
	A	B	C
Volume (per unit)	1 cu yd	2 cu yd	3 cu yd
Value (per unit)	$200	$300	$250
Weight (per unit)	200 lb	250 lb	400 lb

Their containers hold 85 cu yd and a total weight of 12,000 lb. They can insure the container for $11,500. How should they load each container?

Solution Let

x_1 be the number of units of type A cargo

x_2 the number of units of type B cargo

x_3 the number of units of type C cargo

The conditions as to volume, value, and weight stated in equation form are

$$x_1 + 2x_2 + 3x_3 = 85$$
$$200x_1 + 300x_2 + 250x_3 = 11,500$$
$$200x_1 + 250x_2 + 400x_3 = 12,000$$

In matrix form,

$$\begin{bmatrix} 1 & 2 & 3 \\ 200 & 300 & 250 \\ 200 & 250 & 400 \end{bmatrix} \begin{bmatrix} x_1 \\ x_2 \\ x_3 \end{bmatrix} = \begin{bmatrix} 85 \\ 11,500 \\ 12,000 \end{bmatrix}$$

Inverting,

$$\left[\begin{array}{ccc|ccc} 1 & 2 & 3 & 1 & 0 & 0 \\ 200 & 300 & 250 & 0 & 1 & 0 \\ 200 & 250 & 400 & 0 & 0 & 1 \end{array}\right]$$

becomes

$$\left[\begin{array}{ccc|ccc} 1 & 0 & 0 & -\frac{23}{13} & \frac{1}{650} & \frac{4}{325} \\ 0 & 1 & 0 & \frac{12}{13} & \frac{2}{325} & -\frac{7}{650} \\ 0 & 0 & 1 & \frac{4}{13} & -\frac{3}{650} & \frac{1}{325} \end{array}\right]$$

The inverse is

$$\left[\begin{array}{ccc} -\frac{23}{13} & \frac{1}{650} & \frac{4}{325} \\ \frac{12}{13} & \frac{2}{325} & -\frac{7}{650} \\ \frac{4}{13} & -\frac{3}{650} & \frac{1}{325} \end{array}\right]$$

The solution is

$$\begin{bmatrix} x_1 \\ x_2 \\ x_3 \end{bmatrix} = \begin{bmatrix} -\frac{23}{13} & \frac{1}{650} & \frac{4}{325} \\ \frac{12}{13} & \frac{2}{325} & -\frac{7}{650} \\ \frac{4}{13} & -\frac{3}{650} & \frac{1}{325} \end{bmatrix} \begin{bmatrix} 85 \\ 11{,}500 \\ 12{,}000 \end{bmatrix}$$

$$= \begin{bmatrix} \dfrac{-1150(85) + 11{,}500 + 8(12{,}000)}{650} \\[2ex] \dfrac{600(85) + 4(11{,}500) - 7(12{,}000)}{650} \\[2ex] \dfrac{200(85) - 3(11{,}500) + 2(12{,}000)}{650} \end{bmatrix}$$

$$= \begin{bmatrix} 15 \\ 20 \\ 10 \end{bmatrix}$$

The examples are somewhat artificial in that they work out to exact answers. In a later chapter we will consider a more powerful method of solving similar problems where the conditions are stated in terms of linear inequalities. This is called linear programming, and it will be used to solve very realistic problems.

EXERCISES 5.7

1. Two sales divisions of a company attempt to outsell each other on two items. On the first item the first division sells 100 for each salesman, and the second division sells 80 per salesman. A total of 980 units are sold.

On the second item the first division sells 70 per salesman, and the second division sells 80, for a total of 830 units. How many salesmen are there in each division?

2. Two divisions of a company compete on dollar volume. During the first six months one division sells \$80,000 per man, the second sells \$90,000 per man, for a total of \$1,860,000. During the second six-month period the averages per man are reversed, and the total sales amount to \$1,880,000. How many men are in each division?

The Good Health Vitamin Company makes vitamins using time in hours per lb as indicated in the tables in Exercises 3–5. If the available hours per division are as indicated, find the number of pounds of each type of vitamin that should be manufactured under the stated conditions.

3.

	Vitamin		
Division	A	B	E
Processing	1	1	0
Testing	10	3	1
Packaging	4	0	2

Available time: 200 hours of processing
1,000 hours of testing
600 hours of packaging

4.

	Vitamin		
Division	G	C	A
Processing	10	8	0
Testing	5	1	2
Packaging	3	2	1

Available time: 26 hours of processing
13 hours of testing
10 hours of packaging

5.

	Vitamin		
Division	B_1	B_2	B_3
Processing	1	3	1
Testing	2	4	1
Packaging	1	1	1

Available time: 8 hours of processing
11 hours of testing
6 hours of packaging

6. The Rock Solid Hammer Company makes three types of hammers: claw, sledge, and tack. The hours required to manufacture 100 of each type from the company's divisions are given below:

	Claw	Sledge	Tack
Stamping	1	5	1
Finishing	1	1	5
Packaging	5	1	1

How many of each type should be made if 14 hours of stamping, 38 hours of finishing, and 18 hours of packaging are available?

7. Two companies are involved in a business venture with three distinct outcomes. If outcome 1 occurs, company (1) will make $5,000 and company (2) will lose $10,000. If outcome 2 takes place, company (1) will make $10,000 and (2) will lose $20,000. If outcome 3 takes place, (1) will lose $5,000 and (2) will make $15,000. The overall average or expected gain for company (1) is $6,000, while the expected or average overall loss for company (2) is $11,000. What are the probabilities of the three outcomes?

8. Two companies are involved in a deal, with the following table of gains and losses depending on the outcome:

	Outcome 1	Outcome 2	Outcome 3
Company A	+ $3,000	+ $2,000	+ $7,000
Company B	+ $4,000	− $2,000	+ $5,000

The expected value of the investment for company A is $2,850 and for B is − $350. What are the probabilities of each outcome?

9. Three people wish to make some joint investments:

Investment	Person (1)	(2)	(3)	Total Dollar Amt
A	1	$\frac{1}{2}$	$\frac{1}{3}$	$ 9,000
B	$\frac{1}{2}$	$\frac{1}{2}$	1	$11,000
C	1	$\frac{1}{6}$	$\frac{1}{12}$	$ 5,500

The table entries indicate the fraction of each investor's total resources he is willing to invest in the given investment. How much does each invest or have to invest?

10. Three banks each have a certain amount they are willing to invest in selected mutual funds. The table below gives the fraction of their total investment fund each is willing to invest in three funds, together with the total dollar amount that would represent. How much is each bank's investment fund?

| | Bank | | | |
	A	B	C	Total
Hot Fund	$\frac{1}{4}$	$\frac{1}{5}$	1	$ 6,000,000
Growth Fund	$\frac{1}{4}$	1	1	$15,000,000
Safe Fund	$\frac{1}{6}$	$\frac{1}{10}$	1	$ 4,000,000

11. A shipping company wishes to load three types of cargo with the following parameters per unit:

| | Type | | |
	A	B	C
Volume (cu yd)	6	6	5
Value ($)	1,200	600	100
Weight (cwt)	3	3	1

The value of the cargo is $135,000, it takes up 900 cu yd, and it weighs 45,000 lb. How much of each type is loaded?

12. A drugstore is asked to make 5 g of a certain compound with three component ingredients. The first drug is 50% inert solids, the second 20%, and the third 40%. The final mixture must be 33% inert solids. The final mixture must contain 1.45 g $C_4H_6O_2$. The first drug contains 0.2 g/g of $C_4H_6O_2$, the second 0.3 g/g, and the third 0.4 g/g. How much of each of the three component drugs should be used?

13. A mixture of tobacco totaling 120 lb is to be made from three basic types. The value of the mixture is to be $160. Type 1 tobacco is worth $2 per lb; type 2, $0.50 per lb; and type 3, $1 per lb. The mixture must include a total of 300 oz of Yenidje leaf tobacco. Type 1 contains 1 oz per lb; type 2, 4 oz per lb; and type 3, 4 oz per lb. How much of each type of tobacco should be used?

14. A mixture of three drugs totals 14 oz. The following table gives the number of g per oz of 3 components each drug contains, together with the number of g required in the final mixture. What amount of each component drug should be used?

	Drug			Total
	A	B	C	Required
Tetracyclin	6	6	5	79
HCL	12	6	10	134

15. A total of \$30,000 is to be invested at 5%, 7%, and 10%. The total return on the investment is \$2,220. If the amount at 5% is reinvested at 7%, the total return is \$2,420. How much is invested at each rate?

16. A total of 6,000 man-hours are to be divided among three projects. Work on the first project costs \$12 per man-hour, work on the second costs \$14 per man-hour, and work on the third costs \$16 per man-hour. The total cost is \$84,000. If all the men on the third project are put to work on the first project, the cost total is \$76,000. How many man-hours were originally assigned to each project?

5.8 COMPACT NOTATION

While its application to systems of equations alone would make matrix theory worthwhile, there are many other uses for it. As has been previously discussed, the increased availability of electronic data-processing equipment has made matrix methods practical. In this section we will consider a more compact matrix notation, one suitable for computer analysis and programming.

NOTATION

If A is an $n \times m$ matrix, it will be denoted

$$A = [a_{ij}]_{nm}$$

That is,

$$A = [a_{ij}]_{nm} = \begin{bmatrix} a_{11} & a_{12} & \cdots & a_{1m} \\ \vdots & & & \\ a_{n1} & \cdots\cdots\cdots & a_{nm} \end{bmatrix}$$

In other words, $[a_{ij}]_{nm}$ denotes the matrix with elements a_{ij}, where i ranges from 1 to n and j ranges from 1 to m.

EXAMPLE Write $[a_{ij}]_{34}$ in expanded form.

Solution

$$[a_{ij}]_{34} = \begin{bmatrix} a_{11} & a_{12} & a_{13} & a_{14} \\ a_{21} & a_{22} & a_{23} & a_{24} \\ a_{31} & a_{32} & a_{33} & a_{34} \end{bmatrix}$$

EXAMPLE   Write

$$B = \begin{bmatrix} b_{11} & b_{12} \\ b_{21} & b_{22} \\ b_{31} & b_{32} \end{bmatrix}$$

in compact form.

Solution $B = [b_{ij}]_{32}$

Note: The subscripts need not be i, j, m, n; they can use any "dummy index." As a general but not absolute rule, letters in the middle of the alphabet are used as index dummy variables. This is also true in most programming languages used in computers.

EXAMPLE   Write $C = [c_{ke}]_{ij}$ in expanded form.

Solution

$$[c_{ke}]_{ij} = \begin{bmatrix} c_{11} & c_{12} & \cdots & c_{1j} \\ c_{21} & c_{22} & \cdots & c_{2j} \\ \vdots & & & \\ c_{i1} & \cdots\cdots\cdots & & c_{ij} \end{bmatrix}$$

The various definitions of matrix equality, sums, products, etc. can be written using compact notation.

DEFINITION: Matrix Equality

If $A = [a_{ij}]_{nm}$ and $B = [b_{ij}]_{pq}$, then

$$A = B$$

if and only if $n = p$ and $m = q$ and $a_{ij} = b_{ij}$ for all possible values for i and j.

EXAMPLE   What is the meaning of $[a_{ij}]_{34} = [b_{ij}]_{nm}$?

Solution

$$[a_{ij}]_{34} = [b_{ij}]_{nm}$$

if and only if $n = 3$, $m = 4$, and

$$a_{11} = b_{11}, a_{12} = b_{12}, a_{13} = b_{13}, a_{14} = b_{14}$$

$$a_{21} = b_{21}, a_{22} = b_{22}, a_{23} = b_{23}, a_{24} = b_{24}$$

$$a_{31} = b_{31}, a_{32} = b_{32}, a_{33} = b_{33}, a_{34} = b_{34}$$

EXAMPLE If $A = [a_{ij}]_{24}$ and $a_{ij} = (-1)^{i+j}$ for all i and j, find the expanded form of A.

Solution

$$A = [a_{ij}]_{24} = \begin{bmatrix} (-1)^{1+1} & (-1)^{1+2} & (-1)^{1+3} & (-1)^{1+4} \\ (-1)^{2+1} & (-1)^{2+2} & (-1)^{2+3} & (-1)^{2+4} \end{bmatrix}$$

$$= \begin{bmatrix} 1 & -1 & 1 & -1 \\ -1 & 1 & -1 & 1 \end{bmatrix}$$

DEFINITION

If $A = [a_{ij}]_{nm}$, then

$$cA = c[a_{ij}]_{nm} = [ca_{ij}]_{nm}$$

EXAMPLE If $A = [a_{ij}]_{24}$ with $a_{ij} = (-1)^{i+j}$, find $3A$.

Solution

$$3A = [3a_{ij}]_{24} = \begin{bmatrix} 3 & -3 & 3 & -3 \\ -3 & 3 & -3 & 3 \end{bmatrix}$$

DEFINITION: Matrix Addition

$$[a_{ij}]_{nm} + [b_{ij}]_{nm} = [a_{ij} + b_{ij}]_{nm}$$

That is,

$$[a_{ij}]_{nm} + [b_{ij}]_{nm} = [c_{ij}]_{nm}$$

implies that $c_{ij} = a_{ij} + b_{ij}$ for all $i = 1$ to n and all $j = 1$ to m.

EXAMPLE Find $A + B$ if $A = [a_{ij}]_{24}$, $a_{ij} = 2^{i+j}$, $B = [b_{ij}]_{24}$, and $b_{ij} = 3^{i+j}$ for all i, j; $i = 1$ to n, $j = 1$ to m.

Solution

$$A + B = [c_{ij}]_{24}$$

where $$c_{ij} = a_{ij} + b_{ij} = 2^{i+j} + 3^{i+j}$$

Thus

$$C = \begin{bmatrix} 2^2 + 3^2 & 2^3 + 3^3 & 2^4 + 3^4 & 2^5 + 3^5 \\ 2^3 + 3^3 & 2^4 + 3^4 & 2^5 + 3^5 & 2^6 + 3^6 \end{bmatrix}$$

$$= \begin{bmatrix} 13 & 35 & 97 & 275 \\ 35 & 97 & 275 & 793 \end{bmatrix}$$

EXAMPLE If $C = [c_{pq}]_{51}$ and $D = [d_{nm}]_{51}$, find the expanded form of $3C - 2D$.

Solution

$$3C - 2D = \begin{bmatrix} 3c_{11} - 2d_{11} \\ 3c_{21} - 2d_{21} \\ 3c_{31} - 2d_{31} \\ 3c_{41} - 2d_{41} \\ 3c_{51} - 2d_{51} \end{bmatrix}$$

DEFINITION

Let $f(i)$ be an algebraic expression involving i. Then for $n < m$ and n and m integers,

$$\sum_{i=n}^{m} f(i) = f(n) + f(n + 1) + \cdots + f(m - 1) + f(m)$$

That is,

$$\sum_{i=n}^{m} f(i)$$

requires the formation of a sum of terms, the first of which is the expression following the sigma or summation, $\sum$, with i replaced by n. The second term has i replaced by $n + 1$, and so forth, until i is replaced by m.*

EXAMPLE Find

$$\sum_{i=2}^{8} i^2$$

Solution

$$\sum_{i=2}^{8} i^2 = (2)^2 + (3)^2 + (4)^2 + (5)^2 + (6)^2 + (7)^2 + (8)^2$$
$$= 203$$

EXAMPLE Expand

$$\sum_{i=1}^{7} a_i$$

Solution

$$\sum_{i=1}^{7} a_i = a_1 + a_2 + a_3 + a_4 + a_5 + a_6 + a_7$$

The index of the sum need not be i as specifically indicated in the definition.

* A discussion of sigma notation may be found in the chapter of additional topics

EXAMPLE Find

$$\sum_{k=2}^{6} (k + 2)$$

Solution

$$\sum_{k=2}^{6} (k + 2) = (2 + 2) + (3 + 2) + (4 + 2) + (5 + 2) + (6 + 2)$$
$$= 30$$

The terms being summed need not contain specific reference to the index.

EXAMPLE Find

$$\sum_{i=1}^{5} (7)$$

Solution

$$\sum_{i=1}^{5} (7) = \underset{\underset{i=1}{\uparrow}}{7} + \underset{\underset{i=2}{\uparrow}}{7} + \underset{\underset{i=3}{\uparrow}}{7} + \underset{\underset{i=4}{\uparrow}}{7} + \underset{\underset{i=5}{\uparrow}}{7} = 35$$

Numerous theorems and applications of sigma notation include many formulas from statistics and calculus. The reader is referred to books in these areas for such applications. However, the following definition is one application to matrix theory.

DEFINITION

If $A = [a_{ij}]_{nm}$ and $B = [b_{ij}]_{mp}$, then

$$AB = C$$

defines C as

$$C = [c_{ij}]_{np}$$

where

$$c_{ij} = \sum_{k=1}^{m} a_{ik} b_{kj}$$

($i = 1$ to n, $j = 1$ to p).

The reader should notice that this is only a compact form of the previous matrix product definition. From the compact notation, the number of columns of A, m, matches the number of rows of B. Expanding the sigma notation used to define c_{ij}, a typical element of the product matrix,

$$c_{ij} = \sum_{k=1}^{m} a_{ik}b_{kj}$$

$$= a_{i1}b_{1j} + a_{i2}b_{2j} + a_{i3}b_{3j} + \cdots + a_{im}b_{mj}$$

This agrees exactly with the previous definition.

EXAMPLE If $C = [c_{ij}]_{23}$ such that $c_{ij} = (-1)^{i+j}$ and $D = [d_{ij}]_{31}$ such that $d_{ij} = 2^{i-j}$, find CD.

Solution Let $CD = E = [e_{ij}]_{21}$. Then

$$e_{11} = \sum_{k=1}^{3} c_{1k}d_{k1}$$

$$= \sum_{k=1}^{3} (-1)^{1+k}2^{k-1}$$

$$= (-1)^2 2^0 + (-1)^3 2^1 + (-1)^4 2^2$$

$$= 1 - 2 + 4 = 3$$

$$e_{21} = \sum_{k=1}^{3} c_{2k}d_{k1}$$

$$= \sum_{k=1}^{3} (-1)^{2+k}2^{k-1}$$

$$= (-1)^3 2^0 + (-1)^4 2 + (-1)^5 2^2$$

$$= -1 + 2 - 4 = -3$$

Since
$$E = [e_{ij}]_{21} = \begin{bmatrix} e_{11} \\ e_{21} \end{bmatrix}$$

$$E = \begin{bmatrix} 3 \\ -3 \end{bmatrix}$$

It is left as an exercise to expand C and D and verify the multiplication.

The identity matrix can be defined in the compact notation.

DEFINITION

$$I = [\delta_{ij}]_{nn} \quad (n \text{ taken from context})$$

where $\delta_{ij} = 0$ if $i \neq j$ and $\delta_{ij} = 1$ if $i = j$.

EXAMPLE If $A = [a_{ij}]_{33}$, verify that $AI = A$.

Solution $AI = [a_{ij}]_{33}[\delta_{ij}]_{33} = [c_{ij}]_{33}$

where
$$c_{ij} = \sum_{k=1}^{3} a_{ik}\delta_{kj}$$

However,

$$\sum_{k=1}^{3} a_{ik}\delta_{kj} = a_{i1}\delta_{1j} + a_{i2}\delta_{2j} + a_{i3}\delta_{3j}$$

Since the δs are zero unless both subscripts are the same, it is clear that

$$a_{i1}\delta_{1j} + a_{i2}\delta_{2j} + a_{i3}\delta_{3j} = a_{ij}\delta_{jj}$$

because all of the other terms are zero but $a_{ij}\delta_{jj} = a_{ij} \cdot 1 = a_{ij}$.

Thus
$$c_{ij} = a_{ij}$$
or
$$C = A$$

Patterns of matrix inversion can be carried out in compact notation, but that discussion is left as a project for the reader.

EXERCISES 5.8

Write each of Exercises 1–4 in expanded form.

1. $[a_{ij}]_{35}$ **2.** $[b_{ke}]_{14}$

3. $[b_{ke}]_{22}$ if $b_{ke} = \dfrac{k + e}{2}$ for all k and e

4. $[a_{ij}]_{32}$ if $a_{ij} = \dfrac{i^2}{j^2}$ for all i and j

Find some way to write the matrices in Exercises 5–8 in compact notation.

5. $\begin{bmatrix} 1 & 2 & 3 \\ 1 & 2 & 3 \end{bmatrix}$ **6.** $\begin{bmatrix} 1 & 2 & 3 \\ 2 & 3 & 4 \\ 3 & 4 & 5 \end{bmatrix}$

7. $\begin{bmatrix} 1 & 2 & 3 \\ 2 & 4 & 6 \\ 3 & 6 & 9 \end{bmatrix}$ **8.** $\begin{bmatrix} 1 & 4 & 9 & 16 \\ 0 & 1 & 4 & 9 \end{bmatrix}$

In Exercises 9–12,

$$\text{Let } A = [a_{ij}]_{33}, \ a_{ij} = i + j \text{ for all } i \text{ and } j$$
$$B = [b_{ij}]_{33}, \ b_{ij} = i - j \text{ for all } i \text{ and } j$$
$$C = [c_{ij}]_{33}, \ c_{ij} = i \cdot j \text{ for all } i \text{ and } j$$

Find the indicated expressions in expanded form.

9. $A - B$

11. $3(A - C) + 2B$

10. $A + 2B - C$

12. $3A - 4B + I$

Write out and evaluate the sums in Exercises 13–20.

13. $\displaystyle\sum_{i=1}^{10} i$

14. $\displaystyle\sum_{k=1}^{10} k^2$

15. $\displaystyle\sum_{j=3}^{15} (j - 2)$

16. $\displaystyle\sum_{i=1}^{6} (3)$

17. $\displaystyle\sum_{i=1}^{5} (i + i^2)$

18. $\displaystyle\sum_{i=1}^{7} (-1)^i$

19. $\displaystyle\sum_{i=1}^{20} (-1)^i i$

20. $\displaystyle\sum_{k=5}^{50} k$

In Exercises 21–25,

$$\text{Let } A = [a_{ij}]_{33}, \; a_{ij} = i + j$$
$$B = [b_{ij}]_{33}, \; b_{ij} = i - j$$
$$C = [c_{ij}]_{33}, \; c_{ij} = i \cdot j$$

Find the indicated products.

21. AB **22.** BA **23.** CA

24. $A(BC)$ **25.** $(AC)B$

26. Use compact notation to discuss possible equality between $A(BC)$ and $(AB)C$ if A, B, and C are all 4×4 matrices.

27. Use compact notation to discuss possible equality between $c(AB)$ and $(cA)B$ where A and B are $n \times n$ matrices and c is a scalar.

28. If $C = [c_{ij}]_{23}$, $c_{ij} = (-1)^{i+j}$, and $D = [d_{ij}]_{31}$ such that $d_{ij} = 2^{i-j}$, verify by expanding and matrix multiplication that $E = \begin{bmatrix} 3 \\ -3 \end{bmatrix}$ is CD.

REVIEW EXERCISES

Determine values for the literal expressions in Exercises 1–4 so that the indicated equalities hold.

1. $[x^2 + x, y, z] = [2, 3, 4]$

2. $\begin{bmatrix} x + 1 \\ x - 1 \\ x + 3 \end{bmatrix} = \begin{bmatrix} 2 \\ 0 \\ 4 \end{bmatrix}$

3. $\begin{bmatrix} x + 1 \\ x - 1 \\ x + 3 \end{bmatrix} = \begin{bmatrix} 2 \\ 1 \\ 4 \end{bmatrix}$

4. $\begin{bmatrix} x + 2 & y - 1 \\ y + 3 & x - 2 \end{bmatrix} = \begin{bmatrix} 2 & 0 \\ 4 & -2 \end{bmatrix}$

Find the indicated products in Exercises 5–10.

5. $[3 \quad 2 \quad -7 \quad 1] \begin{bmatrix} 4 \\ 2 \\ 6 \\ 3 \end{bmatrix}$

6. $[3 \quad 7 \quad 5] \begin{bmatrix} -1 \\ 2 \\ 3 \end{bmatrix}$

7. $\begin{bmatrix} -1 \\ 2 \\ 3 \end{bmatrix} [3 \quad 7 \quad 5]$

8. $\begin{bmatrix} 4 \\ 2 \\ 6 \\ 3 \end{bmatrix} [3 \quad 2 \quad -7 \quad 1]$

9. $\left(\begin{bmatrix} -1 \\ 4 \\ 6 \end{bmatrix} [2 \quad 3 \quad 7 \quad -5] \right) \begin{bmatrix} 4 \\ 6 \\ 2 \\ -3 \end{bmatrix}$

10. $\left(\begin{bmatrix} -1 \\ 4 \\ 6 \end{bmatrix} [2 \quad 3 \quad 4 \quad -5] \right) \begin{bmatrix} 4 \\ 6 \\ -2 \\ 3 \end{bmatrix}$

In Exercises 11–13, let A be an n × m matrix, B be an m × p matrix, and C be a p × q matrix.

11. Is $A(BC)$ defined, and if so what is it?
12. Is $(AB)C$ defined, and if so what is it?
13. Assuming both are defined, does $A(BC) = (AB)C$?

Find the inverses, if they exist, of the matrices in Exercises 14–20.

14. $\begin{bmatrix} 0 & 2 & 1 \\ 1 & 3 & 0 \\ 3 & -2 & 1 \end{bmatrix}$

15. $\begin{bmatrix} 3 & 0 & 2 \\ 5 & 0 & 6 \\ 3 & 1 & 5 \end{bmatrix}$

16. $\begin{bmatrix} 2 & 0 & 1 \\ 2 & 2 & 2 \\ 2 & 1 & -2 \end{bmatrix}$

17. $\begin{bmatrix} 1 & -5 & 2 \\ 2 & 3 & -1 \\ 2 & -2 & 1 \end{bmatrix}$

18. $\begin{bmatrix} 2 & 1 & 4 & 2 \\ 4 & 3 & 9 & 3 \\ 2 & -2 & -1 & 4 \\ 0 & 3 & -6 & -4 \end{bmatrix}$

19. $\begin{bmatrix} 2 & 0 & 0 & 3 \\ 3 & -6 & 2 & -1 \\ 2 & -4 & -1 & -3 \\ 1 & -5 & -2 & -4 \end{bmatrix}$

20. $\begin{bmatrix} 2 & 0 & 1 & 1 \\ 2 & -2 & -1 & 1 \\ 1 & -1 & 0 & 1 \\ 1 & 0 & -1 & 1 \end{bmatrix}$

Solve the systems of equations in Exercises 21–24.

21. $2x_1 + 4x_2 - 3x_3 = 15$
$\quad x_1 + x_2 - 2x_3 = -1$
$\quad 5x_1 - x_2 - 2x_3 = -1$

22. $3x_1 - x_2 + x_3 = 4$
$\quad 2x_1 + 3x_2 - x_3 = -1$
$\quad 2x_1 - 2x_2 + x_3 = 0$

23. $x_1 - x_2 - x_3 = 1$
$\quad 2x_1 + x_3 + x_4 = 0$
$\quad x_1 - x_2 + x_4 = 1$
$\quad x_1 - x_3 + x_4 = 0$

24. $x_1 + x_2 + x_3 + x_4 = -2$
$\quad x_1 + 2x_2 + 2x_3 + x_4 = 1$
$\quad 3x_1 - x_2 + 5x_3 + 2x_4 = -2$
$\quad -x_1 + 4x_2 - 2x_3 - x_4 = 1$

25. Two investments, one at 5% and one at 6%, total \$85,000. The total return is \$4,390. How much is invested at each rate?

26. Three investments—some at 5%, some at 6%, and some at 8%—total \$144,000. The total return is \$8,820. If the investments at 5% and 8% are interchanged, the return is \$9,300. How much is invested at each rate?

CHAPTER **6**

LINEAR PROGRAMMING

Linear programming is a method of optimization—for instance, the optimization of profits when they can be affected by many factors. The theory was developed by John von Neuman, G. B. Dantzig, and T. C. Koopmans, to name only a few. It was first applied by the U.S. Air Force. An early application was in the Berlin airlift. The simplex method, which is presented in a later section of this chapter, was developed by G. B. Dantzig, and a number of important theorems were proved by von Neuman.

Consider the following problem statement.

PROBLEM: A San Francisco firm has a rush order for its outlet in Hawaii. To meet this order it charters a cargo jet. After making up its rush order it finds that the jet can carry an additional 15,000 kg and/or 100 cu m of cargo. To make use of this space the company plans to use a reserve cash fund of $24,000 to ship amounts of two additional types of items needed in the islands. The various physical and economic parameters for these items are given in the following table.

Product	Volume/ unit (cu m)	Weight/ unit (kg)	Cost/ unit ($)	Profit/ unit ($)
A	2	400	300	310
B	3	250	800	500

How much of each type of item should the firm ship in order to maximize the profit?

This problem sounds somewhat similar to the problems in the unit on applications of equations, yet there are several important differences. First, the problem is one of maximizing a variable profit, not just making the solution fit the various conditions exactly. Second, the conditions are limiting or boundary conditions on what takes place rather than absolute conditions (equality) that must be met. For example, the total volume of the units shipped need only be less than or equal to the maximum the plane can hold, 100 cu m. The total weight shipped need not equal 15,000 kg, it need only be less than or equal to that figure. The company need not spend the entire $24,000 available, only some of it if they wish. The symbolic statement of the problem involves a system of conditions which must be met, together with the "profit function," which is to be made as large as possible. Specifically, if x_1 is the number of units of product A and x_2 is the number of units of product B to be shipped, the problem can be stated in the following terms:

$$2x_1 + 3x_2 \leq 100 \qquad \text{(Volume condition)} \qquad (1)$$
$$400x_1 + 250x_2 \leq 15{,}000 \qquad \text{(Weight condition)} \qquad (2)$$
$$300x_1 + 800x_2 \leq 24{,}000 \qquad \text{(Cost condition)} \qquad (3)$$
$$x_1 \geq 0 \qquad (4)$$
$$x_2 \geq 0 \qquad (5)$$
$$\text{Profit} = P = 310x_1 + 500x_2 \ (\$) \qquad (6)$$

Equation (1) states that the total volume of the additional cargo is less than or equal to 100 cu m. Equation (2) states that the total weight of new cargo must be less than or equal to 15,000 kg. Equation (3) describes the condition placed on total cost. Equations (4) and (5) remind us that we cannot ship negative numbers of units, and (6) describes the profit to be maximized.

This then is the type of problem one tries to solve using the methods of linear programming. We will try to maximize a *linear expression* in the variables (in this case profit) while at the same time meeting various conditions on those variables, conditions stated in terms of a system of *linear inequalities*—thus the name *linear programming*.

In the following sections of this chapter we will develop methods for solving this specific problem, together with procedures that could be used on similar problems. We will begin by developing some background about linear inequalities and their solutions, then we will move to a graphical procedure for solving the simplest types of linear programming (LP) problems, those involving two variables. Using the insights this will provide, we will then extend the concepts to more complex cases and the development of nongeometric procedures for solution.

6.1 OPEN INEQUALITIES IN TWO VARIABLES

The expression

$$2x_1 + 3x_2 \leq 100 \tag{1}$$

is an example of an open inequality in two variables. We have used sub-scripted variables rather than the more familiar x and y in hopes that any procedures we develop or discover can be extended to n, or an unspecified number of variables. Like its counterpart, the open equation in two variables, (1) has infinitely many solutions, or pairs of values that satisfy it. Specifically, the solution set, or collection of all solutions to (1), is a set of ordered pairs. Thus in set notation the solution set to (1) is

$$\{(x_1, x_2)| \ 2x_1 + 3x_2 \leq 100\}$$

We are assuming that the reader is familiar with the symbol $\leq$ (read "less than or equal to") and the other inequality symbols: $<$ (less than), $\geq$ (greater than or equal to), and $>$ (greater than). To understand the nature of this set of ordered pairs, we might consider graphing all the pairs of this set on a rectangular coordinate plane such as Figure 6.1.1.

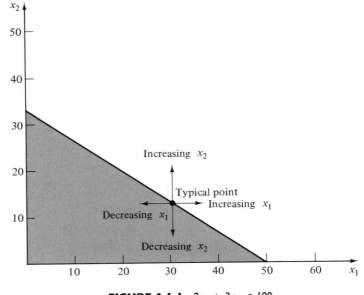

FIGURE 6.1.1 $2x_1 + 3x_2 \leq 100$.

The solid line in the figure is the graph of the linear equation

$$2x_1 + 3x_2 = 100$$

(Note the use of x_1, x_2, coordinates.) One can then proceed to argue that all of the points such that

$$2x_1 + 3x_2 < 100$$

are on the same side of this line, or in the same "half plane." One intuitive argument might go like this: Consider a typical point on this line as shown in Figure 6.1.1, i.e., a point where

$$2x_1 + 3x_2 = 100$$

Suppose from this point one "moves" to the right of this point. Points to the right will have larger x_1 values, without changing x_2. This will increase the value of $2x_1 + 3x_2$, making it larger—larger than its value on the line, i.e., 100. Clearly, for every point "to the right" of the line, $2x_1 + 3x_2 > 100$, and in a similar manner, for every point "to the left," $2x_1 + 3x_2 < 100$. Therefore, the graph of

$$\{(x_1, x_2)|\ 2x_1 + 3x_2 \le 100\}$$

is the set of points to the left of the line, together with the line itself. This region has been shaded, to indicate that it represents the solution set.

Since we developed matrix notation in the previous chapter, it is useful to extend it to handle inequalities.

DEFINITION

Let A and B be two n-dimensional vectors (either column or row vectors). Then

$$A < B \text{ if and only if } a_i < b_i \text{ for all } i = 1 \text{ to } i = n$$
$$A > B \text{ if and only if } a_i > b_i \text{ for all } i$$
$$A \ge B \text{ if and only if } a_i \ge b_i \text{ for all } i$$

and

$$A \le B \text{ if and only if } a_i \le b_i \text{ for all } i$$

EXAMPLE   Write (1) above in matrix notation.

Solution   If we let

$$A = [2 \quad 3] \quad \text{and} \quad X = \begin{bmatrix} x_1 \\ x_2 \end{bmatrix}$$

(1) becomes

$$A \cdot X \le 100$$

EXAMPLE   Find the graph of the solution set to

$$[400 \quad 250]\begin{bmatrix} x_1 \\ x_2 \end{bmatrix} \leq 15{,}000.$$

Solution $[400 \quad 250]\begin{bmatrix} x_1 \\ x_2 \end{bmatrix} \leq 15{,}000$ can be written

$$400x_1 + 250x_2 \leq 15{,}000$$

We can then construct the graph of the solution set for this inequality by considering the graph of

$$400x_1 + 250x_2 = 15{,}000$$

The graph of this linear equation can be found by finding several specific points. We do this by assigning any value to one of the variables and calculating the corresponding value of the second variable. Thus if we let $x_1 = 0$, then

$$400(0) + 250x_2 = 15{,}000$$
$$250x_2 = 15{,}000$$
$$x_2 = 60$$

The point with coordinates $(0, 60)$ is on the graph. If we let $x_2 = 0$, then

$$400x_1 + 0 = 15{,}000$$
$$x_1 = 37.5$$

The graph contains the "point" $(37.5, 0)$ (actually the point with coordinates $(37.5, 0)$, but we will use *point* for short). These points have been plotted on an x_1, x_2 coordinate system in Figure 6.1.2, and then they have been connected with a straight line. Which side of the boundary line then contains the remaining points of the solution to the inequality? The easy way to find out is to pick a point at random on either side of the line. If the coordinates of that point satisfy the inequality

$$400x_1 + 250x_2 \leq 15{,}000$$

that is the solution side. If it fails to "work," the other side or half plane is the required point set. Specifically, consider $(0, 0)$. Replacing x_1 by 0 and x_2 by 0,

$$400(0) + 250(0) \leq 15{,}000$$
$$0 \leq 15{,}000$$

This is true, thus the side containing $(0, 0)$ forms the remaining graph of the solution. This side has been shaded in Figure 6.1.2.

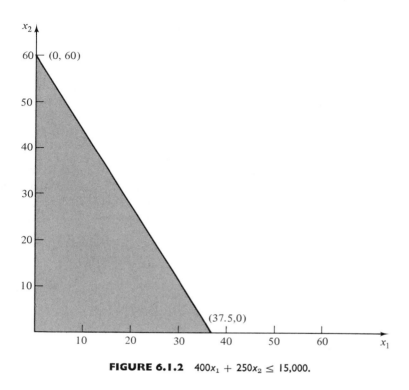

FIGURE 6.1.2 $400x_1 + 250x_2 \le 15{,}000.$

The reader might note that both of the examples above could be considered in a single matrix inequality and their solution graphs placed on a single coordinate axis.

EXAMPLE   Write the two inequalities

$$2x_1 + 3x_2 \le 100$$

$$400x_1 + 250x_2 \le 15{,}000$$

in a single matrix inequality and graph their solution sets on a single coordinate system.

Solution   The two inequalities can be written

$$\begin{bmatrix} 2 & 3 \\ 400 & 250 \end{bmatrix} \begin{bmatrix} x_1 \\ x_2 \end{bmatrix} \leq \begin{bmatrix} 100 \\ 15{,}000 \end{bmatrix}$$

By applying matrix multiplication, the left side of this inequality can be found as

$$\begin{bmatrix} 2x_1 + 3x_2 \\ 400x_1 + 250x_2 \end{bmatrix} \leq \begin{bmatrix} 100 \\ 15{,}000 \end{bmatrix}$$

which by definition is true if and only if

$$2x_1 + 3x_2 \leq 100$$

and
$$400x_1 + 250x_2 \leq 15{,}000$$

The graphs of Figures 6.1.1 and 6.1.2 can be drawn on a single coordinate system. This has been done in Figure 6.1.3. Note that the graph of the solution to the matrix inequality is the portion of 6.1.3 with shading in both directions—the portion with points whose coordinates meet both inequality conditions.

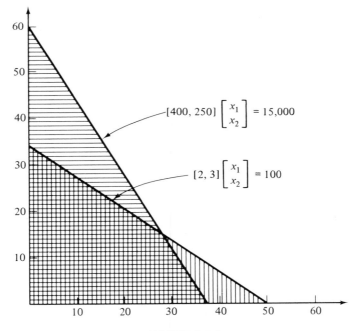

$$[400, 250] \begin{bmatrix} x_1 \\ x_2 \end{bmatrix} = 15{,}000$$

$$[2, 3] \begin{bmatrix} x_1 \\ x_2 \end{bmatrix} = 100$$

FIGURE 6.1.3

EXAMPLE Extend the matrix notation to provide a single complete statement of the system of inequalities required for a solution of the problem proposed in the introduction to this chapter.

Solution The system of inequalities was

$$2x_1 + 3x_2 \leq 100$$
$$400x_1 + 250x_2 \leq 15{,}000$$
$$300x_1 + 800x_2 \leq 24{,}000$$
$$x_1 \geq 0$$
$$x_2 \geq 0$$

To use matrix notation the conditions

$$x_1 \geq 0 \quad \text{and} \quad x_2 \geq 0$$

will have to be put in a $\leq$ form. Since $x_1 \geq 0$ implies $-x_1 \leq 0$ and $x_2 \geq 0$ implies $-x_2 \leq 0$, we can consider the system of conditions as

$$2x_1 + 3x_2 \leq 100$$
$$400x_1 + 250x_2 \leq 15{,}000$$
$$300x_1 + 800x_2 \leq 24{,}000$$
$$-x_1 \leq 0$$
$$-x_2 \leq 0$$

In matrix form,

$$
\begin{bmatrix}
2 & 3 \\
400 & 250 \\
300 & 800 \\
-1 & 0 \\
0 & -1
\end{bmatrix}
\begin{bmatrix} x_1 \\ x_2 \end{bmatrix}
\leq
\begin{bmatrix}
100 \\
15{,}000 \\
24{,}000 \\
0 \\
0
\end{bmatrix}
$$

To complete the graph of the solution set we need only note in which half plane

$$300x_1 + 800x_2 < 24{,}000$$
$$-x_1 < 0$$
$$-x_2 < 0$$

and superimpose these conditions on the previous graphs. This has been done in Figure 6.1.4. The shaded portion of that figure corresponds to the solution set of the matrix inequality.

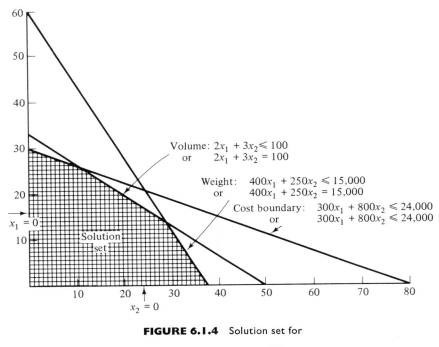

FIGURE 6.1.4   Solution set for

$$\begin{bmatrix} 2 & 3 \\ 400 & 250 \\ 300 & 800 \\ -1 & 0 \\ 0 & -1 \end{bmatrix} \begin{bmatrix} x_1 \\ x_2 \end{bmatrix} \leq \begin{bmatrix} 100 \\ 15,000 \\ 24,000 \\ 0 \\ 0 \end{bmatrix}$$

The following definitions will provide the necessary terminology to make a complete and compact analysis.

DEFINITIONS

The points in a plane on one side of a line form an **open half space.** The points in a plane on one side of a line, together with the points on the line itself form a **closed half plane.**

The intersection of any number of closed half planes forms a **polyhedral convex set.**

In terms of polyhedral convex sets, each of the shaded portions of Figures 6.1.1 to 6.1.4 forms a polyhedral convex set. These terms will be extended to consider problems in more than two dimensions. The set is called *polyhedral convex* because it is a special case of a *convex set,* which is any set of points such that any line segment drawn connecting two points in the set contains only points which are also in the set. Note that if A is any $n \times 2$ matrix and B is any $n \times 1$ matrix, the solutions to the matrix inequality

$$A \begin{bmatrix} x_1 \\ x_2 \end{bmatrix} \leq B$$

will form a polyhedral convex set, and each such set represents the solution of an inequality of this type.

EXERCISES 6.1

In Exercises 1–15, for each $n \times 2$ matrix A, with $X = \begin{bmatrix} x_1 \\ x_2 \end{bmatrix}$ and B an $n \times 1$ matrix, find the graph that represents the solution to the matrix inequality $AX \leq B$.

1. $A = [2 \quad 7]$, $B = [16] = 16$

2. $A = [4 \quad -3]$, $B = [20] = 20$

3. $A = \begin{bmatrix} 2 & 7 \\ 4 & -3 \end{bmatrix}$, $B = \begin{bmatrix} 16 \\ 20 \end{bmatrix}$

4. $A = \begin{bmatrix} 1 & 3 \\ 3 & 2 \end{bmatrix}$, $B = \begin{bmatrix} 6 \\ 1 \end{bmatrix}$

5. $A = \begin{bmatrix} 2 & 5 \\ 1 & 1 \end{bmatrix}$, $B = \begin{bmatrix} 10 \\ 1 \end{bmatrix}$

6. $A = \begin{bmatrix} 3 & 4 \\ 2 & -1 \end{bmatrix}$, $B = \begin{bmatrix} 120 \\ 40 \end{bmatrix}$

7. $A = \begin{bmatrix} 4 & 5 \\ -2 & -1 \\ 0 & -1 \end{bmatrix}$ $B = \begin{bmatrix} 20 \\ -1 \\ 0 \end{bmatrix}$

8. $A = \begin{bmatrix} 2 & 1 \\ 1 & 6 \\ -1 & 0 \\ 0 & -1 \end{bmatrix}$, $B = \begin{bmatrix} 6 \\ 6 \\ 0 \\ 0 \end{bmatrix}$

9. $A = \begin{bmatrix} 6 & 5 \\ 1 & 2 \\ 2 & 1 \\ 0 & -1 \\ -1 & 0 \end{bmatrix}$, $B = \begin{bmatrix} 30 \\ 10 \\ 8 \\ 0 \\ 0 \end{bmatrix}$

10. $A = \begin{bmatrix} 3 & 1 \\ 1 & 3 \\ 1 & 1 \\ 0 & -1 \\ -1 & 0 \end{bmatrix}$, $B = \begin{bmatrix} 15 \\ 15 \\ 7 \\ 0 \\ 0 \end{bmatrix}$

11. $A = \begin{bmatrix} 17 & 12 \\ 1 & 1 \\ 13 & 25 \\ 0 & -1 \\ -1 & 0 \end{bmatrix}$, $B = \begin{bmatrix} 204 \\ 14 \\ 325 \\ 0 \\ 0 \end{bmatrix}$

12. $A = \begin{bmatrix} -1 & 1 \\ 1 & -1 \\ 1 & 1 \\ -1 & -1 \end{bmatrix}$, $B = \begin{bmatrix} 5 \\ 5 \\ 5 \\ 5 \end{bmatrix}$

13. $A = \begin{bmatrix} 7 & 4 \\ -1 & 0 \\ -1 & -2 \end{bmatrix}$, $B = \begin{bmatrix} 1 \\ 0 \\ -10 \end{bmatrix}$

14. $A = \begin{bmatrix} -9 & -5 \\ 3 & 5 \\ 0 & -1 \end{bmatrix}$, $B = \begin{bmatrix} -45 \\ 45 \\ 0 \end{bmatrix}$

15. $A = \begin{bmatrix} 1 & 1 \\ 1 & 0 \\ -1 & 1 \\ 0 & -1 \\ -1 & -1 \\ -1 & 0 \\ 1 & -1 \\ 0 & 1 \end{bmatrix}$, $B = \begin{bmatrix} 25 \\ 15 \\ 10 \\ 0 \\ -5 \\ 0 \\ 10 \\ 15 \end{bmatrix}$

Translate each statement in Exercises 16–20 into a matrix inequality and construct the graph showing the solution set of each.

16. A dealer has 10 each of two models of television sets. He plans to ship 15 or less of these to another dealer.

17. A company plans to build two types of multiple family homes. One type of unit provides for 6 families and the other for 4 families. They must provide housing for at least 60 families. By contract the company must employ at least 10 heating/cooling specialists, but the 6-unit types can make use of only 2 such specialists at a time, and the 4-unit houses use only one specialist at a time. The 4-family units have a number of special electronic features and require 5 electronic technicians at a time, while the 6-family units need only 2. There are a total of 50 electronic technicians available.

18. Two divisions of a company are to work together to make two types of items. Division I has 1,944 man-hours available, while division II has 2,080 man-hours. A unit of product A requires 144 man-hours from division I and 104 man-hours from division II. A unit of product B requires 104 man-hours from division I and 160 man-hours from division II.

19. A drug company wishes to make a new compound from two experimental drugs. The final compound should contain not less than 5 or more than 15 units total of the two drugs, and in addition, the difference between the amount of each drug used cannot be more than 5 units.

20. A paint plant plans to mix two types of white base to produce a special paint. In the final mixture the number of units of type II base cannot exceed that of type I by more than 5 units. On the other hand, there can be as much as 15 units more type I base than type II base. Type I base contains 3 g of lead per unit, and type II base contains 4 g per unit. The final mixture must contain less than 60 g of lead. A unit of type I is

200 gallons of liquid, and a unit of type II is 100 gallons of liquid; the company needs at least 100 gallons of the final mix.

6.2 LINEAR PROGRAMMING

We are now in a position to solve the problem posed in the introduction to this chapter. Its solution will permit us to establish a pattern which can be used on similar problems. Recall that the problem was to find the maximum value of the expression

$$P = 310x_1 + 500x_2$$

under the conditions

$$\begin{bmatrix} 2 & 3 \\ 400 & 250 \\ 300 & 800 \\ -1 & 0 \\ 0 & -1 \end{bmatrix} \begin{bmatrix} x_1 \\ x_2 \end{bmatrix} \leq \begin{bmatrix} 100 \\ 15{,}000 \\ 24{,}000 \\ 0 \\ 0 \end{bmatrix}$$

Figure 6.2.1 shows the graphical solution to the matrix inequality. To meet the various conditions we must select a point within the solution set. But where? A better question is where not? First, the point where P is a maximum cannot be in the interior. To see this consider any point (a, b) within the solution. Such a point is shown in Figure 6.2.1. At that point

$$P = 310a + 500b$$

Clearly, we can make P larger by increasing a, moving to the right, yet we would still be within the solution set. Since this argument applies to any point in the interior, we can conclude that an interior point cannot yield a maximum. Our conclusion is that the maximum of P must occur on the boundary of the solution set. But where on the boundary? Again we can conclude where it is not. It cannot occur between "corners," or intersections of the boundary lines. To see this assume, for example, that the maximum occurs at a point (c, d) on the line $2x_1 + 3x_2 = 100$. Since (c, d) is on this line, we have

$$P = 310c + 500d \quad \text{and} \quad 2c + 3d = 100$$

Using the second equation, we solve for d:

$$3d = 100 - 2c$$

$$d = \frac{100 - 2c}{3}$$

This result can be inserted into the expression for P. Specifically,

$$P = 310c + 500\left(\frac{100 - 2c}{3}\right)$$

$$= 310c + \frac{50,000}{3} - \frac{1,000c}{3}$$

$$= \frac{-70c}{3} + \frac{50,000}{3}$$

This is the value of P for any point on the line.

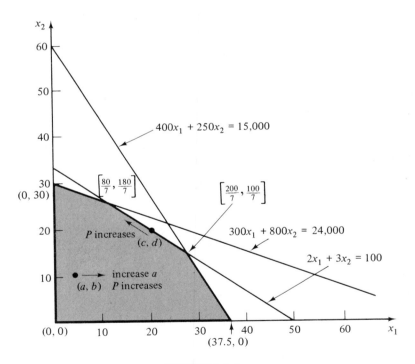

FIGURE 6.2.1

If we study this expression we see that we can increase P by moving upward along the line—i.e., by making c decrease—yet remaining on the line. A similar analysis shows that the maximum value of P cannot occur between corners on any of the border lines. Our conclusion is that the maximum of P must occur at a corner. Notice that there is a finite number of corners—specifically, 5. The corners are:

$$(0, 0)$$
$$(0, 30)$$
$$(37.5, 0)$$
$$\left(\tfrac{80}{7}, \tfrac{180}{7}\right)$$
$$\left(\tfrac{200}{7}, \tfrac{100}{7}\right)$$

At $(0, 0)$ $P = 310(0) + 500(0) = 0$

At $(0, 30)$ $P = 310(0) + 500(30) = 15{,}000$

At $(37.5, 0)$ $P = 310(37.5) + 500(0) = 11{,}625$

At $\left(\tfrac{80}{7}, \tfrac{180}{7}\right)$ $P = 310(\tfrac{80}{7}) + 500(\tfrac{180}{7})$

$$= \frac{24{,}800 + 90{,}000}{7}$$

$$= \frac{114{,}800}{7} = 16{,}400$$

At $\left(\tfrac{200}{7}, \tfrac{100}{7}\right)$ $P = 310(\tfrac{200}{7}) + 500(\tfrac{100}{7})$

$$= \frac{62{,}000 + 50{,}000}{7}$$

$$= \frac{112{,}000}{7} = 16{,}000$$

The maximum value of P occurs at $\left(\tfrac{80}{7}, \tfrac{180}{7}\right)$ and is $\dfrac{114{,}800}{7} = \$16{,}400$.

Translating back to the original problem, the company will make a maximum profit if it ships $\tfrac{80}{7}$ units of A (actually 11 units of A) and $\tfrac{180}{7}$ units of B (actually 25 units of B).

We should modify the corners to round them off to the nearest whole number of units. Thus $\left(\tfrac{80}{7}, \tfrac{180}{7}\right)$ rounds off to $(11, 25)$, with

$$P = 310(11) + 500(25) = \$15{,}910$$

and $\left(\tfrac{200}{7}, \tfrac{100}{7}\right)$ rounds off to $(28, 14)$, with

$$P = 310(28) + 500(14) = \$15{,}680$$

The rounding off here must be considered because a whole number of units must be shipped. This is not the case in our next example.

The most straightforward problems of the type being considered here are those stated in matrix inequality form rather than in word form. Specifically, they are of the type: Determine the maximum or minimum value of $P =$ linear expressions in x_1, x_2. $AX \le B$, and A, X, and B are appropriate matrices.

EXAMPLE Find the minimum value of $P = 6x_1 - 4x_2$ under the conditions that

$$\begin{bmatrix} -1 & -6 \\ 1 & 1 \\ -5 & -1 \end{bmatrix} \begin{bmatrix} x_1 \\ x_2 \end{bmatrix} \le \begin{bmatrix} -6 \\ 3 \\ -5 \end{bmatrix}$$

Solution The graphical analysis is shown in Figure 6.2.2.

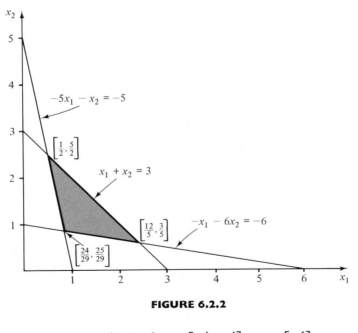

FIGURE 6.2.2

$$
\begin{aligned}
-x_1 - 6x_2 &\le -6 \\
x_1 + x_2 &\le 3 \\
-5x_1 - x_2 &\le -5
\end{aligned}
\quad \text{or} \quad
\begin{bmatrix} -1 & -6 \\ 1 & 1 \\ -5 & -1 \end{bmatrix} \begin{bmatrix} x_1 \\ x_2 \end{bmatrix} \le \begin{bmatrix} -6 \\ 3 \\ -5 \end{bmatrix}
$$

The corners or intersections of the borders are

$$\left(\tfrac{24}{29}, \tfrac{25}{29}\right), \left(\tfrac{12}{5}, \tfrac{3}{5}\right) \quad \text{and} \quad \left(\tfrac{1}{2}, \tfrac{5}{2}\right)$$

At $\left(\tfrac{24}{29}, \tfrac{25}{29}\right)$ $P = 6\left(\tfrac{24}{29}\right) - 4\left(\tfrac{25}{29}\right) = \tfrac{44}{29}$

At $\left(\tfrac{12}{5}, \tfrac{3}{5}\right)$ $P = 6\left(\tfrac{12}{5}\right) - 4\left(\tfrac{3}{5}\right) = \tfrac{60}{5} = 12$

At $\left(\tfrac{1}{2}, \tfrac{5}{2}\right)$ $P = 6\left(\tfrac{1}{2}\right) - 4\left(\tfrac{5}{2}\right) = -7$

The minimum value of P is -7 at $\left(\tfrac{1}{2}, \tfrac{5}{2}\right)$.

EXAMPLE A druggist has to make up a special mixture of vitamins A, B, and C. He has two stock vitamin compounds on hand, which have the following characteristics:

	Type	
	I	II
Vitamin A (units/g)	19	30
Vitamin B (units/g)	33	20
Vitamin C (units/g)	23	28

The new mixture cannot contain more than 114 units of vitamin A or 110 units of vitamin B, and the amount of compound I cannot exceed that of compound II by more than 1 g. How should the druggist make the mixture to yield a maximum amount of vitamin C?

Solution If x_1 is used to symbolize the number of g of compound I in the final mixture and the number of g of compound II is x_2, then the expression to be maximized is

$$P = 23x_1 + 28x_2$$

Considering the compound that cannot exceed 114 units of vitamin A,

$$19x_1 + 30x_2 \leq 114$$

The condition on the amount of vitamin B is

$$33x_1 + 20x_2 \leq 110$$

The condition that there can be no more than 1 g more of compound I than of compound II becomes

$$x_1 - x_2 \leq 1$$

Since there must be positive amounts of each drug type

$$x_1 \geq 0$$
$$x_2 \geq 0$$

Summarizing these data, we have to maximize

$$P = 23x_1 + 28x_2$$

while

$$19x_1 + 30x_2 \le 114$$
$$33x_1 + 20x_2 \le 110$$
$$x_1 - x_2 \le 1$$
$$x_2 \ge 0$$
$$x_1 \ge 0$$

Writing the last two statements in "less than" form,

$$0(x_1) - x_2 \le 0$$
$$-x_1 + 0x_2 \le 0$$

In matrix form, these statements are

$$\begin{bmatrix} 19 & 30 \\ 33 & 20 \\ 1 & -1 \\ 0 & -1 \\ -1 & 0 \end{bmatrix} \begin{bmatrix} x_1 \\ x_2 \end{bmatrix} \le \begin{bmatrix} 114 \\ 110 \\ 1 \\ 0 \\ 0 \end{bmatrix}$$

Figure 6.2.3 shows the graphical analysis of these conditions.

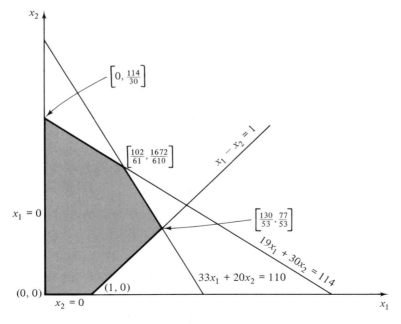

FIGURE 6.2.3

Taking the equations of the boundaries in pairs, we can find the corners:

$$\left.\begin{array}{l} x_1 = 0 \\ 19x_1 + 30x_2 = 114 \end{array}\right\} \text{ together}$$

yield $(0, \frac{114}{30})$.

$$\left.\begin{array}{l} x_2 = 0 \\ x_1 - x_2 = 1 \end{array}\right\}$$

yield $(1, 0)$.

$$\left.\begin{array}{l} x_1 - x_2 = 1 \\ 33x_1 + 20x_2 = 110 \end{array}\right\}$$

yield $(\frac{130}{53}, \frac{77}{53})$.

$$\left.\begin{array}{l} 19x_1 + 30x_2 = 114 \\ 33x_1 + 20x_2 = 110 \end{array}\right\}$$

yield $(\frac{102}{61}, \frac{1672}{610})$.

$$\left.\begin{array}{l} x_1 = 0 \\ x_2 = 0 \end{array}\right\}$$

yield $(0, 0)$.

The corners are

$$(0, 0), (\tfrac{102}{61}, \tfrac{1672}{610}), (\tfrac{130}{53}, \tfrac{77}{53}), (1, 0), (0, \tfrac{114}{30})$$

Evaluating P at each of these points,

At $(0, 0)$ $\quad\quad\quad P = 23(0) + 28(0) = 0$

At $(\frac{102}{61}, \frac{1672}{610})$ $\quad P = 23(\frac{1020}{610}) + 28(\frac{1672}{610}) = \frac{70276}{610} \approx 115.2$

At $(\frac{130}{53}, \frac{77}{53})$ $\quad\, P = 23(\frac{130}{53}) + 28(\frac{77}{53}) = \frac{5146}{53} \approx 97.1$

At $(1, 0)$ $\quad\quad\quad P = 23(1) + 28(0) = 23$

At $(0, \frac{114}{30})$ $\quad\,\, P = 23(0) + 28(\frac{114}{30}) = \frac{3192}{30} = 106.4$

The maximum value of P is $\frac{70276}{610}$ at $(\frac{102}{61}, \frac{1672}{610})$. This means that the druggist should use $\frac{102}{61} \approx 1.672$ g of compound I and $\frac{1672}{610} \approx 2.741$ g of compound II.

EXERCISES 6.2

In Exercises 1–9, find the maximum or minimum value of P as indicated under the described conditions.

1. Find the largest value of $P = 3x_1 + 4x_2$ provided

$$\begin{bmatrix} 4 & 5 \\ -2 & -1 \\ 0 & -1 \end{bmatrix} \begin{bmatrix} x_1 \\ x_2 \end{bmatrix} \le \begin{bmatrix} 20 \\ -1 \\ 0 \end{bmatrix}$$

2. Find the smallest value of $P = 4x_1 + 3x_2$ provided

$$\begin{bmatrix} 2 & 1 \\ 1 & 6 \\ -1 & 0 \\ 0 & -1 \end{bmatrix} \begin{bmatrix} x_1 \\ x_2 \end{bmatrix} \leq \begin{bmatrix} 6 \\ 6 \\ 0 \\ 0 \end{bmatrix}$$

3. Find the smallest value of $P = 6x_1 - 4x_2$ provided

$$\begin{bmatrix} 17 & 12 \\ 1 & 1 \\ 13 & 25 \\ 0 & -1 \\ -1 & 0 \end{bmatrix} \begin{bmatrix} x_1 \\ x_2 \end{bmatrix} \leq \begin{bmatrix} 204 \\ 14 \\ 325 \\ 0 \\ 0 \end{bmatrix}$$

4. Find the largest value of $P = 6x_1 - 6x_2$ provided

$$\begin{bmatrix} 3 & 1 \\ 1 & 3 \\ 1 & 1 \\ 0 & -1 \\ -1 & 0 \end{bmatrix} \begin{bmatrix} x_1 \\ x_2 \end{bmatrix} \leq \begin{bmatrix} 15 \\ 15 \\ 7 \\ 0 \\ 0 \end{bmatrix}$$

5. Find the largest value of $P = 16x_1 + 4x_2$ provided

$$\begin{bmatrix} 6 & 5 \\ 1 & 2 \\ 2 & 1 \\ 0 & -1 \\ -1 & 0 \end{bmatrix} \begin{bmatrix} x_1 \\ x_2 \end{bmatrix} \leq \begin{bmatrix} 30 \\ 10 \\ 8 \\ 0 \\ 0 \end{bmatrix}$$

6. Find the largest value of $P = 12x_1 - 4x_2$ provided

$$\begin{bmatrix} -1 & 1 \\ 1 & -1 \\ 1 & 1 \\ -1 & -1 \end{bmatrix} \begin{bmatrix} x_1 \\ x_2 \end{bmatrix} \leq \begin{bmatrix} 5 \\ 5 \\ 15 \\ -5 \end{bmatrix}$$

7. Find the difference between the largest and smallest values of $P = 4x_1 - 7x_2$ provided

$$\begin{bmatrix} 5 & 4 \\ 4 & 5 \\ -1 & 0 \\ 0 & -1 \end{bmatrix} \begin{bmatrix} x_1 \\ x_2 \end{bmatrix} \leq \begin{bmatrix} 20 \\ 20 \\ 0 \\ 0 \end{bmatrix}$$

8. Find the difference between the largest and smallest values of $P = 7x_1 + 3x_2$ provided

$$\begin{bmatrix} -9 & -5 \\ 3 & 5 \\ 0 & -1 \end{bmatrix} \begin{bmatrix} x_1 \\ x_2 \end{bmatrix} \leq \begin{bmatrix} -45 \\ 45 \\ 0 \end{bmatrix}$$

9. Find the difference between the largest and smallest values of $P = 3x_1 - 7x_2$ provided

$$\begin{bmatrix} 1 & 1 \\ 1 & 0 \\ -1 & 1 \\ 0 & -1 \\ -1 & -1 \\ -1 & 0 \\ 1 & -1 \\ 0 & 1 \end{bmatrix} \begin{bmatrix} x_1 \\ x_2 \end{bmatrix} \leq \begin{bmatrix} 25 \\ 15 \\ 10 \\ 0 \\ -5 \\ 0 \\ 10 \\ 15 \end{bmatrix}$$

10. If the dealer in Exercise 6.1.16 makes \$50 on the first type of television and \$63 on the second type, how many of each type should he ship to maximize his profit?

11. The builder in Exercise 6.1.17 makes \$11,600 on each six-family unit and \$8,900 on each four-family unit. How can he maximize his profit under the described conditions?

12. If in the case described in Exercise 6.1.18 a man-hour in division I produces an average profit of \$.75 and that of division II yields \$.81, what should the company do to maximize its profit?

13. If the new compound described in Exercise 6.1.19 is to be as cheap as possible and the first experimental drug costs \$18 per unit and the second costs \$12.50 per unit, how should the new compound be made?

14. The paint mixture described in Exercise 6.1.20 must contain a maximum amount of "radar blue." Paint type I contains .60 oz per unit, and type II contains .54 oz per unit. How should the mixture be made?

15. A company wishes to keep supplies of two basic chemicals in stock. It has storage space for 15 cu yd of material. Chemical A weighs 1 ton per cu yd, and chemical B weighs 2 tons per cu yd. The storage weight limit is 20 tons. How much of each type of material should be stocked to maximize profit if the company makes \$450 per ton on type A and \$360 per ton on type B?

16. A maker of food supplements is asked to make up a special food for the space program. It is to be made from a mixture of two stock types the maker has on hand. The final mixture cannot contain more than 170

cal. Type A contains 30 cal/oz, and type B contains 50 cal/oz. The mixture must contain at least 100 units of vitamin E. Type A contains 10 units of E/oz and costs $5.25/oz. Type B contains 30 units of E/oz and costs $6.00/oz. How should the mixture be made to keep the cost as low as possible?

6.3 MORE ABOUT LINEAR PROGRAMMING

The graphical method of solution works reasonably well in the two-variable problems considered in the previous sections of this chapter. One might consider the simple linear programming problem below involving three variables.

Problem: The XYZ Company wishes to ship three types of units to a dealer by cargo jet. The weight and economic parameters are given in the following table:

	Product Type		
	X	Y	Z
Weight (kg)	1,200	1,500	2,000
Profit ($)	400	700	600

If the jet can carry 6,000 kg of material and volume is no problem, how many units of each type should be shipped to maximize profit?

To solve the problem using geometric methods would require extending our ability to graph to three dimensions. Instead of the two perpendicular coordinate axes previously considered, let us consider three such axes, as in Figure 6.3.1. Points in space are tagged or identified with an ordered triple of numbers (x_1, x_2, x_3), as shown in the figure.

To actually solve the problem we would have to determine what kind of geometric conditions correspond to the algebraic conditions we can place on the problem. That is, we would have to determine where the *feasible points* were, those points having coordinates corresponding to actual numerical values that meet the conditions of the problem. Once this has been done, we would have to study the nature of the boundaries of the set of all feasible points and then determine how to move along these boundaries to the solution point or next feasible point. The first step is to translate the words in which the problem is stated into appropriate mathematical sentences. Specifically, if we let

x_1 equal the number of units of product X

x_2 equal the number of units of product Y

and x_3 equal the number of units of product Z

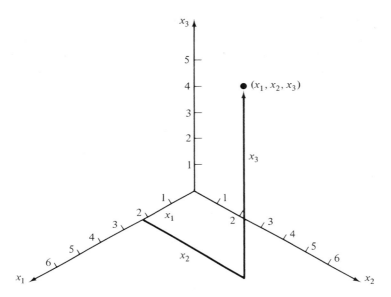

FIGURE 6.3.! Rectangular coordinates in three-dimensional space.

the weight condition becomes

$$1,200x_1 + 1,500x_2 + 2,000x_3 \leq 6,000$$

The profit is given by

$$P = 400x_1 + 700x_2 + 600x_3$$

with

$$x_1 \geq 0, x_2 \geq 0, x_3 \geq 0$$

In matrix form, the problem is to maximize

$$P = 400x_1 + 700x_2 + 600x_3$$

provided

$$\begin{bmatrix} 1,200 & 1,500 & 2,000 \\ -1 & 0 & 0 \\ 0 & -1 & 0 \\ 0 & 0 & -1 \end{bmatrix} \begin{bmatrix} x_1 \\ x_2 \\ x_3 \end{bmatrix} \leq \begin{bmatrix} 6,000 \\ 0 \\ 0 \\ 0 \end{bmatrix}$$

The problem can be done geometrically. It can be shown, using three-dimensional geometry, that

$$1,200x_1 + 1,500x_2 + 2,000x_3 = 6,000$$

$$x_1 = 0$$

$$x_2 = 0$$

and

$$x_3 = 0$$

each define plane surfaces in three-dimensional space. Further, the four planes enclose a volume in which all four conditions are met, i.e., the set of feasible points. Further, we could show that this solid has four corners with coordinates $(0, 0, 0)$, $(5, 0, 0)$, $(0, 4, 0)$, and $(0, 0, 3)$. We now evaluate P at each point:

At $(0, 0, 0)$ $P = 0$

At $(5, 0, 0)$ $P = 400(5) + 700(0) + 600(0) = 2,000$

At $(0, 4, 0)$ $P = 400(0) + 700(4) + 600(0) = 2,800$

At $(0, 0, 3)$ $P = 400(0) + 700(0) + 600(3) = 1,800$

The maximum profit occurs at $(0, 4, 0)$, indicating that XYZ Company should ship 4 units of product Y.

You may not be willing to accept this result. How are we sure of the geometry? How does one find the corners? And so forth. We have not explained how because even if we did this geometric method solution would only work in a very few cases with very simple three-dimensional geometry. In other words, the "geometric attack" is a dead end.

To solve problems involving more than two variables we have a very powerful algorithm which will yield the solution, provided a solution exists. Recall that an algorithm is a mechanical process which after the application of a finite number of steps will yield the solution. This particular algorithm is known as the *simplex method*. It was first made available to the general user only in 1951. We will consider two basic types of problems, one involving the maximum value of a linear expression and the other seeking the minimum value. We will begin with the maximum value problem.

The Simplex Method

The simplex algorithm is becoming popular because it does not depend on graphing and because it is readily adaptable to computers. Some of the theorems about the simplex method are very abstract and will not be proved here. However, the method is not difficult to apply. The basic theory is best demonstrated by an example.

EXAMPLE   Find the maximum value of $z = 5x_1 + 4x_2$ given that

$$\begin{cases} x_1 + x_2 \leq 4 \\ 2x_1 + x_2 \leq 7 \\ \quad\quad x_1 \geq 0 \\ \quad\quad x_2 \geq 0 \end{cases}$$

Solution   Figure 6.3.2 shows the graph of the constraints and the corner points.

At $(0, 0)$	$z = 0$
At $(0, 4)$	$z = 16$
At $(3, 1)$	$z = 15 + 4 = 19$
At $(\frac{7}{2}, 0)$	$z = \frac{35}{2} = 17\frac{1}{2}$

Therefore, $z = 19$ is maximum and occurs when $x_1 = 3$ and $x_2 = 1$.

The simplex method requires that the variables $x_1, x_2, x_3, \ldots, x_n$ all be nonnegative. Therefore, the constraints $x_1 \geq 0$ and $x_2 \geq 0$ from the above example will be assumed but will not be directly used. The other two constraints, $x_1 + x_2 \leq 4$ and $2x_1 + x_2 \leq 7$, will be transformed to equalities by the introduction of *slack variables*. One definition of $a \leq b$ is $a \leq b$ if and only if there exists $h \geq 0$ such that $a + h = b$. For the inequality $a \leq b$, h is the **slack variable.** Then in the example, if $x_1 + x_2 \leq 4$, there is an x_3 such that $x_3 \geq 0$ and $x_1 + x_2 + x_3 = 4$. Also, if $2x_1 + x_2 \leq 7$, then there is an x_4 such that $x_4 \geq 0$ and $2x_1 + x_2 + x_4 = 7$. Therefore, the example is equivalent to finding a solution to the system of equations

$$
\begin{aligned}
x_1 + x_2 \; + x_3 \qquad\qquad\quad &= 4 \\
2x_1 + x_2 \; + \qquad x_4 \qquad &= 7 \\
-5x_1 - 4x_2 \; + \qquad\qquad z \; &= 0
\end{aligned}
\tag{1}
$$

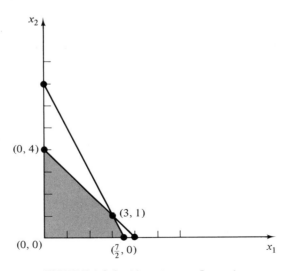

FIGURE 6.3.2   Maximize $z = 5x_1 + 4x_2$.

which will maximize z under the additional conditions that x_1, x_2, x_3, and x_4 all be nonnegative. In matrix form, (1) becomes

$$
\begin{array}{cc|ccc|c}
x_1 & x_2 & x_3 & x_4 & z & \\
1 & 1 & 1 & 0 & 0 & 4 \\
② & 1 & 0 & 1 & 0 & 7 \\
\hline
-5 & -4 & 0 & 0 & 1 & 0
\end{array}
\tag{2}
$$

Matrix (2) is called the **simplex tableau**. It may seem that the slack variables, x_3 and x_4, are unnecessary since we know that the solutions to linear programming problems are on the borders—in fact, at the corners of regions. This particular example could be solved without using the slack variables. However, it will be shown that they are necessary in other problems.

An obvious solution to (1) is $x_1 = 0$, $x_2 = 0$, $x_3 = 4$, $x_4 = 7$, and $z = 0$. This solution will not maximize z since from the last row of the simplex tableau, $z = 5x_1 + 4x_2 + 0x_3 + 0x_4$, and z could be increased by increasing either x_1 or x_2. z will increase more rapidly if x_1 is increased since for each unit increase of x_1, z increases 5 units, while for each unit increase of x_2, z increases only 4 units. Therefore, we will let $x_2 = 0$ for the time being and will increase x_1. Note that the variable to be increased, x_1, is in the column of (2) that contains the most negative number in the last row, -5. There are limits to the amount that x_1 can be increased while x_2 is held at 0. The first row implies that x_1 cannot exceed $\frac{4}{1} = 4$ without making the slack variable negative, which is not allowed. From inspecting the second row, x_1 cannot exceed $\frac{7}{2}$. But in order to satisfy both constraints, x_1 cannot exceed the smaller of the two values 4 and $\frac{7}{2}$. So let $x_1 = \frac{7}{2}$. If $x_1 = \frac{7}{2}$ and $x_2 = 0$, then $z = 5 \cdot \frac{7}{2} = \frac{35}{2}$.

The results that have been achieved so far can be shown in matrix form by using the second row of (2) to eliminate the other coefficients of x_1. This will be done using the ⓪, which is called the **pivot**. First divide the second row by 2.

$$
\begin{array}{cc|ccc|c}
x_1 & x_2 & x_3 & x_4 & z & \\
1 & 1 & 1 & 0 & 0 & 4 \\
1 & \frac{1}{2} & 0 & \frac{1}{2} & 0 & \frac{7}{2} \\
\hline
-5 & -4 & 0 & 0 & 1 & 0
\end{array}
\tag{3}
$$

Then multiply the second row by -1 and add it to the first row. Finally multiply the second row by $+5$ and add it to the third row, creating (4):

$$\begin{array}{c} \begin{array}{ccccc} x_1 & x_2 & x_3 & x_4 & z \end{array} \\ \left[\begin{array}{ccccc|c} 0 & \boxed{\tfrac{1}{2}} & 1 & -\tfrac{1}{2} & 0 & \tfrac{1}{2} \\ 1 & \tfrac{1}{2} & 0 & \tfrac{1}{2} & 0 & \tfrac{7}{2} \\ \hline 0 & -\tfrac{3}{2} & 0 & \tfrac{5}{2} & 1 & \tfrac{3 \cdot 5}{2} \end{array} \right] \end{array} \qquad (4)$$

Matrix (4) shows the value of x_1 in the rightmost column as $\tfrac{7}{2}$ and the current value of z as $\tfrac{3 \cdot 5}{2}$. Matrix (4) has the further advantage of making z independent of x_1. That is, $z = 0x_1 + \tfrac{3}{2}x_2 + 0x_3 - \tfrac{5}{2}x_4 + \tfrac{3 \cdot 5}{2}$. Now z will increase if x_2 is increased. That may mean that x_1 will have to be changed from its present value of $\tfrac{7}{2}$, but a change in x_1 will not affect z. Note in (4) that x_2, the variable to be increased, is in the column of (4) that contains the most negative value of the last row—in fact, the only negative value in the last row. To decide how much x_2 can be increased and still comply with the constraints of the problem, we momentarily ignore x_1 and divide each of the entries above the most negative, $-\tfrac{3}{2}$, into the extreme right-hand entry of its row: $\tfrac{1}{2}/\tfrac{1}{2} = 1$ and $\tfrac{7}{2}/\tfrac{1}{2} = 7$. The least quotient, which is 1, identifies the pivot, which is circled in (4). What is implied is in the first row x_2 can be increased to 1, while in the second it can be increased to 7. But to satisfy both constraints simultaneously, we must limit x_2 to the smaller value, 1. If $z = 0x_1 + \tfrac{3}{2}x_2 + 0x_3 - \tfrac{5}{2}x_4 + \tfrac{3 \cdot 5}{2}$ and if $x_2 = 1$ and x_4 is chosen to be 0, then $z = 0 + \tfrac{3}{2} + 0 + 0 + \tfrac{3 \cdot 5}{2} = \tfrac{3 \cdot 8}{2} = 19$, and 19 is the maximum value z will reach under the conditions stated.

But the same results can be obtained in matrix form. Use the pivot in (4) to eliminate the other coefficients of x_2. Multiplying the first row by 2 gives

$$\begin{array}{c} \begin{array}{ccccc} x_1 & x_2 & x_3 & x_4 & z \end{array} \\ \left[\begin{array}{ccccc|c} 0 & \boxed{1} & 2 & -1 & 0 & 1 \\ 1 & \tfrac{1}{2} & 0 & \tfrac{1}{2} & 0 & \tfrac{7}{2} \\ \hline 0 & -\tfrac{3}{2} & 0 & \tfrac{5}{2} & 1 & \tfrac{3 \cdot 5}{2} \end{array} \right] \end{array} \qquad (5)$$

If the first row is multiplied by $-\tfrac{1}{2}$ and added to the second row, then is multiplied by $\tfrac{3}{2}$ and added to the third row, (6) results:

$$\begin{array}{c} \begin{array}{ccccc} x_1 & x_2 & x_3 & x_4 & z \end{array} \\ \left[\begin{array}{ccccc|c} 0 & 1 & 2 & -1 & 0 & 1 \\ 1 & 0 & -1 & 1 & 0 & 3 \\ \hline 0 & 0 & 3 & 1 & 1 & \tfrac{3 \cdot 8}{2} = 19 \end{array} \right] \end{array} \qquad (6)$$

From (6), $z = 0x_1 + 0x_2 - 3x_3 - x_4 + 19$. z will be maximum when x_3 and x_4 are both 0. The maximum value of z is 19 and is found in the lower right-hand corner of (6). The values $x_1 = 3$ and $x_2 = 1$, which maximize z, are in the rightmost column of (6), above 19.

The simplex algorithm is a set of rules telling us how to create matrices (1) through (6) above. These rules are based upon the type of reasoning that was presented in the example. The process is made more efficient by omitting writing the variables above each matrix and by eliminating the column for z, the function to be maximized.

Consider a linear programming problem of the following type

$$a_{11}x_1 + a_{12}x_2 + \cdots + a_{1n}x_n \le b_1$$
$$a_{21}x_1 + a_{22}x_2 + \cdots + a_{2n}x_n \le b_2$$
$$\vdots$$
$$a_{m1}x_1 + a_{m2}x_2 + \cdots + a_{mn}x_n \le b_m$$

with the additional condition that

$$x_1 \ge 0, x_2 \ge 0, \ldots, x_n \ge 0 \quad \text{and} \quad b_1 \ge 0, b_2 \ge 0, \ldots, b_n \ge 0$$

where we are trying to maximize

$$P = c_1x_1 + c_2x_2 + \cdots + c_nx_n$$

We can convert this to a problem involving linear equations rather than linear inequalities by introducing a set of m non-negative variables:

$$x_{n+1}, x_{n+2}, \ldots, x_{n+m}$$

x_{n+1} is to be assigned a value such that

$$a_{11}x_1 + a_{12}x_2 + \cdots + a_{1n}x_n \le b_1$$

becomes

$$a_{11}x_1 + a_{12}x_2 + \cdots + a_{1n}x_n + x_{n+1} = b_1$$

In a similar way x_{n+2} is used to convert

$$a_{21}x_1 + a_{22}x_2 + \cdots + a_{2n}x_n \le b_2$$

to

$$a_{21}x_1 + a_{22}x_2 + \cdots + a_{2n}x_n + x_{n+2} = b_2$$

and so forth. These variables are called the *slack variables*.

The system of inequalities becomes

$$a_{11}x_1 + a_{12}x_2 + \cdots + a_{1n}x_n + x_{n+1} + 0 \cdot x_{n+2} + 0 \cdot x_{n+3} + \cdots + 0 \cdot x_{n+m}$$
$$= b_1$$

$$a_{21}x_1 + a_{22}x_2 + \cdots + a_{2n}x_n + 0 \cdot x_{n+1} + x_{n+2} + 0 \cdot x_{n+3} + \cdots + 0 \cdot x_{n+m}$$
$$= b_2$$
$$\vdots$$
$$a_{m1}x_1 + a_{m2}x_2 + \cdots + a_{mn}x_n + 0 \cdot x_{n+1} + \cdots + x_{n+m}$$
$$= b_m$$

In matrix form, considering only the coefficients, we write

$$\begin{bmatrix} a_{11} & a_{12} & a_{13} & \cdots & a_{1n} & 1 & 0 & \cdots & 0 & b_1 \\ a_{21} & a_{22} & \cdots\cdots\cdots\cdots & a_{2n} & 0 & 1 & \cdots & 0 & b_2 \\ \vdots & & & & & & & & \\ a_{m1} & a_{m2} & \cdots\cdots\cdots\cdots & a_{mn} & 0 & 0 & \cdots & 1 & b_m \end{bmatrix}$$

To apply the simplex method we extend this matrix to the initial simplex tableau by adding another row to it:

$$\begin{bmatrix} a_{11} & a_{12} & \cdots & a_{1n} & 1 & 0 & \cdots & 0 & b_1 \\ a_{21} & \cdots\cdots\cdots\cdots\cdots & a_{2n} & 0 & 1 & \cdots & 0 & b_2 \\ \vdots & & & & & & & & \vdots \\ -c_1 & -c_2 & \cdots & -c_n & 0 & 0 & \cdots & 0 & 0 \end{bmatrix}$$

Note that the final row consists of the negatives of the coefficients of the linear expression whose maximum we wish to find. Under the columns corresponding to the slack variables we have placed zeros. Also, a zero has been placed under the column of constants.

The Simplex Algorithm

The method requires the repeating of the following steps over and over on the simplex tableau until one reaches the **terminal tableau.**

STEP 1: Determine which value of the values $-c_1, -c_2, \ldots, -c_n$, the first n numbers in the final row, is the most negative number—i.e., of the *negative* numbers in this portion of the final row, the one with the *largest absolute value*. The column containing this value will be called the **pivot column.** If there are no negative values, the terminal tableau has been reached. If several values are most negative—i.e., equal in absolute value—we may choose any one.

STEP 2: Each *positive* number of the pivot column is divided into its corresponding number of the right-hand column of the simplex tableau. The row whose value yields the *smallest* quotient by this process is designated the **pivot row.** The number in the pivot row and pivot column is called the **pivot** and is circled for identification.

STEP 3: The pivot row is "tagged" with the number of the pivot column—i.e., if the pivot is in column 2, row 4, row number 4 is tagged 2 for future identification. (This is a device for getting the solution in the correct order.) Matrix row operations are then used to complete steps 4 and 5.

STEP 4: The pivot element is made 1 by dividing all the elements in the pivot row by the original value of the pivot.

STEP 5: Using row operations (specifically, recall that one can modify any row by adding any multiple of one row to the corresponding elements of another row), the remaining numbers in the *pivot column* are "zeroed," i.e., made zero.

STEP 6: If after completing step 5 any negative numbers remain among the first n numbers of the final row of the simplex tableau, one returns to step 1. If no negative *indicators*—i.e., numbers in the first n positions of the last row of the tableau—remain, the terminal tableau has been reached, and the solution to the linear programming problem can be read as below.

Interpreting the Terminal Tableau

1. The maximum value of P is the number in the terminal tableau in its last row and last column.

2. The corner of the feasible space that yields this maximum value is found by setting up an ordered n-tuple

$$(\quad _1, \quad _2, \quad _3, \ldots, \quad _n)$$

Notice that each position of the n-tuple has a number tag from 1 to n to identify it. In each "slot" we insert the last entry in the row of the terminal tableau tagged with that number. If no row is so tagged, then a zero is placed in that slot.

To clarify the process we can consider as an example one of the two-dimensional problems we have already solved using the graphical method.

EXAMPLE   Use the simplex method to solve the linear programming problem: Maximize $P = 310x_1 + 500x_2$ provided

$$\begin{bmatrix} 2 & 3 \\ 400 & 250 \\ 300 & 800 \\ -1 & 0 \\ 0 & -1 \end{bmatrix} \begin{bmatrix} x_1 \\ x_2 \end{bmatrix} \leq \begin{bmatrix} 100 \\ 15,000 \\ 24,000 \\ 0 \\ 0 \end{bmatrix}$$

Solution   This problem is the sample problem considered in previous sections. The final two rows of the coefficient matrix are simply statements that $x_1 \geq 0$ and $x_2 \geq 0$. Since these are not required in the simplex algorithm, they may be omitted here. Then the actual system of inequalities is

$$2x_1 + 3x_2 \leq 100$$
$$400x_1 + 250x_2 \leq 15,000$$
$$300x_1 + 800x_2 \leq 24,000$$

and $$P = 310x_1 + 500x_2$$

The initial simplex tableau is

$$\left[\begin{array}{ccc|ccc|c}
2 & 3 & 1 & 0 & 0 & 100 \\
400 & 250 & 0 & 1 & 0 & 15,000 \\
300 & 800 & 0 & 0 & 1 & 24,000 \\
\hline
-310 & -500 & 0 & 0 & 0 & 0
\end{array}\right]$$

Applying step 1, -500 is the most negative entry; thus the second column becomes the pivot column.

Pivot row "2" $\rightarrow$
$\uparrow$
Tag

$$\left[\begin{array}{ccc|ccc|c}
2 & 3 & 1 & 0 & 0 & 100 \\
400 & 250 & 0 & 1 & 0 & 15,000 \\
300 & \text{\textcircled{800}} & 0 & 0 & 1 & 24,000 \\
\hline
-310 & -500 & 0 & 0 & 0 & 0
\end{array}\right]
\begin{array}{l}
100 \div 3 \approx 33.3 \\
15,000 \div 250 = 60 \\
24,000 \div 800 = 30
\end{array}$$

$\uparrow$
Pivot column

According to step 2, we then divide each number in the pivot column into the corresponding number in the final column. We find

$$100 \div 3 = 33\tfrac{1}{3}$$

$$15,000 \div 250 = 60$$

$$24,000 \div 800 = 30$$

The smallest quotient is 30, in the third row. This makes the third row the pivot row. From step 3, this row is tagged with the number 2. Following the directions of step 4, each element of this row is divided by 800. The tableau becomes

$$2\left[\begin{array}{ccc|ccc|c}
2 & 3 & 1 & 0 & 0 & 100 \\
400 & 250 & 0 & 1 & 0 & 15,000 \\
\frac{3}{8} & \text{\textcircled{1}} & 0 & 0 & \frac{1}{800} & 30 \\
\hline
-310 & -500 & 0 & 0 & 0 & 0
\end{array}\right]$$

Using row 3 and following the directions of step 5, we adjust the remaining elements of the second column to zero:

$$2\left[\begin{array}{ccc|ccc|c}
\frac{7}{8} & 0 & 1 & 0 & -\frac{3}{800} & 10 \\
\frac{2450}{8} & 0 & 0 & 1 & -\frac{25}{80} & 7500 \\
\frac{3}{8} & 1 & 0 & 0 & \frac{1}{800} & 30 \\
\hline
-\frac{245}{2} & 0 & 0 & 0 & \frac{5}{8} & 15,000
\end{array}\right]$$

Since there is still a negative indicator ($-\frac{245}{2}$ in column 1), we must start over with step 1 of the simplex algorithm:

$$
\begin{array}{c}
\text{Pivot row 1} \rightarrow \\
\uparrow \\
\text{Tag} \\
2 \\
\\
\end{array}
\left[
\begin{array}{ccc|cc|c}
\boxed{\tfrac{7}{8}} & 0 & 1 & 0 & -\tfrac{3}{800} & 10 \\
\tfrac{2450}{8} & 0 & 0 & 1 & -\tfrac{25}{80} & 7500 \\
\tfrac{3}{8} & 1 & 0 & 0 & \tfrac{1}{800} & 30 \\
\hline
-\tfrac{245}{2} & 0 & 0 & 0 & \tfrac{5}{8} & 15{,}000
\end{array}
\right]
$$

$$
\begin{array}{c}
\uparrow \\
\text{Pivot column}
\end{array}
$$

Since
$$10 \div \tfrac{7}{8} \approx 11.4$$
$$7500 \div \tfrac{2450}{8} \approx 24.5$$

and
$$30 \div \tfrac{3}{8} = 80$$

row 1 becomes the pivot row.
We divide row 1 by $\tfrac{7}{8}$, converting the pivot to 1. The tableau becomes

$$
\begin{array}{c}
1 \\
\\
2 \\
\\
\end{array}
\left[
\begin{array}{ccc|cc|c}
1 & 0 & \tfrac{8}{7} & 0 & -\tfrac{3}{700} & \tfrac{80}{7} \\
\tfrac{2450}{8} & 0 & 0 & 1 & -\tfrac{25}{80} & 7500 \\
\tfrac{3}{8} & 1 & 0 & 0 & \tfrac{1}{800} & 30 \\
\hline
-\tfrac{245}{2} & 0 & 0 & 0 & \tfrac{5}{8} & 15{,}000
\end{array}
\right]
$$

Using row 1 to adjust the remaining elements of column 1 to zero, we have

$$
\begin{array}{c}
1 \\
\\
2 \\
\\
\end{array}
\left[
\begin{array}{cc|cc|cc}
1 & 0 & \tfrac{8}{7} & 0 & -\tfrac{3}{700} & \tfrac{80}{7} \\
0 & 0 & -\tfrac{2450}{7} & 1 & 1 & 4000 \\
0 & 1 & -\tfrac{3}{7} & 0 & \tfrac{1}{350} & \tfrac{180}{7} \\
\hline
0 & 0 & 140 & 0 & \tfrac{1}{10} & 16{,}400
\end{array}
\right]
$$

Since there are no remaining negative indicators, we have reached the terminal tableau. We can now read the answer to the problem. $P = 310x_1 + 500x_2$ has a maximum value of 16,400. Notice that the final value in the row tagged 1 is 80/7. This means that the maximum value of P occurs at

$$
\underset{\substack{1 \qquad 2}}{\left(\tfrac{80}{7}, \tfrac{180}{7}\right)}
$$

This agrees with the solution we found by graphical method.

One way to clarify the simplex method is a *flow diagram*. This is shown in Figure 6.3.3.

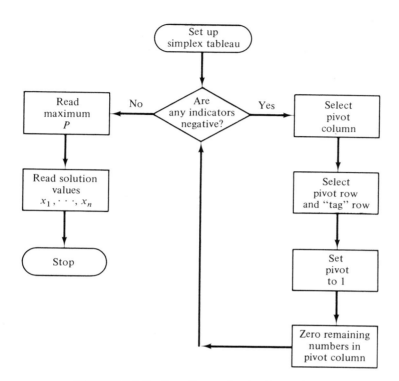

FIGURE 6.3.3 Flow diagram for simplex method.

EXAMPLE   Find the maximum value of $P = 23x_1 + 28x_2$ provided

$$19x_1 + 30x_2 \leq 114$$
$$33x_1 + 20x_2 \leq 110$$
$$x_1 - x_2 \leq 1$$

Solution   The graphical solution to this problem is shown in Figure 6.2.3. Applying the simplex method, the initial tableau is

$$
\text{Pivot row 2} \rightarrow
\begin{bmatrix}
19 & 30 & 1 & 0 & 0 & 114 \\
33 & 20 & 0 & 1 & 0 & 110 \\
1 & -1 & 0 & 0 & 1 & 1 \\
\hline
-23 & -28 & 0 & 0 & 0 & 0
\end{bmatrix}
\begin{matrix}
\frac{114}{30} = 3.8 \\
\frac{110}{20} = 5.5 \\
 \\

\end{matrix}
$$

$$\underset{\text{Pivot column}}{\uparrow}$$

Adjusting the pivot to 1,

$$2 \begin{bmatrix} \frac{19}{30} & \textcircled{1} & \frac{1}{30} & 0 & 0 & \frac{19}{5} \\ 33 & 20 & 0 & 1 & 0 & 110 \\ 1 & -1 & 0 & 0 & 1 & 1 \\ \hline -23 & -28 & 0 & 0 & 0 & 0 \end{bmatrix}$$

Adjusting the remaining elements in the pivot column to zero and selecting a new pivot,

$$\begin{matrix} 2 \\ 1 \\ {} \\ {} \end{matrix} \begin{bmatrix} \frac{19}{30} & 1 & \frac{1}{30} & 0 & 0 & \frac{19}{5} \\ \textcircled{\tfrac{61}{3}} & 0 & -\frac{2}{3} & 1 & 0 & 34 \\ \frac{49}{30} & 0 & \frac{1}{30} & 0 & 1 & \frac{24}{5} \\ \hline -\frac{79}{15} & 0 & \frac{14}{15} & 0 & 0 & \frac{532}{5} \end{bmatrix} \quad \begin{matrix} \frac{19}{5} \div \frac{19}{30} = 6 \\ 34 \div \frac{61}{3} \approx 1.67 \\ \frac{24}{5} \div \frac{49}{30} \approx 3 \\ {} \end{matrix}$$

Adjusting the new pivot to 1,

$$\begin{matrix} 2 \\ 1 \\ {} \\ {} \end{matrix} \begin{bmatrix} \frac{19}{30} & 1 & \frac{1}{30} & 0 & 0 & \frac{19}{5} \\ 1 & 0 & -\frac{2}{61} & \frac{3}{61} & 0 & \frac{102}{61} \\ \frac{49}{30} & 0 & \frac{1}{30} & 0 & 1 & \frac{24}{5} \\ \hline -\frac{79}{15} & 0 & \frac{14}{15} & 0 & 0 & \frac{532}{5} \end{bmatrix}$$

Adjusting the remaining elements in the pivot column to zero,

$$\begin{matrix} 2 \\ 1 \\ {} \\ {} \end{matrix} \begin{bmatrix} 0 & 1 & \frac{33}{610} & -\frac{19}{610} & 0 & \frac{836}{305} \\ 1 & 0 & -\frac{2}{61} & \frac{3}{61} & 0 & \frac{102}{61} \\ 0 & 0 & \frac{53}{610} & -\frac{49}{610} & 1 & \frac{631}{305} \\ \hline 0 & 0 & \frac{232}{305} & \frac{79}{305} & 0 & \frac{35138}{305} \end{bmatrix} \begin{matrix} = x_2 \\ = x_1 \\ {} \\ {} \end{matrix}$$

Since there are no negative indicators, this is the terminal tableau. The maximum value of P is $\frac{35138}{305} = \frac{70276}{610}$ at $x_1 = \frac{102}{61}$, $x_2 = \frac{836}{305} = \frac{1672}{610}$. This agrees with the result found graphically.

This section has considered only problems involving two variables. This means that the graphical method of solution could be used as a check on the process. The next section will extend the process to three or more variables. Problems involving minimum values of a linear expression will be considered in a later section. A note of warning: Should it occur that in steps 1 or 2 the column with the most negative indicator fails to have a positive entry, or the pivot lies in a row which has already been used, the linear program will not yield a solution. The examples and problems considered here avoid such cases, but the reader is warned that they do exist.

EXERCISES 6.3

Use the simplex algorithm to solve the linear programming problems in Exercises 1–12. In every case $x_1 \geq 0$ and $x_2 \geq 0$.

1. Maximize $P = 6x_1 + 4x_2$ provided

$$\begin{bmatrix} 3 & 5 \\ 6 & 5 \end{bmatrix} \begin{bmatrix} x_1 \\ x_2 \end{bmatrix} \leq \begin{bmatrix} 45 \\ 60 \end{bmatrix}$$

2. Maximize $P = 2x_1 + 5x_2$ provided

$$\begin{bmatrix} 3 & 5 \\ 6 & 5 \end{bmatrix} \begin{bmatrix} x_1 \\ x_2 \end{bmatrix} \leq \begin{bmatrix} 45 \\ 60 \end{bmatrix}$$

3. Maximize $P = 5x_1 + 6x_2$ provided

$$\begin{bmatrix} 3 & 5 \\ 6 & 5 \end{bmatrix} \begin{bmatrix} x_1 \\ x_2 \end{bmatrix} \leq \begin{bmatrix} 45 \\ 60 \end{bmatrix}$$

4. Maximize $P = 4x_1 + 3x_2$ provided

$$\begin{bmatrix} 2 & 1 \\ 1 & 6 \end{bmatrix} \begin{bmatrix} x_1 \\ x_2 \end{bmatrix} \leq \begin{bmatrix} 6 \\ 6 \end{bmatrix}$$

5. Maximize $P = 6x_1 - 4x_2$ provided

$$\begin{bmatrix} 17 & 12 \\ 1 & 1 \\ 13 & 25 \end{bmatrix} \begin{bmatrix} x_1 \\ x_2 \end{bmatrix} \leq \begin{bmatrix} 204 \\ 14 \\ 325 \end{bmatrix}$$

6. Maximize $P = -2x_1 + 5x_2$ provided

$$\begin{bmatrix} 17 & 12 \\ 1 & 1 \\ 13 & 25 \end{bmatrix} \begin{bmatrix} x_1 \\ x_2 \end{bmatrix} \leq \begin{bmatrix} 204 \\ 14 \\ 325 \end{bmatrix}$$

7. Find the largest value of $P = 6x_1 - 6x_2$ provided

$$\begin{bmatrix} 3 & 1 \\ 1 & 3 \\ 1 & 1 \end{bmatrix} \begin{bmatrix} x_1 \\ x_2 \end{bmatrix} \leq \begin{bmatrix} 15 \\ 15 \\ 7 \end{bmatrix}$$

8. Find the largest value of $P = 16x_1 + 4x_2$ provided

$$\begin{bmatrix} 6 & 5 \\ 1 & 2 \\ 2 & 1 \end{bmatrix} \begin{bmatrix} x_1 \\ x_2 \end{bmatrix} \leq \begin{bmatrix} 30 \\ 10 \\ 8 \end{bmatrix}$$

9. Find the largest value of $P = 6x_1 + 4x_2$ provided

$$\begin{bmatrix} 3 & 4 \\ 5 & 3 \\ 1 & 0 \end{bmatrix} \begin{bmatrix} x_1 \\ x_2 \end{bmatrix} \leq \begin{bmatrix} 36 \\ 45 \\ 7 \end{bmatrix}$$

10. Find the largest value of $P = 8x_1 - 4x_2$ provided

$$3x_1 + 4x_2 \leq 36$$
$$5x_1 + 3x_2 \leq 45$$
$$x_1 \qquad \leq 7$$

11. Find the largest value of $P = 14x_1 + 30x_2$ provided

$$x_1 + 3x_2 \leq 15$$
$$x_1 + 2x_2 \leq 10$$
$$x_1 + x_2 \leq 6$$

12. Maximize $P = 16x_1 + 40x_2$ provided

$$x_1 + 2x_2 \leq 14$$
$$3x_1 + 2x_2 \leq 1$$
$$x_1 + x_2 \leq 1$$

13. The XYZ Company plans to ship two kinds of dry material in individual plastic bags in a container which will hold .6 cu m and can hold 10 kg. The materials have the following parameters:

	Volume (cu m)	Weight (kg/cu m)	Profit (per unit)
A	0.2	1	$30
B	0.3	4	$70

How should they load containers to maximize profit?

14. The Your Health Food Supplement Company plans to mix two kinds of food supplements with the following characteristics:

	Type I	Type II
Fat (cal/g)	8	13
Carbohydrates (cal/g)	16	7
Vitamin A (units/g)	450	500

How should they make the mixture to maximize the vitamin A content if the final mixture cannot contain more than 104 cal of fat and 112 cal of carbohydrates?

15. A garden products firm plans to package fertilizer with a maximum amount of nitrogen. The product is to be made from two stock mixtures with the following characteristics:

	A	*B*
Bacteria (ct/kg)	450	800
Sulfur (g/kg)	70	55
Nitrogen (g/kg)	850	900

The bacteria count of the final mixture cannot exceed 1,440, and there may be no more than 154 g sulfur. How should they make up the new product?

6.4 THE SIMPLEX METHOD APPLIED TO PROBLEMS WITH THREE OR MORE VARIABLES

The power of the simplex method comes into play when considering problems involving more than two variables. The three-variable problem "solved" geometrically in the last section can illustrate the process.

EXAMPLE   Find the maximum value of $P = 400x_1 + 700x_2 + 600x_3$ provided

$$1,200x_1 + 1,500x_2 + 2,000x_3 \leq 6,000$$
$$x_1 \geq 0, x_2 \geq 0 \quad \text{and} \quad x_3 \geq 0$$

Solution   The initial simplex tableau is

$$\begin{bmatrix} 1,200 & \boxed{1,500} & 2,000 & 1 & 6,000 \\ \hline -400 & -700 & -600 & 0 & 0 \end{bmatrix}$$

The second column is the pivot column, with 1,500 the pivot. We then divide the first row by 1,500, tagging it with a 2:

$$2 \begin{bmatrix} \frac{4}{5} & 1 & \frac{4}{3} & \frac{1}{1500} & 4 \\ \hline -400 & -700 & -600 & 0 & 0 \end{bmatrix}$$

Zeroing the only remaining element of column 2, we have

$$2 \begin{bmatrix} \frac{4}{5} & 1 & \frac{4}{3} & \frac{1}{1500} & 4 \\ \hline 160 & 0 & \frac{1000}{3} & \frac{7}{15} & 2,800 \end{bmatrix}$$

This is the terminal tableau. Thus the maximum value of P occurs at $(0, 4, 0)$ and is 2,800. This agrees with the previous result.

EXAMPLE   Find the maximum value of $P = 7x_1 + 4x_2 + 8x_2$ provided

$$x_1 + 2x_2 + x_3 \ \leq 16 \quad (x_1 \geq 0, x_2 \geq 0, \text{ and } x_3 \geq 0)$$
$$4x_1 + 2x_2 + 2x_3 \leq 34$$

Solution   The simplex tableau is

$$3 \begin{bmatrix} 1 & 2 & \textcircled{1} & 1 & 0 & 16 \\ 4 & 2 & 2 & 0 & 1 & 34 \\ \hline -7 & -4 & -8 & 0 & 0 & 0 \end{bmatrix} \quad \begin{matrix} \frac{16}{1} = 16 \\ \frac{34}{2} = 17 \end{matrix}$$

With the pivot in column 3, zeroing, we have

$$3 \begin{bmatrix} 1 & 2 & 1 & 1 & 0 & 16 \\ 2 & -2 & 0 & -2 & 1 & 2 \\ \hline 1 & 12 & 0 & 8 & 0 & 128 \end{bmatrix}$$

The maximum value of P is 128 at $x_1 = 0$, $x_2 = 0$, $x_3 = 16$.

EXAMPLE   Find the maximum value of $P = 4x_1 + 2x_2 - x_3$ provided

$$x_1 + x_2 \qquad \leq 3$$
$$x_1 + \qquad x_3 \leq 5$$
$$x_2 + x_3 \leq 2$$

Solution   The simplex tableau is

$$1 \begin{bmatrix} \textcircled{1} & 1 & 0 & 1 & 0 & 0 & 3 \\ 1 & 0 & 1 & 0 & 1 & 0 & 5 \\ 0 & 1 & 1 & 0 & 0 & 1 & 2 \\ \hline -4 & -2 & +1 & 0 & 0 & 0 & 0 \end{bmatrix}$$

The pivot is in column 1: zeroing, we have

$$1 \begin{bmatrix} 1 & 1 & 0 & 1 & 0 & 0 & 3 \\ 0 & -1 & 1 & -1 & 1 & 0 & 2 \\ 0 & 1 & 1 & 0 & 0 & 1 & 2 \\ \hline 0 & 2 & 1 & 4 & 0 & 0 & 12 \end{bmatrix}$$

The maximum value of P is 12 at $(x_1, x_2, x_3) = (3, 0, 0)$.

EXAMPLE   Find the maximum value of $P = 5x_1 - 3x_2 + 6x_3$ with $x_1 \geq 0, x_2 \geq 0, x_3 \geq 0$, and

$$x_1 + x_2 + x_3 \leq 10$$
$$x_1 \leq 7$$
$$x_2 \leq 4$$
$$x_3 \leq 7$$

Solution   The simplex tableau is

$$
\begin{array}{c}
\\
\\
\\
3\\

\end{array}
\left[
\begin{array}{ccc|cccc|c}
1 & 1 & 1 & 1 & 0 & 0 & 0 & 10 \\
1 & 0 & 0 & 0 & 1 & 0 & 0 & 7 \\
0 & 1 & 0 & 0 & 0 & 1 & 0 & 4 \\
0 & 0 & \textcircled{1} & 0 & 0 & 0 & 1 & 7 \\
\hline
-5 & 3 & -6 & 0 & 0 & 0 & 0 & 0
\end{array}
\right]
\begin{array}{l}
10 \div 1 = 10 \\
\\
\\
7 \div 1 = 7
\end{array}
$$

$$
\begin{array}{c}
1\\
\\
\\
3\\

\end{array}
\left[
\begin{array}{ccc|cccc|c}
\textcircled{1} & 1 & 0 & 1 & 0 & 0 & -1 & 3 \\
1 & 0 & 0 & 0 & 1 & 0 & 0 & 7 \\
0 & 1 & 0 & 0 & 0 & 1 & 0 & 4 \\
0 & 0 & 1 & 0 & 0 & 0 & 1 & 7 \\
\hline
-5 & +3 & 0 & 0 & 0 & 0 & 6 & 42
\end{array}
\right]
\begin{array}{l}
3 \div 1 = 3 \\
7 \div 1 = 7 \\
\\
\\
\end{array}
$$

$$
\begin{array}{c}
1\\
\\
\\
3\\

\end{array}
\left[
\begin{array}{ccc|cccc|c}
1 & 1 & 0 & 1 & 0 & 0 & -1 & 3 \\
0 & -1 & 0 & -1 & 1 & 0 & 1 & 4 \\
0 & 1 & 0 & 0 & 0 & 1 & 0 & 4 \\
0 & 0 & 1 & 0 & 0 & 0 & 1 & 7 \\
\hline
0 & 8 & 0 & 5 & 0 & 0 & 1 & 57
\end{array}
\right]
$$

The maximum value of P is 57 at $x_1 = 3$, $x_2 = 0$, $x_3 = 7$, i.e., $(3, 0, 7)$.

EXAMPLE   The Low-Fat Milk Company wishes to add a food supplement to its dry milk food mix. It has three types of supplements on hand with the following characteristics:

	Type A	B	C
Bacteria count (tens/cc)	1	3	1
Vitamin A (100 units/cc)	2	2	1
Weight (g/cc)	3	1	4
Vitamin B$_1$ (100 units/cc)	2	4	3

The bacteria count of the final additive cannot exceed 60. The weight of the final additive cannot exceed 7 g. The amount of vitamin A cannot exceed 500 units. How should the supplement be constructed to maximize the amount of vitamin B_1 added?

Solution   If we let

x_1 equal the number of cc of supplement A

x_2 equal the number of cc of supplement B

x_3 equal the number of cc of supplement C

The symbolic statement of the problem becomes:

Maximize $P = 2x_1 + 4x_2 + 3x_3$

provided $x_1 \geq 0,\ x_2 \geq 0,\ x_3 \geq 0$

and

$$x_1 + 3x_2 + x_3 \leq 6$$
$$2x_1 + 2x_2 + x_3 \leq 5$$
$$3x_1 + x_2 + 4x_3 \leq 7$$

Applying the simplex method,

$$
2\begin{bmatrix}
1 & ③ & 1 & 1 & 0 & 0 & 6 \\
2 & 2 & 1 & 0 & 1 & 0 & 5 \\
3 & 1 & 4 & 0 & 0 & 1 & 7 \\
\hline
-2 & -4 & -3 & 0 & 0 & 0 & 0
\end{bmatrix}
\begin{array}{l}
\frac{6}{3} = 2 \\
\frac{5}{2} = 2\frac{1}{2} \\
\frac{7}{1} = 7 \\
\\
\end{array}
$$

$$
2\begin{bmatrix}
\frac{1}{3} & 1 & \frac{1}{3} & \frac{1}{3} & 0 & 0 & 2 \\
2 & 2 & 1 & 0 & 1 & 0 & 5 \\
3 & 1 & 4 & 0 & 0 & 1 & 7 \\
\hline
-2 & -4 & -3 & 0 & 0 & 0 & 0
\end{bmatrix}
$$

$$
\begin{array}{c}
2 \\
\\
3 \\
\\
\end{array}
\begin{bmatrix}
\frac{1}{3} & 1 & \frac{1}{3} & \frac{1}{3} & 0 & 0 & 2 \\
\frac{4}{3} & 0 & \frac{1}{3} & -\frac{2}{3} & 1 & 0 & 1 \\
\frac{8}{3} & 0 & ⑪\frac{11}{3} & -\frac{1}{3} & 0 & 1 & 5 \\
\hline
-\frac{2}{3} & 0 & -\frac{5}{3} & \frac{4}{3} & 0 & 0 & 8
\end{bmatrix}
\begin{array}{l}
2 \div \frac{1}{3} = 6 \\
1 \div \frac{1}{3} = 3 \\
5 \div \frac{11}{3} = \frac{15}{11} \\
\\
\end{array}
$$

$$
\begin{array}{c}
2 \\
\\
3 \\
\\
\end{array}
\begin{bmatrix}
\frac{1}{3} & 1 & \frac{1}{3} & \frac{1}{3} & 0 & 0 & 2 \\
\frac{4}{3} & 0 & \frac{1}{3} & -\frac{2}{3} & 1 & 0 & 1 \\
\frac{8}{11} & 0 & 1 & -\frac{1}{11} & 0 & \frac{3}{11} & \frac{15}{11} \\
\hline
-\frac{2}{3} & 0 & -\frac{5}{3} & \frac{4}{3} & 0 & 0 & 8
\end{bmatrix}
$$

$$2 \begin{bmatrix} \frac{1}{11} & 1 & 0 & \frac{4}{11} & 0 & -\frac{1}{11} & \frac{17}{11} \\ \frac{12}{11} & 0 & 0 & -\frac{7}{11} & 1 & -\frac{1}{11} & \frac{6}{11} \\ \frac{8}{11} & 0 & 1 & -\frac{1}{11} & 0 & \frac{3}{11} & \frac{15}{11} \\ \hline -\frac{6}{11} & 0 & 0 & \frac{13}{11} & 0 & \frac{5}{11} & \frac{113}{11} \end{bmatrix}$$
3 label on third row.

The maximum value of P is $\frac{113}{11}$ at $(0, \frac{17}{11}, \frac{15}{11})$, i.e., $x_1 = 0$, $x_2 = \frac{17}{11}$, and $x_3 = \frac{15}{11}$.

You may have noticed two things. First, the "middle" of the simplex tableau, that part relating to the slack variables, is not needed in the process of solving the problems thus far considered. That is, the three entries $\frac{13}{11}$, 0, and $\frac{5}{11}$ in the above example do not appear to have anything to do with the problem solution. Second, no minimum problems have been considered. We will see how these two items are related in the next section.

EXERCISES 6.4

Find the maximum value of the given P in Exercises 1–11. Assume the additional condition that $x_1 \geq 0$, $x_2 \geq 0$, etc.

1. $P = 3x_1 + 4x_2 + 6x_3$ provided
$$12x_1 + 14x_2 + 6x_3 \leq 8$$

2. $P = 6x_1 + 8x_2 - 3x_3$ provided
$$12x_1 + 5x_2 + x_3 \leq 10$$

3. $P = 10x_1 + 15x_2 + 20x_3$ provided
$$2x_1 + 4x_2 + 3x_3 \leq 20$$
$$3x_1 + x_2 \leq 4$$

4. $P = 9x_1 + 7x_2 + 4x_3$ provided
$$x_1 + x_2 + 3x_3 \leq 8$$
$$2x_1 + 4x_2 + x_3 \leq 7$$

5. $P = 3x_1 + 9x_2 + 2x_3$ provided
$$3x_1 + 7x_2 + x_3 \leq 1$$
$$8x_1 + 6x_2 + 5x_3 \leq 2$$

6. $P = 9x_1 + 4x_3$ provided
$$x_1 + 7x_2 \leq 9$$
$$x_2 + 7x_3 \leq 9$$

7. $P = 7x_1 + 3x_2 + 9x_3$ provided

$$5x_1 + 8x_2 + 2x_3 \leq 2$$
$$8x_2 + 7x_3 \leq 3$$
$$3x_1 + 7x_2 + 8x_3 \leq 6$$

8. $P = 7x_1 + x_2 + 5x_3$ provided

$$x_1 + 3x_2 + 6x_3 \leq 12$$
$$7x_1 + x_2 + 5x_3 \leq 20$$
$$5x_1 + 5x_2 + 5x_3 \leq 25$$

9. $P = 6x_1 + 4x_2 - 7x_3$ provided

$$3x_1 + 9x_2 + 6x_3 \leq 12$$
$$9x_1 + 7x_3 \leq 30$$
$$2x_1 + 8x_2 \leq 24$$

10. $P = 3x_1 + 2x_2 + 4x_3 + 5x_4$ provided

$$x_1 + 2x_2 + 3x_3 \leq 6$$
$$x_2 + 2x_3 + x_4 \leq 8$$
$$2x_1 + x_3 + 4x_4 \leq 12$$
$$x_1 + x_2 + x_3 + x_4 \leq 4$$

11. $P = x_1 + 2x_2 + x_3 + 2x_4$ provided

$$x_1 + x_2 + x_3 + x_4 \leq 20$$
$$5x_1 + 4x_2 + 3x_3 + x_4 \leq 120$$
$$x_1 + 4x_2 + 3x_3 + 2x_4 \leq 72$$
$$6x_1 + 4x_2 + 3x_3 + x_4 \leq 124$$

12. The XYZ Shipping Company plans to ship material by barge which will hold 12,000 kg or 1,800 cu m. They plan to ship three types of items with characteristics as shown below:

	Type		
	A	B	C
Weight ($\times$ 1,000 kg) per unit	1	2	1
Volume ($\times$ 100 cu m) per unit	2	4	6
Profit (per unit)	$200	$700	$300

How should the barge be loaded for maximum profit?

13. Answer Exercise 12 provided the barge will hold 20,000 kg, with a maximum volume of 4,500 cu m, if the units have the following characteristics:

	Type A	B	C
Weight ($\times$ 1,000 kg) per unit	10	5	15
Volume ($\times$ 100 cu m) per unit	10	10	20
Profit (per unit)	$700	$300	$400

14. The ZTQ Drug Company is asked to make up 1 kg or less of a new drug compound with three existing compounds. The existing drugs have the following components:

	Type I	II	III
Units of Vitamin A ($\times$ 100) per kg	2	4	5
Units of Vitamin B_1 ($\times$ 100) per kg	6	4	12
Units of vitamin E per kg	200	400	600

How should they make the new drug if it cannot contain more than 10,000 units of vitamin A or 12,000 units of vitamin B_1 and must contain a maximum amount of vitamin E?

15. A blood sample contains less than a 1,200 count of antigen A, less than 2,000 count of antigen B, and less than 1,000 count of antigen C. It also contains a maximum count of antigen D. The sample is known to contain three types of blood, each with a characteristic count of each antigen.

Antigen/unit	Type I	II	III
A	20	60	30
B	50	80	20
C	60	60	50
D	350	450	200

How much of each blood type is present?

16. An electronics firm plans to make a special switching circuit for a computer from stock "plug-in" units. Unit type I weighs 20 g/unit, is 10 cc in volume, costs $5/unit and contains 1,000 switching circuits. Unit type II weighs 30 g/unit, is 50 cc in volume, and costs $5/unit. It contains 650 switching circuits. Unit type III has 800 switching circuits, weighs 40 g/unit, is 40 cc in volume, and costs $10/unit. If the total weight of the final unit must be less than 120 g, if it must be less than 90 cc in volume, and if it must cost less than $50, how should it be made to maximize the number of switching circuits?

17. A company is asked to make up 100 kg or less of a mixture using any or all of three types of cattle feed. The new mixture must contain less than 100 g sulfur and 30 g tetracycline. It must contain a maximum amount of food solids. Food type I contains 15 g/kg sulfur, 5 g/kg tetracycline, and 120 g/kg food solids. Food type II contains 20 g/kg sulfur, 4 g/kg tetracycline, and 75 g/kg food solids. Food type III contains 10 g/kg sulfur, 6 g/kg tetracycline, and 80 g/kg food solids. How should the mixture be made?

6.5   THE MINIMUM PROBLEM AND DUALITY

Consider the problem of finding the minimum value of $Q = 6y_1 + 4y_2$ provided $y_1 \geq 0$, $y_2 \geq 0$, and

$$y_1 + 3y_2 \geq 5$$
$$y_1 + y_2 \geq 3$$
$$3y_1 + y_2 \geq 5$$

Since this is a minimum problem, with the constraints requiring combinations of y_1 and y_2 to be greater than specific values, the simplex method would not seem to apply. The problem can be done using the graphical method. The solution space is shown in Figure 6.5.1. The possible solution points are (0, 5), (1, 2), (2, 1), and (5, 0). We can evaluate Q at each possible solution point:

At (5, 0) $Q = 6(5) + 4(0) = 30$

At (2, 1) $Q = 6(2) + 4(1) = 16$

At (1, 2) $Q = 6(1) + 4(2) = 14$

At (0, 5) $Q = 6(0) + 4(5) = 20$

The minimum value of Q is 14 at $y_1 = 1$, $y_2 = 2$. The simplex method can be applied by using the dual concept.

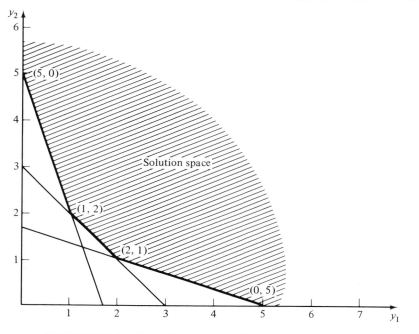

FIGURE 6.5.1   The minimum value of $P = 6y_1 + 4y_2$, provided

$$\begin{bmatrix} 1 & 3 \\ 1 & 1 \\ 3 & 1 \end{bmatrix} \begin{bmatrix} y_1 \\ y_2 \end{bmatrix} \geq \begin{bmatrix} 5 \\ 3 \\ 5 \end{bmatrix}$$

DEFINITION

The **dual** of the linear programming problem of maximizing $P = c_1x_1 + c_2x_2 + \cdots + c_nx_n$ provided

$$a_{11}x_1 + a_{12}x_2 + \cdots + a_{1n}x_n \leq b_1$$
$$a_{21}x_1 + a_{22}x_2 + \cdots + a_{2n}x_n \leq b_2$$
$$\vdots$$
$$a_{m1}x_1 + a_{m2}x_2 + \cdots + a_{mn}x_n \leq b_m$$

is the problem of finding the minimum value of

$$Q = b_1y_1 + b_2y_2 + \cdots + b_my_m$$

subject to the conditions that

$$a_{11}y_1 + a_{21}y_2 + a_{31}y_3 + \cdots + a_{m1}y_m \geq c_1$$
$$a_{12}y_1 + a_{22}y_2 + a_{32}y_3 + \cdots + a_{m2}y_m \geq c_2$$
$$a_{13}y_1 + a_{23}y_2 + a_{33}y_3 + \cdots + a_{m3}y_m \geq c_3$$
$$\vdots$$
$$a_{1n}y_1 + a_{2n}y_2 + \qquad \cdots + a_{mn}y_m \geq c_n$$

In other words, the dual of a linear programming problem is found by making the rows into columns; the coefficients of the linear expression become the bounding conditions, and those of the boundary conditions become the coefficients of the new linear expression.

The first system is also the dual of the second. All x_i, y_i, b_i, and c_i are assumed to be positive.

EXAMPLE   Find the dual of the sample problem considered above and solve.

Solution   The dual is: Maximize $P = 5x_1 + 3x_2 + 5x_3$ provided

$$x_1 + x_2 + 3x_3 \leq 6 \tag{1}$$

$$3x_1 + x_2 + x_3 \leq 4$$

In matrix form, the dual of finding the maximum of $P = c_1x_1 + c_2x_2 + \cdots + c_nx_n$ provided

$$\begin{bmatrix} a_{11} & a_{12} & \cdots & a_{1n} \\ a_{21} & & & a_{2n} \\ \vdots & & & \\ a_{m1} & & & a_{mn} \end{bmatrix} \begin{bmatrix} x_1 \\ \vdots \\ x_n \end{bmatrix} \leq \begin{bmatrix} b_1 \\ \vdots \\ b_m \end{bmatrix}$$

is: Find the minimum of $Q = b_1y_1 + b_2y_2 + \cdots + b_my_m$ provided

$$\begin{bmatrix} a_{11} & a_{21} & \cdots & a_{m1} \\ a_{12} & & & \\ \vdots & & & \\ a_{1n} & & & a_{mn} \end{bmatrix} \begin{bmatrix} y_1 \\ \vdots \\ y_m \end{bmatrix} \geq \begin{bmatrix} c_1 \\ \vdots \\ c_n \end{bmatrix}$$

Note the size of each matrix. The matrix of coefficients in the maximum problem is an $m \times n$ matrix; that in the minimum problem is an $n \times m$ matrix. This is clear because the one problem has as many rows as its dual has columns, and vice versa.

Why is the dual concept important? Let us "solve" the dual of the minimum problem with which we began the section. The dual is stated in (1). This problem can be done using the simplex method. The tableau is

$$\begin{bmatrix} 1 & 1 & 3 & 1 & 0 & 6 \\ 3 & 1 & 1 & 0 & 1 & 4 \\ \hline -5 & -3 & -5 & 0 & 0 & 0 \end{bmatrix}$$

Since there are two most negative indicators, we will arbitrarily choose the right-hand -5 as the most negative. Then circling the pivot and dividing by 3, we have

$$\begin{bmatrix} \frac{1}{3} & \frac{1}{3} & \textcircled{1} & \frac{1}{3} & 0 & 2 \\ 3 & 1 & 1 & 0 & 1 & 4 \\ \hline -5 & -3 & -5 & 0 & 0 & 0 \end{bmatrix}$$

There is no need to tag rows as before because such tags will not be needed to read the answer. Working in the usual way, this becomes

$$\begin{bmatrix} \frac{1}{3} & \frac{1}{3} & 1 & \frac{1}{3} & 0 & 2 \\ \textcircled{\frac{8}{3}} & \frac{2}{3} & 0 & -\frac{1}{3} & 1 & 2 \\ \hline -\frac{10}{3} & -\frac{4}{3} & 0 & \frac{5}{3} & 0 & 10 \end{bmatrix} \begin{array}{l} \frac{2}{1/3} = 6 \\ \frac{2}{8/3} = \frac{6}{8} = \frac{3}{4} \end{array}$$

Selecting a new pivot and multiplying by $\frac{3}{8}$,

$$\begin{bmatrix} \frac{1}{3} & \frac{1}{3} & 1 & \frac{1}{3} & 0 & 2 \\ \textcircled{1} & \frac{1}{4} & 0 & -\frac{1}{8} & \frac{3}{8} & \frac{3}{4} \\ \hline -\frac{10}{3} & -\frac{4}{3} & 0 & \frac{5}{3} & 0 & 10 \end{bmatrix}$$

$$\begin{bmatrix} 0 & \frac{1}{4} & 1 & \frac{3}{8} & -\frac{1}{8} & \frac{7}{4} \\ 1 & \frac{1}{4} & 0 & -\frac{1}{8} & -\frac{3}{8} & \frac{3}{4} \\ \hline 0 & -\frac{1}{2} & 0 & \frac{5}{4} & \frac{5}{4} & \frac{25}{2} \end{bmatrix}$$

Since there is still a negative indicator $(-\frac{1}{2})$, we select yet another pivot and continue:

$$\begin{bmatrix} 0 & \frac{1}{4} & 1 & \frac{3}{8} & -\frac{1}{8} & \frac{7}{4} \\ 4 & \textcircled{1} & 0 & -\frac{1}{2} & \frac{3}{2} & 3 \\ \hline 0 & -\frac{1}{2} & 0 & \frac{5}{4} & \frac{5}{4} & \frac{25}{2} \end{bmatrix}$$

$$\begin{bmatrix} -1 & 0 & 1 & \frac{1}{2} & -\frac{1}{2} & 1 \\ 4 & 1 & 0 & -\frac{1}{2} & \frac{3}{2} & 3 \\ \hline 2 & 0 & 0 & 1 & 2 & 14 \end{bmatrix}$$

This is the terminal tableau, but note the unusual result. The maximum value of P is equal to the minimum value of Q, which was found by graphing to be 14. Further, the values 1 and 2 which appear in the previously meaningless middle of the tableau correspond to the values of y_1 (1) and y_2 (2), which yield the minimum value of Q. The fact is that this is always the case. That is, we can always solve a minimum problem by (1) finding its dual, (2) solving the dual, and (3) observing that the maximum of P and the minimum of Q have the same value and further, that the values under the slack variable columns will correspond to the solution point of the minimum problem. This result was proved by John von Neuman in 1947. It is called the *Fundamental Theorem of Linear Programming*.

> **THEOREM** **The Fundamental Theorem of Linear Programming**
>
> If a linear programming problem and its dual both have a feasible point, then they have a best feasible point, and the values of both linear expressions involved are equal there.

This result, as with the validity of the simplex method, will not be proved here.

EXAMPLE   Find the minimum value of

$$Q = 2y_1 + 3y_2 + 4y_3$$

if

$$2y_1 + y_2 + 3y_3 \geq 12$$
$$y_1 + 2y_2 + y_3 \geq 18$$

$$y_1 \geq 0, y_2 \geq 0, y_3 \geq 0$$

Solution   The dual of this problem is: Maximize $P = 12x_1 + 18x_2$ provided

$$2x_1 + x_2 \leq 2$$
$$x_1 + 2x_2 \leq 3$$
$$3x_1 + x_2 \leq 4$$

The simplex tableau is

$$\begin{bmatrix} 2 & 1 & 1 & 0 & 0 & 2 \\ 1 & ② & 0 & 1 & 0 & 3 \\ 3 & 1 & 0 & 0 & 1 & 4 \\ \hline -12 & -18 & 0 & 0 & 0 & 0 \end{bmatrix}$$

Processing this in the usual manner,

$$\begin{bmatrix} 2 & 1 & 1 & 0 & 0 & 2 \\ \frac{1}{2} & ① & 0 & \frac{1}{2} & 0 & \frac{3}{2} \\ 3 & 1 & 0 & 0 & 1 & 4 \\ \hline -12 & -18 & 0 & 0 & 0 & 0 \end{bmatrix}$$

$$\begin{bmatrix} \textcircled{$\frac{3}{2}$} & 0 & 1 & -\frac{1}{2} & 0 & \frac{1}{2} \\ \frac{1}{2} & 1 & 0 & \frac{1}{2} & 0 & \frac{3}{2} \\ \frac{5}{2} & 0 & 0 & -\frac{1}{2} & 1 & \frac{5}{2} \\ \hline -3 & 0 & 0 & 9 & 0 & 27 \end{bmatrix}$$

$$\begin{bmatrix} \textcircled{1} & 0 & \frac{2}{3} & -\frac{1}{3} & 0 & \frac{1}{3} \\ \frac{1}{2} & 1 & 0 & \frac{1}{2} & 0 & \frac{3}{2} \\ \frac{5}{2} & 0 & 0 & -\frac{1}{2} & 1 & \frac{5}{2} \\ \hline -3 & 0 & 0 & 9 & 0 & 27 \end{bmatrix}$$

$$\begin{bmatrix} 1 & 0 & \frac{2}{3} & -\frac{1}{3} & 0 & \frac{1}{3} \\ 0 & 1 & -\frac{1}{3} & \frac{2}{3} & 0 & \frac{4}{3} \\ 0 & 0 & -\frac{5}{3} & \frac{1}{3} & 1 & \frac{5}{3} \\ \hline 0 & 0 & 2 & 8 & 0 & 28 \end{bmatrix}$$

This is the terminal tableau. Thus the minimum value of Q is 28, and it occurs at $(2, 8, 0)$, i.e., $y_1 = 2$, $y_2 = 8$, $y_3 = 0$.

Not all linear programming problems have solutions; indeed, most of them do not. We have examined the simplex method only in an elementary form. The algorithm can be extended in some cases, and you are referred to books on this topic for such extensions.

EXERCISES 6.5

Find the duals of each of the linear programming problems in Exercises 1–10 and solve the problem with $y_1 \geq 0$, $y_2 \geq 0$, etc. (Warning: when using the dual, simplifying by dividing out common factors may influence the answer.)

1. Find the minimum value of $Q = 3y_1 + 4y_2$ provided

$$3y_1 + y_2 \geq 3$$
$$y_1 + 2y_2 \geq 2$$

2. Find the minimum value of $Q = 4y_1 + 7y_2$ provided

$$2y_1 + 5y_2 \geq 10$$
$$5y_1 + 4y_2 \geq 20$$

3. Find the minimum value of $Q = 3y_1 + 4y_2 + 5y_3$ provided

$$2y_1 + y_2 + 3y_3 \geq 12$$

4. Find the minimum value of $Q = y_1 + y_2 + y_3$ provided

$$2y_1 + 3y_2 + 6y_3 \geq 24$$

5. Find the minimum value of $Q = y_1 + 3y_2 + 4y_3$ provided

$$2y_1 + y_2 + 4y_3 \geq 18$$
$$y_1 + 2y_2 + 3y_3 \geq 24$$

6. Find the minimum value of $Q = 5y_1 + 6y_2 + 8y_3$ provided

$$5y_1 + 10y_2 + 20y_3 \geq 100$$
$$y_1 + 2y_2 + y_3 \geq 20$$

7. Find the minimum value of $Q = 3y_1 + 7y_2 + 16y_3$ provided

$$y_1 + 2y_2 \qquad\quad \geq \ 8$$
$$3y_2 + \ y_3 \geq \ 9$$
$$y_1 + \qquad 4y_3 \geq 12$$

8. Find the minimum value of $Q = 2y_1 + 4y_2 + 2y_3$ provided

$$y_1 \qquad\qquad \geq \ 6$$
$$y_1 + y_2 \qquad \geq \ 8$$
$$y_2 + y_3 \geq 10$$

9. Find the minimum value of $Q = y_1 + 2y_2 + y_3 + 2y_4$ provided

$$y_1 + 2y_2 + 3y_3 \qquad\qquad \geq \ 6$$
$$2y_2 + \ y_3 + \ y_4 \geq 12$$
$$3y_1 + \ y_2 + \qquad\quad y_4 \geq 24$$

10. Find the minimum value of $Q = 6y_1 + 4y_2 + 3y_3 + 2y_4$ provided

$$y_1 + y_2 + \ y_3 + \ y_4 \geq \ 4$$
$$2y_1 + \qquad\quad y_3 \qquad\quad \geq \ 6$$
$$y_2 + 3y_3 + 2y_4 \geq \ 8$$
$$y_1 + \qquad\qquad\quad y_4 \geq 12$$

11. The XYZ Food Company is asked to make up a compound with a minimum amount of fat but with at least 6 g of protein and 8 g of carbohydrates. The compound is to be made from food type A, which contains 2 g/oz protein, 1 g/oz carbohydrates, and 3 g/oz fat, and food type B, which contains 1 g/oz protein, 4 g/oz carbohydrates, and 8 g/oz fat. How should the new food be made?

12. One food supplement contains 3 g/oz fat and 2 g/oz protein. A second contains 4 g/oz fat and 6 g/oz protein. What is the smallest mixture of these supplements that contains at least 12 g of fat and 15 g of protein?

13. The Brand X Paint Company must make at least 1 gal of a special paint. It plans to use three stock types it has on hand with the following characteristics:

	Type		
	I	*II*	*III*
Basic white (oz/gal)	2	4	3
Brightener (units/gal)	2	1	2
Cost (per gal)	$3	$2	$1

The final mixture must contain at least 12 oz of basic white and at least 6 units of the brightener. How should the paint be made to minimize cost?

14. The XYZ Card Company buys cards from three suppliers. Card supplier *A* provides cases at $50 per case. Each case provides 100 type I cards, 200 type II cards, and 800 type III cards. Supplier *B* charges $60 per case. Each case contains 100 type I cards, 100 type II cards, and 400 type III cards. Supplier *C* charges $80 a case. Each case contains 200 type I cards, 400 type II cards, and 400 type III cards. XYZ needs at least 4,000 type I cards, 2,500 type II cards, and 2,400 type III cards. How should they order to minimize costs?

15. An aggregate company needs at least 7 tons of small crushed rock, 48 tons of large crushed rock, and 10 tons of sand. It ships in raw material from three sources. Source A sends carloads which contain 1 ton of small rock, 6 tons of large rock, and 1 ton of sand each for $1,100. Source B sends carloads which contain 1 ton of small rock, 4 tons of large rock, and 2 tons of sand for $1,400 per carload. Source C sends carloads which contain 1 ton of small rock, 8 tons of large rock, and 3 tons of sand for $1,600 per carload. How should the company order to minimize costs?

REVIEW EXERCISES

In Exercises 1–6, find the maximum value for P using the graphical method.

1. $P = 10x_1 + 5x_2$; $\begin{bmatrix} 1 & 1 \\ 7 & 15 \\ -1 & 0 \\ 0 & -1 \end{bmatrix} \begin{bmatrix} x_1 \\ x_2 \end{bmatrix} \leq \begin{bmatrix} 11 \\ 105 \\ 0 \\ 0 \end{bmatrix}$

2. $P = 12x_1 + 7x_2;$ $\begin{bmatrix} 1 & 1 \\ 1 & 2 \\ -1 & 0 \\ 0 & -1 \end{bmatrix} \begin{bmatrix} x_1 \\ x_2 \end{bmatrix} \leq \begin{bmatrix} 10 \\ 16 \\ 0 \\ 0 \end{bmatrix}$

3. $P = 12x_1 + 6x_2;$ $\begin{bmatrix} 8 & 25 \\ 1 & 1 \\ 9 & 19 \\ -1 & 0 \\ 0 & -1 \end{bmatrix} \begin{bmatrix} x_1 \\ x_2 \end{bmatrix} \leq \begin{bmatrix} 200 \\ 14 \\ 171 \\ 0 \\ 0 \end{bmatrix}$

4. $P = 3x_1 + 7x_2;$ $\begin{bmatrix} 3 & 2 \\ 2 & -1 \\ 14 & 5 \\ -1 & 0 \\ 0 & -1 \end{bmatrix} \begin{bmatrix} x_1 \\ x_2 \end{bmatrix} \leq \begin{bmatrix} 30 \\ 0 \\ 70 \\ 0 \\ 0 \end{bmatrix}$

5. $P = 3x_1 + 7x_2;$ $\begin{bmatrix} 1 & 1 \\ 1 & 1 \\ -1 & 1 \\ -1 & 0 \\ 0 & -1 \end{bmatrix} \begin{bmatrix} x_1 \\ x_2 \end{bmatrix} \leq \begin{bmatrix} 2 \\ 3 \\ 0 \\ 0 \\ 0 \end{bmatrix}$

6. $P = 8x_1 + 4x_2;$ $\begin{bmatrix} 1 & 1 \\ -1 & 1 \\ -1 & 0 \\ 0 & -1 \\ 0 & 1 \end{bmatrix} \begin{bmatrix} x_1 \\ x_2 \end{bmatrix} \leq \begin{bmatrix} 2 \\ 0 \\ 0 \\ 0 \\ \frac{1}{2} \end{bmatrix}$

In Exercises 7–10, find the minimum value of Q by using the graphical method.

7. $Q = 3y_1 + 4y_2;$ $\begin{bmatrix} 1 & 0 \\ 0 & 1 \\ 1 & 1 \\ 3 & 1 \end{bmatrix} \begin{bmatrix} y_1 \\ y_2 \end{bmatrix} \geq \begin{bmatrix} 0 \\ 0 \\ 10 \\ 15 \end{bmatrix}$

8. $Q = 5y_1 + 4y_2;$ $\begin{bmatrix} 10 & 3 \\ 1 & 1 \\ 1 & 5 \\ 1 & 0 \\ 0 & 1 \end{bmatrix} \begin{bmatrix} y_1 \\ y_2 \end{bmatrix} \geq \begin{bmatrix} 30 \\ 5 \\ 10 \\ 0 \\ 0 \end{bmatrix}$

9. $Q = 7y_1 + 8y_2;$ $\begin{bmatrix} 12 & 5 \\ 7 & 10 \\ 2 & 10 \\ 1 & 0 \\ 0 & 1 \end{bmatrix} \begin{bmatrix} y_1 \\ y_2 \end{bmatrix} \geq \begin{bmatrix} 60 \\ 70 \\ 30 \\ 0 \\ 0 \end{bmatrix}$

10. $Q = 12y_1 + 15y_2;$ $\begin{bmatrix} -1 & 1 \\ 1 & 1 \\ 14 & 5 \\ 1 & 0 \\ 0 & 1 \end{bmatrix} \begin{bmatrix} y_1 \\ y_2 \end{bmatrix} \geq \begin{bmatrix} 0 \\ 10 \\ 70 \\ 0 \\ 0 \end{bmatrix}$

In Exercises 11–16, find the maximum value of P; $x_1 \geq 0$, $x_2 \geq 0$, $x_3 \geq 0$.

11. $P = 7x_1 + 3x_2 + 4x_3$ provided
$$2x_1 + 6x_2 + 2x_3 \leq 12$$

12. $P = 8x_1 + 10x_2 + 20x_3$ provided
$$5x_1 + 14x_2 + 12x_3 \leq 60$$

13. $P = 6x_1 + 4x_2 + 5x_3$ provided
$$2x_1 + 7x_2 + 3x_3 \leq 42$$
$$x_1 + 5x_2 + x_3 \leq 20$$

14. $P = 12x_1 + 7x_2 + 10x_3$ provided
$$3x_1 + 4x_2 + 7x_3 \leq 56$$
$$x_1 + x_2 + 3x_3 \leq 18$$

15. $P = 7x_1 + 4x_2 + 3x_3$ provided
$$2x_1 + 4x_2 + 3x_3 \leq 12$$
$$5x_1 + 5x_2 + 10x_3 \leq 30$$
$$2x_1 + x_2 + 2x_3 \leq 10$$

16. $P = 6x_1 + 5x_2 + 4x_3$ provided
$$x_1 + 2x_2 + x_3 \leq 12$$
$$5x_1 + 10x_2 + x_3 \leq 30$$
$$2x_1 + 6x_2 + 8x_3 \leq 24$$

Find the minimum value of Q in Exercises 17–20.

17. $Q = 3y_1 + 4y_2 + 6y_3$ provided
$$y_1 + 4y_2 + y_3 \geq 6$$
$$y_1 + y_2 + 2y_3 \geq 4$$

18. $Q = 10y_1 + 6y_2 + 8y_3$ provided

$$y_1 + 2y_2 + 3y_3 \geq 12$$
$$5y_1 + 10y_2 + 5y_3 \geq 20$$

19. $Q = 6y_1 + 4y_2$ provided

$$3y_1 + y_2 \geq 6$$
$$2y_1 + 8y_2 \geq 8$$
$$5y_1 + 5y_2 \geq 15$$

20. $Q = 8y_1 + 4y_2$ provided

$$3y_1 + 6y_2 \geq 18$$
$$5y_1 + y_2 \geq 20$$

Exercises 21–23 are based on the following table describing types of baby food (figures per kg).

Type	Protein g	Fat g	Carbohy-drates g	Fiber g	Ash g	Calories	Cost	Profit
Oatmeal	15	8	65	1	3	3,950	$1.20	0.60
Barley	10	4	75	2	2	3,700	$1.10	0.58
Rice	7	6	74	1	5	3,800	$1.00	0.56

21. How should the barley and rice be combined to produce a mixed cereal with no more than 300 g of carbohydrates and no more than 8 g of fiber to maximize profit?

22. How should the three types be combined to produce a mixture with less than 400 g of carbohydrates and less than 24 g of fat but with a maximum amount of protein?

23. How should oatmeal and rice be combined at minimum cost to produce a mixture of at least 15 g ash and 120 g protein?

CHAPTER 7

GAME THEORY

Many of the concepts of finite mathematics come together in the *theory of games* to form a powerful and very new application of mathematics. As its name implies, game theory examines mathematical models for games and gamelike activities. Naturally game theory applies to regular games such as football, baseball, bridge, or poker, but its real goal is far more basic and important. The main body of game theory is involved with the analysis of situations of conflict. As such, it is a dynamic subject, attempting to structure models with which the parties involved can select courses of action that will optimize the possibilities of reaching a desired goal. Such "games" can and in real applications often do include problems related to economic, political, and military situations. In these contexts the "games" become important, and the players are companies, political parties, generals, or even nations. Although we will examine some specific games, the details of the game are not the real factor in game theory. In fact, a complete analysis of a specific game is possible only in the simplest of cases. Because of this, we will be more interested in the general approach to and the general structure of our games and how our analysis can be used to gain insight into real-world problems.

7.1 THE NATURE OF GAMES

Game theory pursues the selection of the "best" strategy as viewed from each of the possible moves of the players and the rewards and penalties that follow.

EXAMPLE Consider the following "game" between the two major political parties. Suppose that the two parties are engaged in a close race for a United States Senate seat in California. Each has $100,000 to invest in a last-minute television campaign in one of the two major TV areas, Los Angeles or San Francisco. Based on past experience, both parties know that each $1,000 they spend more than their opposition in Los Angeles will produce 1,000 votes, while a $1,000 difference in San Francisco will produce only 600 more votes. Describe the various moves and vote "payoffs" in each case.

Solution Each party has three possible moves—it can spend its $100,000 in Los Angeles or in San Francisco, or it can save the $100,000. If party R spends its $100,000 in Los Angeles and so does party D, no one wins extra votes. If party R spends its money in Los Angeles and party D spends it in San Francisco, party R wins an extra 40,000 votes. (It gets 100,000 extra in Los Angeles, while party D gets 60,000 extra in San Francisco, a difference of 40,000 votes.) If party R goes with Los Angeles and party D saves its $100,000, R wins an extra 100,000 votes. If R goes to San Francisco and D to Los Angeles, R loses 40,000 votes. If both go to San Francisco, no one wins extra votes. If R goes to San Francisco and D sits on its hands, R wins 60,000 votes. If R saves its money and D spends in Los Angeles, D wins 100,000 extra votes; if D spends in San Francisco, it wins 60,000 votes; and if D also saves its money, neither wins extra votes. A table can be used to summarize this information.

R's Action	D's Action		
	Spend in LA	Spend in SF	Save money
Spend in LA	- 0 -	R + 40,000 votes	R + 100,000 votes
Spend in SF	D + 40,000 votes	- 0 -	R + 60,000 votes
Save money	D + 100,000 votes	D + 60,000 votes	- 0 -

The example illustrates the key features of a "game" required for game theory analysis.

Because we must limit our discussion to some reasonable size, we will restrict it to two-person games—games with only two players. For reasons which will be clear below, these players are called R and C. Whatever the specific nature of the game involved, it will be assumed that it has the following features:

ASSUMPTION 1: Each player will have the choice of a number of moves. For convenience we will assume that R has m moves, identified as $R_1, R_2, \ldots, R_m$. Player C will have n moves, identified as $C_1, \ldots, C_n$.

ASSUMPTION 2: The game can be reduced to the following basic form: Each player selects a move. The rules of the game then determine the winner and the player's reward or payoff. Further, these payoffs can be assigned a numerical value.

For convenience the requirements of Assumption 2 can be met in the form of payoff matrices. We can state the specifics of a game in terms of two $m \times n$ matrices. We can view the game as player R sees it by considering an $m \times n$ matrix as below:

$$
A = \begin{array}{c} \\ R_1 \\ R_2 \\ \vdots \\ R_m \end{array}
\begin{array}{cccc} C_1 & C_2 & \cdots & C_n \\ \left[\begin{array}{cccc} a_{11} & a_{12} & \cdots & a_{1n} \\ a_{21} & & & \\ \vdots & & & \\ a_{m1} & \cdots\cdots\cdots & & a_{mn} \end{array}\right] \end{array}
$$

Each row of the matrix corresponds to one of player R's m possible moves. They have been labeled R_1 to R_m for identification. Each column corresponds to player C's n possible moves; thus the columns are labeled C_1 to C_n. The actual entries in the matrix correspond to the payoffs involved. Thus a_{ij}, the entry in the ith row and jth column, is the payoff if player R selects move R_i and player C selects move C_j. We shall agree that positive values for a_{ij}, indicate that player R wins an amount $|a_{ij}|$, and negative values indicate that he loses an amount $|a_{ij}|$.

There is a second payoff matrix

$$
\overline{A} = \begin{array}{c} \\ R_1 \\ \vdots \\ R_m \end{array}
\begin{array}{ccc} C_1 & \cdots\cdots & C_n \\ \left[\begin{array}{ccc} \bar{a}_{11} & \cdots\cdots & \bar{a}_{1n} \\ \vdots & & \\ \bar{a}_{m1} & \cdots\cdots & \bar{a}_{mn} \end{array}\right] \end{array}
$$

giving similar information about the game from the point of view of player C. Again we shall agree that positive values of $\bar{a}_{ij}$ correspond to "player C loses," and negative values of $\bar{a}_{ij}$ correspond to C winning an amount $|\bar{a}_{ij}|$. The reason for this seemingly awkward notation for C will become clear below when we consider zero-sum games.

EXAMPLE   Consider the "rock, scissors, and paper" game we have all played. Each player makes either a fist (rock), two fingers (scissors), or an open hand (paper) behind his or her back. At a signal each player exposes her or his hand. The winner is determined by:

"Rock breaks scissors"

"Scissors cut paper"

"Paper covers rock"

Assume that each player bets $1 and construct the payoff matrices for this game.

Solution   For player R the matrix is

$$
\begin{array}{c}
 \\
R_1 \ (rock) \\
R_2 \ (scissors) \\
R_3 \ (paper)
\end{array}
\begin{array}{ccc}
C_1 \ (rock) & C_2 \ (scissors) & C_3 \ (paper) \\
\left[\begin{array}{ccc}
0 & +1 & -1 \\
-1 & 0 & +1 \\
+1 & -1 & 0
\end{array} \right]
\end{array}
$$

For player C the payoff matrix (remember, a negative entry means a winning value for C) is

$$
\begin{array}{c}
 \\
R_1 \ (rock) \\
R_2 \ (scissors) \\
R_3 \ (paper)
\end{array}
\begin{array}{ccc}
C_1 \ (rock) & C_2 \ (scissors) & C_3 \ (paper) \\
\left[\begin{array}{ccc}
0 & +1 & -1 \\
-1 & 0 & +1 \\
+1 & -1 & 0
\end{array} \right]
\end{array}
$$

This example is a special case. The two payoff matrices are identical. That is, player R wins exactly what player C loses, and vice versa. This type of game is called a **zero-sum game**. The example below illustrates a **non-zero-sum game**.

EXAMPLE   A real estate developer has enough funds to buy one of two pieces of land. For identification we shall call these site I and site II. He knows that a major company will build a plant on one of three parcels of land. These are directly on site I, on a site adjacent to site I, and on site II. If the developer buys site I and the plant locates there, he will make $1,000,000. If the plant locates adjacent to site I, he will make $500,000. If the plant locates at site II, he will lose $250,000. On the other hand, if he buys site II and the plant locates at site I, he will lose $500,000. If it locates adjacent to site I, he will lose $100,000. If it locates at site II, he will make $250,000. The outlook of the company building the plant is somewhat different.

If the developer has bought site I, it will cost the company an extra $250,000 to locate there. If they locate adjacent to site I, they will break even. If they locate at site II, they will save $250,000. On the other hand, if the developer has bought site II and the new industry locates at site I, it will save \$500,000. It can locate adjacent to site I at no additional cost or at site II for a loss of \$250,000. Set up this information in a game theory matrix format.

Solution The game matrix for the land developer is

	C_1 (locate at I)	C_2 (locate adjacent to I)	C_3 (locate at II)
R_1 (buy site I)	1,000,000	500,000	−250,000
R_2 (buy site II)	−500,000	−100,000	250,000

For the industry the payoff matrix is

	C_1 (locate at I)	C_2 (locate adjacent to I)	C_3 (locate at II)
R_1 (buy site I)	+250,000	0	−250,000
R_2 (buy site II)	−500,000	0	+250,000

For the most part, non-zero-sum games are too complicated for our discussion, but it is worthwhile to point out that they do exist. For example, we might consider a game in which all the entries in the R matrix are positive and all those in the C matrix negative. This would be a positive-sum game, where both parties would win something (remember, in the C matrix player C "wins" with negative entries). We could also consider a negative-sum game, one in which both parties lose (perhaps a game theory model of war—are there any winners?).

The reader will notice that if we restrict our attention to zero-sum games, we can describe the game with a single matrix. In this case the same matrix describes the payoffs for both players; thus there is a single game matrix. Conversely, any $m \times n$ matrix corresponds to some two-person, zero-sum game. If we assume that any game is played over and over, we might ask how R should play to maximize his or her winning and how C should play to minimize her or his losses (the minimum loss for C would be winning). It is not necessary for both players to be real persons.

EXAMPLE A stock broker plans to invest either in gold or silver stocks. He knows that the federal government plans to take one of three courses of action. First, the government might raise the price of gold. In this case the broker will make \$10,000 if he has invested in gold stock and will lose \$5,000 if he has invested in silver stock. The government might raise the price of silver, in which case he loses \$4,000 if he has invested in gold and makes \$7,500 if he has invested in silver. The government might take no action, in which case the broker will lose \$1,000 no matter which investment he makes. Construct a game matrix for these possibilities.

Solution   The game matrix is given as

	C_1 (government supports gold)	C_2 (government supports silver)	C_3 (no government action)
R_1 (invest in gold)	10,000	−4,000	−1,000
R_2 (invest in silver)	−5,000	7,500	−1,000

Considering this example, who is player C? It is not the government but rather the stock market. In a sense this is a "one-against-the-system" game. We shall see later that a suitable plan or "strategy" can be constructed provided we can make reasonable numerical estimates describing the situation.

The simple act of putting the game in matrix form can give the players an edge in selecting the best move. This is illustrated in the following example.

EXAMPLE   In a certain game the payoff matrix is

	C_1	C_2	C_3
R_1	0	1	3
R_2	−2	0	5
R_3	−6	−7	0

Assume that both players wish to maximize their winnings. Is there a "best" move for each player?

Solution   Consider the game from R's point of view. He would like to win his maximum amount, 5, by selecting row 2 (move R_2). The problem is that he knows C can see this and thus might choose column 1 (move C_1), in which case R would lose 2. Because of C, R must look at the worst thing that might happen in each case. He then lists the greatest loss for himself in each row. From his point of view the payoff matrix appears to be

	C_1	C_2	C_3	Row minimum
R_1	0	1	3	0
R_2	−2	0	5	−2
R_3	−6	−7	0	−7
	0	1	5	Column maximum

From R's view, if he selects row 1 (move R_1), he will at least break even no matter what action C takes.

C's point of view is similar; he would like to select column 2 (move C_2) and would like R to select row 3 (move R_3). In this case he would win 7. But it is not reasonable to assume that R will go along. Rather, C must look at the worst thing that can happen for each choice. C then lists the maximum value of each column at the bottom of each column. C notes that by selecting column 1 (move C_1), he can assure that he will at least break even and will make a profit if R goofs up. Thus R's best move is R_1, and C's best move is C_1. The net effect if both players make their best move is a zero payoff. Later we shall see this as an example of a "fair" game.

EXERCISES 7.1

Determine the game matrix or matrices for each of Exercises 1–6. Also, determine if the game is zero-sum.

1. Each player marks *heads* or *tails* on a sheet of paper. If they match, player R wins \$1 from player C. If they fail to match, player C wins \$1 from player R.

2. Players R and C play roulette without a table. Player R selects to bet either red, black, or green. Without seeing R's choice, C selects red, black, or green. If C selects red or black and they match what R selected, R wins \$1 from C. If C selects green and that matches what R selected, R wins \$18 from C. If the two choices fail to match, C wins \$1 from R.

3. The quarterback of a pro football team must select either a run or a pass play. The defensive captain of the other team will select either a 6-man line or a blitz (special pass rush). If the blitz is called against the pass, there will be a 5-yard loss. If the blitz is called against a running play, there will be a 7-yard gain. The 6-man line can hold a run to no gain, but the 6-man line will allow a 15-yard gain if it is used against the pass.

4. Suppose that the quarterback of Exercise 3 plans to call one of three plays—a run, a short pass, or a long pass. Against the blitz the run makes 7 yards, the short pass gains 10 yards, and the long pass loses 6 yards. Against a 6-man rush the run loses 1 yard, the short pass gains 2 yards, and the long pass makes 35 yards.

5. Two land developers plan to bid on two tracts of land. Developer R knows he will not outbid developer C. If he bids on the same tract as C, he will lose \$1,000. If he bids on a different tract, he will make \$1,500. Developer C will make \$1,000 on either tract.

6. An appliance dealer plans to stock one of two kinds of stereos. He has a special customer who plans to buy one or the other of the two types. If his special customer buys the one he has in stock, the dealer makes \$400. If he has stocked the wrong one, he will make only \$200. The customer will save \$100 if he gets the one in stock and will break even (i.e., not save anything) otherwise.

In Exercises 7–12, for each zero-sum game whose matrix is given, determine the maximum amount R can win and the maximum amount C can win. Also determine if there seems to be a best strategy for R and C.

7.

$$\begin{array}{cc} & C_1 \quad C_2 \\ \begin{array}{c} R_1 \\ R_2 \end{array} & \left[\begin{array}{cc} 1 & -1 \\ -1 & 1 \end{array}\right] \end{array}$$

8.

$$\begin{array}{cc} & C_1 \quad C_2 \\ \begin{array}{c} R_1 \\ R_2 \end{array} & \left[\begin{array}{cc} 1 & 0 \\ -1 & 0 \end{array}\right] \end{array}$$

9.

$$\begin{array}{cc} & C_1 \quad C_2 \quad C_3 \\ \begin{array}{c} R_1 \\ R_2 \end{array} & \left[\begin{array}{ccc} 5 & 0 & -5 \\ 4 & -3 & 2 \end{array}\right] \end{array}$$

10.

$$\begin{array}{cc} & C_1 \quad C_2 \\ \begin{array}{c} R_1 \\ R_2 \\ R_3 \end{array} & \left[\begin{array}{cc} 10 & 0 \\ -4 & 1 \\ -3 & 4 \end{array}\right] \end{array}$$

11.

$$\begin{array}{cc} & C_1 \quad C_2 \quad C_3 \\ \begin{array}{c} R_1 \\ R_2 \\ R_3 \end{array} & \left[\begin{array}{ccc} 0 & 4 & 5 \\ -4 & 0 & 6 \\ -5 & -6 & 0 \end{array}\right] \end{array}$$

12.

$$\begin{array}{cc} & C_1 \quad C_2 \quad C_3 \\ \begin{array}{c} R_1 \\ R_2 \\ R_3 \end{array} & \left[\begin{array}{ccc} 1 & -1 & 0 \\ -1 & 0 & 1 \\ -1 & 2 & -1 \end{array}\right] \end{array}$$

Exercises 13–16 are based on the following non-zero-sum game:

$$\begin{array}{cc} R\text{'s matrix} & C\text{'s matrix} \\ \begin{array}{cc} & C_1 \quad C_2 \\ \begin{array}{c} R_1 \\ R_2 \end{array} & \left[\begin{array}{cc} +1 & -2 \\ +2 & -1 \end{array}\right] \end{array} & \begin{array}{cc} & C_1 \quad C_2 \\ \begin{array}{c} R_1 \\ R_2 \end{array} & \left[\begin{array}{cc} -1 & -2 \\ +2 & +1 \end{array}\right] \end{array} \end{array}$$

13. Is there a pair of moves, one for R and one for C, such that both R and C are winners?

14. Is there a pair of moves such that C and R both lose?

15. If R and C are real-life opponents, what would be the outcome R would most like to see?

16. If R and C are real-life opponents, what outcome would C most like to see?

7.2 STRATEGY AND EXPECTED VALUE

From this point on in our discussion, we will consider only zero-sum games. Any future reference to a game will mean a zero-sum game. Our next step will be to consider the problem of selecting an appropriate strategy.

DEFINITION

A **strategy** for player R will refer to a specific plan to be used by R in choosing his or her moves if the game is repeated a number of times. A strategy for player C is a specific plan for her or his moves.

The definition of *strategy* does not place any restrictions on the procedure for selection. The players may select a strategy which requires them to make the same move each time or one involving a random pattern of moves. In any event, we will assume that any strategy for R can be reduced to a set of m probabilities $p_1, p_2, \ldots, p_m$, where p_i gives the fraction of the time or probability R will use move R_i.

EXAMPLE Consider the vote game described in section 7.1. The table describing the outcome becomes the game matrix. If we let party R become player R and party D become player C, we have a zero-sum game with this payoff matrix:

$$\begin{bmatrix} 0 & 40{,}000 & 100{,}000 \\ -40{,}000 & 0 & 60{,}000 \\ -100{,}000 & -60{,}000 & 0 \end{bmatrix}$$

Define some strategy for players R and C.

Solution One strategy for R might be

$$p_1 = \tfrac{1}{3}, \; p_2 = \tfrac{1}{4}, \; p_3 = \tfrac{5}{12}$$

indicating that R plans a $\tfrac{1}{3}$ probability of using move 1 (spending money in Los Angeles), a $\tfrac{1}{4}$ probability of using move 2 (spending money in San Francisco), and a $\tfrac{5}{12}$ probability of using move 3 (saving his money). A possible strategy for C is

$$q_1 = \tfrac{1}{2}, \; q_2 = \tfrac{1}{2}, \; q_3 = 0$$

indicating that $\tfrac{1}{2}$ the time C will spend in Los Angeles and $\tfrac{1}{2}$ the time he will spend in San Francisco.

For compact notation we can view that a specific strategy corresponds to a row vector:

$$P = [p_1 \quad p_2 \quad \cdots \quad p_m]$$

where $p_i \geq 0 \quad \text{for all} \quad i = 1 \quad \text{to} \quad i = m$

and $p_1 + p_2 + \cdots + p_m = 1$

In the same way any strategy for C corresponds to an n-dimensional column vector Q such that

$$Q = \begin{bmatrix} q_1 \\ q_2 \\ \vdots \\ q_n \end{bmatrix}$$

where $q_j \geq 0 \quad \text{for all} \quad j = 1 \quad \text{to} \quad j = n$

and $q_1 + q_2 + \cdots + q_n = 1$

DEFINITIONS

$P = [p_1 \ \cdots \ p_m]$ represents a **pure strategy** for R if for some i $p_i = 1$ and all other values of p are zero.

$$Q = \begin{bmatrix} q_1 \\ \vdots \\ q_n \end{bmatrix}$$

is a **pure strategy** for C if $q_j = 1$ for some specific j and $q_j = 0$ for all other values of j.

Any strategy for R or C that is not a pure strategy is called a **mixed strategy**.

Notice that as yet nothing has been said about a *best* strategy. That is to come. The concept of a matrix representation of a game together with vectors defining strategies leads to the following concept. Suppose we consider the vector-matrix product

$$PAQ$$

where A is the payoff matrix for some game and P and Q are strategies. Notice that

$$PAQ = [p_1 \ \cdots \ p_m] \begin{bmatrix} a_{11} & \cdots & a_{1n} \\ \cdot & & \\ a_{m1} & \cdots & a_{mn} \end{bmatrix} \begin{bmatrix} q_1 \\ \cdot \\ q_n \end{bmatrix}$$

This is a product of a $1 \times m$ matrix times an $m \times n$ matrix; the result is a $1 \times n$ matrix. This in turn is multiplied by an $n \times 1$ matrix, resulting in a scalar. Specifically,

$$PA = [p_1 \ \cdots \ p_m] \begin{bmatrix} a_{11} & \cdots & a_{1n} \\ \cdot & & \\ a_{m1} & \cdots & a_{mn} \end{bmatrix}$$

$$= [p_1 a_{11} + p_2 a_{21} + p_3 a_{31} + \cdots + p_m a_{m1} \quad p_1 a_{12} + p_2 a_{22}$$
$$+ p_3 a_{32} + \cdots + p_m a_{m2} \quad \cdots \quad p_1 a_{1n} + p_2 a_{2n} + \cdots + p_m a_{mn}]$$

Then $(PA)Q = PA \cdot \begin{bmatrix} q_1 \\ \cdot \\ q_n \end{bmatrix}$

$$= (p_1 a_{11} + p_2 a_{21} + \cdots + p_m a_{m1})q_1$$
$$+ (p_1 a_{12} + p_2 a_{22} + \cdots + p_m a_{m2})q_2 + \cdots$$
$$+ (p_1 a_{1n} + p_2 a_{2n} + \cdots + p_m a_{mn})q_n$$
$$= p_1 a_{11} q_1 + p_2 a_{21} q_1 + p_3 a_{31} q_1 + \cdots + p_m a_{m1} q_1 + \cdots$$
$$+ p_1 a_{1n} q_n + p_2 a_{2n} q_n + \cdots + p_m a_{mn} q_n$$

The reader can verify that this result is also found as $P(AQ)$.

EXAMPLE Consider the "rock, scissors, and paper" game of the previous section. Suppose that player R chooses to be a "rock" $\frac{1}{2}$ of the time and paper $\frac{1}{2}$ of the time. Player C chooses to be rock, paper, and scissors $\frac{1}{3}$ of the time each. Form the strategy vectors P and Q and find PAQ.

Solution

$$P = [\tfrac{1}{2} \quad 0 \quad \tfrac{1}{2}]$$

$$A = \begin{bmatrix} 0 & 1 & -1 \\ -1 & 0 & 1 \\ 1 & -1 & 0 \end{bmatrix}$$

$$Q = \begin{bmatrix} \tfrac{1}{3} \\ \tfrac{1}{3} \\ \tfrac{1}{3} \end{bmatrix}$$

$$PA = [\tfrac{1}{2} \quad 0 \quad \tfrac{1}{2}] \begin{bmatrix} 0 & 1 & -1 \\ -1 & 0 & 1 \\ 1 & -1 & 0 \end{bmatrix}$$

$$= [\tfrac{1}{2}\cdot 0 + 0(-1) + \tfrac{1}{2}\cdot 1 \quad \tfrac{1}{2}\cdot 1 + 0\cdot 0 + \tfrac{1}{2}(-1) \quad \tfrac{1}{2}(-1) + 0\cdot 1 + \tfrac{1}{2}\cdot 0]$$

$$= [\tfrac{1}{2} \quad 0 \quad -\tfrac{1}{2}]$$

$$PAQ = [\tfrac{1}{2} \quad 0 \quad -\tfrac{1}{2}] \begin{bmatrix} \tfrac{1}{3} \\ \tfrac{1}{3} \\ \tfrac{1}{3} \end{bmatrix} = \tfrac{1}{2}(\tfrac{1}{3}) + 0(\tfrac{1}{3}) + (-\tfrac{1}{2})(\tfrac{1}{3}) = 0$$

What is the point of the product PAQ? Consider the individual terms that make it up. A typical one is

$$p_i a_{ij} q_j$$

where i is a number between 1 and m and j is a number between 1 and n. We can rewrite this as

$$p_i q_j a_{ij}$$

Now p_i and q_j are probabilities. Specifically, the product

$$p_i q_j$$

is the probability that R will choose row i and C will choose column j. a_{ij} is the amount R will win *if R chooses R_i and C chooses column j.* Thus $p_i q_j a_{ij}$ is a probability times an amount R will win (or lose if $a_{ij} < 0$). The overall product PAQ is a sum of such products. We have encountered this type of sum before; it is an expected value! PAQ is the expected value of the game from R's point of view.

EXAMPLE In the example above on the "rock, scissors, and paper" game $PAQ = 0$. What does this mean?

Solution $PAQ = 0$ implies that on the average R will break even; she or he has an expected value of zero.

EXAMPLE Suppose that for the "rock, scissors, and paper" game above C discovers R's plan and changes her or his strategy to

$$Q = \begin{bmatrix} 0 \\ 0 \\ 1 \end{bmatrix}$$

Find the expected value of the game.

Solution As before, $PA = [\frac{1}{2} \quad 0 \quad -\frac{1}{2}]$. Then

$$PAQ = [\frac{1}{2} \quad 0 \quad -\frac{1}{2}] \begin{bmatrix} 0 \\ 0 \\ 1 \end{bmatrix}$$

$$= \frac{1}{2}(0) + 0(0) + (-\frac{1}{2})1 = -\frac{1}{2}$$

Clever C, he or she now expects to win $\frac{1}{2}$ a dollar on the average.

EXAMPLE A dealer in pork futures plans to buy one of two lots from a farmer. The farmer can give extra food to one of the lots. The economic results are summarized below:

	Farmer	
Dealer	Feed 1	Feed 2
Buy 1	+1,000	+2,000
Buy 2	−1,000	+1,000

Positive entries indicate increased profit for the dealer. This can be viewed as a two-person game with matrix

$$\begin{bmatrix} 1 & 2 \\ -1 & 1 \end{bmatrix}$$

where the entries represent thousands of dollars. If R uses a strategy $[\frac{3}{4} \quad \frac{1}{4}]$ and C uses $\begin{bmatrix} \frac{2}{3} \\ \frac{1}{3} \end{bmatrix}$, find PAQ.

Solution

$$PA = \begin{bmatrix} \tfrac{3}{4} & \tfrac{1}{4} \end{bmatrix} \begin{bmatrix} 1 & 2 \\ -1 & 1 \end{bmatrix}$$

$$= \begin{bmatrix} \tfrac{3}{4}(1) + \tfrac{1}{4}(-1) & \tfrac{3}{4}(2) + \tfrac{1}{4}(1) \end{bmatrix}$$

$$= \begin{bmatrix} \tfrac{1}{2} & \tfrac{7}{4} \end{bmatrix}$$

$$PAQ = \begin{bmatrix} \tfrac{1}{2} & \tfrac{7}{4} \end{bmatrix} \begin{bmatrix} \tfrac{2}{3} \\ \tfrac{1}{3} \end{bmatrix} = \tfrac{1}{2}(\tfrac{2}{3}) + \tfrac{7}{4}(\tfrac{1}{3})$$

$$= \tfrac{2}{6} + \tfrac{7}{12}$$

$$= \tfrac{11}{12} = 0.91667$$

The expected value of the game for P is $\tfrac{11}{12}$. That is, the dealer will average a net gain of \$916.67.

DEFINITION

For a given game with payoff matrix A and strategies P and Q,

$$EV(P, Q) = PAQ$$

will be called the **expected value of the game.**

EXERCISES 7.2

In Exercises 1–10, for each game with game matrix A and strategies P and Q, find EV(P, Q).

1. $A = \begin{bmatrix} 1 & -1 \\ -1 & 1 \end{bmatrix}$, $P = \begin{bmatrix} \tfrac{1}{2} & \tfrac{1}{2} \end{bmatrix}$, $Q = \begin{bmatrix} \tfrac{1}{2} \\ \tfrac{1}{2} \end{bmatrix}$

2. $A = \begin{bmatrix} 1 & -1 \\ -1 & 1 \end{bmatrix}$, $P = \begin{bmatrix} 1 & 0 \end{bmatrix}$, $Q = \begin{bmatrix} 0 \\ 1 \end{bmatrix}$

3. $A = \begin{bmatrix} 1 & -1 \\ -1 & 1 \end{bmatrix}$, $P = \begin{bmatrix} 0 & 1 \end{bmatrix}$, $Q = \begin{bmatrix} 0 \\ 1 \end{bmatrix}$

4. $A = \begin{bmatrix} 1 & -1 \\ -1 & 1 \end{bmatrix}$, $P = \begin{bmatrix} 0 & 1 \end{bmatrix}$, $Q = \begin{bmatrix} 1 \\ 0 \end{bmatrix}$

5. $A = \begin{bmatrix} 3 & 2 \\ -1 & 2 \end{bmatrix}$, $P = \begin{bmatrix} \tfrac{3}{4} & \tfrac{1}{4} \end{bmatrix}$, $Q = \begin{bmatrix} 1 \\ 0 \end{bmatrix}$

6. $A = \begin{bmatrix} 1 & -2 \\ -2 & 3 \end{bmatrix}$, $P = \begin{bmatrix} \tfrac{1}{3} & \tfrac{2}{3} \end{bmatrix}$, $Q = \begin{bmatrix} \tfrac{3}{4} \\ \tfrac{1}{4} \end{bmatrix}$

7. $A = \begin{bmatrix} 1 & 2 & 1 \\ -2 & 0 & 3 \\ -1 & -3 & -1 \end{bmatrix}$, $P = [\frac{1}{4} \ \frac{1}{2} \ \frac{1}{4}]$, $Q = \begin{bmatrix} \frac{1}{4} \\ \frac{1}{2} \\ \frac{1}{4} \end{bmatrix}$

8. $A = \begin{bmatrix} 1 & 2 & 1 \\ -2 & 0 & 3 \\ -1 & -3 & -1 \end{bmatrix}$, $P = [\frac{1}{3} \ \frac{1}{3} \ \frac{1}{3}]$, $Q = \begin{bmatrix} \frac{2}{3} \\ 0 \\ \frac{1}{3} \end{bmatrix}$

9. $A = \begin{bmatrix} 0 & 2 & 1 & 4 \\ -2 & 0 & 2 & 3 \\ -1 & -2 & 0 & 4 \\ -4 & 3 & -4 & 0 \end{bmatrix}$, $P = [\frac{1}{4} \ \frac{1}{4} \ \frac{1}{4} \ \frac{1}{4}]$, $Q = \begin{bmatrix} \frac{1}{3} \\ \frac{1}{3} \\ 0 \\ \frac{1}{3} \end{bmatrix}$

10. $A = \begin{bmatrix} 0 & 2 & 1 & 4 \\ -2 & 0 & 2 & 3 \\ -1 & -2 & 0 & 4 \\ -4 & 3 & -4 & 0 \end{bmatrix}$, $P = [\frac{1}{2} \ \frac{1}{2} \ 0 \ 0]$, $Q = \begin{bmatrix} \frac{1}{2} \\ \frac{1}{2} \\ 0 \\ 0 \end{bmatrix}$

11. Determine the expected value of a game

$$A = \begin{bmatrix} a_{11} & \cdots & a_{1n} \\ \vdots & & \\ a_{m1} & \cdots & a_{mn} \end{bmatrix}$$

if both R and C select *pure* strategies.

12. For the vote game used as an example in this section, what is the value of PAQ if

$$P = [\frac{1}{2} \ \frac{1}{2} \ 0], \quad Q = \begin{bmatrix} 1 \\ 0 \\ 0 \end{bmatrix}$$

Exercises 13 and 14 are based on the following data:

A farmer can choose to plant wheat or corn. His profit will depend on which crop the government plans to support. The economic results are described in the following table, given in terms of the farmer's profit (in dollars):

Farmer Plants	Government Supports		
	Wheat	Corn	Neither
Wheat	5,000	4,000	3,000
Corn	3,000	4,000	2,000

13. Assume that 50% of the time he chooses to plant wheat and 50% of the time corn, while one year out of three the government supports wheat, one year out of three it supports corn, and one year out of three it supports neither. Construct strategy vectors describing the problem.

14. For the vectors in Exercise 13, what is the "value" of PAQ?

Exercises 15 and 16 are based on the following data:

A pro quarterback is given three plays to call—a run, a short pass, and a long pass—by his coach. He is told to call each $\frac{1}{3}$ of the time. The defensive captain of the other team can call a blitz (special pass rush) or use his regular defense. He is told by his coach to call the blitz $\frac{1}{4}$ of the time. Against a run the blitz will usually yield a 5-yard gain; against a short pass it yields a gain of 15 yards; and against a long pass it usually holds the other team to no gain. The regular defense can hold a run to 1 yard and a short pass to 3 yards, but on the average the regular defense loses 20 yards to the long pass.

15. Construct the game matrix and strategy vectors for this game.

16. What is the expected value of this game?

7.3 THE VALUE OF A GAME

The next step in the analysis of games is to inquire into the existence of a "best" strategy for the players. For R the best strategy, $\bar{P}$, is such that his winnings are as large as possible, no matter what strategy C chooses. From C's point of view, the optimum strategy, $\bar{Q}$, will result in the lowest losses for C regardless of which strategy R chooses.

DEFINITIONS

If for a game with payoff matrix A, out of all the possible values for the strategy vectors P and Q there exist specific vectors $\bar{P}$ and $\bar{Q}$ and a real number v such that

$$\bar{P}AQ \geq v \text{ for all possible choices for } Q$$

and

$$PA\bar{Q} \leq v \text{ for all possible choices for } P$$

then $\bar{P}$ and $\bar{Q}$ are called **optimum** or **best strategies** for the game, and v is called the **value of the game.**

The definition does not say how to find values for $\overline{P}$ and $\overline{Q}$, nor does it claim that such values must exist or that they are unique. It only provides a test for such values and specifies the properties they must have if they do exist. We will see that based on these properties we can develop an understanding of their nature sufficient to construct methods for finding them. The question of their existence is considered in the Fundamental Theorem of Game Theory.

FUNDAMENTAL THEOREM OF GAME THEORY

For every matrix game A there exist strategies $\overline{P}$ and $\overline{Q}$ and a real number v such that

$$\overline{P}AQ \geq v \text{ for all possible choices for } Q$$

and

$$PA\overline{Q} \leq v \text{ for all possible choices for } P$$

Nothing in the Fundamental Theorem of Game Theory implies that $\overline{P}$ and $\overline{Q}$ are unique, only that they exist.

DEFINITION

If the value of the game is zero, then the game is called a **fair game**.

While we are not going to prove the Fundamental Theorem of Game Theory, we can and will prove some important theorems about $\overline{P}$ and $\overline{Q}$. This will seem a little abstract at times, but these theorems will be key factors in the development of methods we can use to find optimum strategies, which we will consider in subsequent sections.

THEOREM

$$EV(\overline{P}, \overline{Q}) = v$$

Proof: By the definition, $\overline{P}$ is such that

$$\overline{P}AQ \geq v \text{ for all possible } Q \text{ including } \overline{Q}$$

Thus $\overline{P}A\overline{Q} \geq v$, which is equivalent to $v \leq \overline{P}A\overline{Q}$.

By the definition of $\bar{Q}$,

$$PA\bar{Q} \leq v \quad \text{for all possible} \quad P \quad \text{including} \quad \bar{P}$$

Thus

$$\bar{P}A\bar{Q} \leq v$$

Therefore,

$$v \leq \bar{P}A\bar{Q} \quad \text{and} \quad \bar{P}A\bar{Q} \leq v$$

The only way this can happen is if

$$\bar{P}A\bar{Q} = v$$

By definition,

$$EV(\bar{P}, \bar{Q}) = \bar{P}A\bar{Q} = v$$

THEOREM

Let A be the $m \times n$ payoff matrix of a game with optimum strategies $\bar{P}$ and $\bar{Q}$. Let v be any number and let

$$V = [v \quad \cdots \quad v] \text{ be an } n\text{-dimensional row vector}$$

and

$$V^T = \begin{bmatrix} v \\ \vdots \\ v \end{bmatrix} \text{ be an } m\text{-dimensional column vector}$$

Then v is the value of the game if and only if

$$\bar{P}A \geq V \quad \text{and} \quad A\bar{Q} \leq V^T$$

Proof: First, let us prove that if v is the value of the game

$$\bar{P}A \geq V \quad \text{and} \quad A\bar{Q} \leq V^T$$

Let $\bar{P} = [\bar{p}_1 \quad \cdots \quad \bar{p}_m]$ since $\bar{P}AQ \geq v$ for all Q. Let

$$Q = \begin{bmatrix} 1 \\ 0 \\ \vdots \\ 0 \end{bmatrix}$$

a pure strategy. Then

$$\bar{P}AQ \geq v$$

implies

$$\bar{p}_1 a_{11} + \bar{p}_2 a_{21} + \cdots + \bar{p}_m a_{m1} \geq v$$

If we let

$$Q = \begin{bmatrix} 0 \\ 1 \\ 0 \\ \vdots \\ 0 \end{bmatrix}$$

$\bar{P}AQ \geq v$ implies

$$\bar{p}_1 a_{12} + \bar{p}_2 a_{22} + \cdots + \bar{p}_m a_{m2} \geq v$$

In general, if we let

$$Q = \begin{bmatrix} 0 \\ \vdots \\ 0 \\ 1 \\ 0 \\ \vdots \\ 0 \end{bmatrix}$$

a pure strategy with a 1 in the ith row, $\bar{P}AQ \geq v$ implies

$$\bar{p}_1 a_{1i} + \bar{p}_2 a_{2i} + \cdots + \bar{p}_m a_{mi} \geq v$$

However, expressions of this type are the components of the row vector

$$\bar{P}A$$

Thus component by component each one of $\bar{P}A$ is greater than v. In vector notation,

$$\bar{P}A \geq V$$

The proof that $A\bar{Q} \leq V^T$ is similar. Now we must prove that if

$$\bar{P}A \geq V \quad \text{and} \quad A\bar{Q} \leq V^T$$

v is the value of the game. Consider

$$\bar{Q} = \begin{bmatrix} \bar{q}_1 \\ \vdots \\ \bar{q}_n \end{bmatrix}$$

$$V\bar{Q} = \begin{bmatrix} v & \cdots & v \end{bmatrix} \begin{bmatrix} \bar{q}_1 \\ \vdots \\ \bar{q}_n \end{bmatrix} = v\bar{q}_1 + v\bar{q}_2 + \cdots + v\bar{q}_n$$

$$= v(\bar{q}_1 + \bar{q}_2 + \cdots + \bar{q}_n) = v \cdot 1 = v$$

In a similar manner,

$$\bar{P}V^T = v$$

Thus since

$$\bar{P}A \geq V$$
$$\bar{P}A\bar{Q} \geq V\bar{Q} = v$$

which is equivalent to

$$v \leq \bar{P}A\bar{Q}$$

Since

$$A\bar{Q} \le V^T$$
$$\bar{P}A\bar{Q} \le \bar{P}V^T = v$$

or

$$v \le \bar{P}A\bar{Q} \le v$$

but only if

$$\bar{P}A\bar{Q} = v$$

Thus v is the value of the game.

THEOREM

The value of a game is unique.

Proof: Assume that a game with matrix A has two values v and u. Then let $\bar{P}_1$ and $\bar{Q}_1$ be the optimum strategies associated with game value v and let $\bar{P}_2$ and $\bar{Q}_2$ be those associated with u. Applying the various definitions, we have:

$$\bar{P}_1AQ \ge v \quad \text{for any } Q \tag{1}$$
$$PA\bar{Q}_1 \le v \quad \text{for any } P \tag{2}$$
$$\bar{P}_2AQ \ge u \quad \text{for any } Q \tag{3}$$
$$PA\bar{Q}_2 \le u \quad \text{for any } P \tag{4}$$

If we replace Q in (1) by $\bar{Q}_2$, we have

$$\bar{P}_1A\bar{Q}_2 \ge v \quad \text{or} \quad v \le \bar{P}_1A\bar{Q}_2$$

If we replace P in (4) by $\bar{P}_1$, we have

$$\bar{P}_1A\bar{Q}_2 \le u$$

Thus

$$v \le \bar{P}_1A\bar{Q}_2 \le u \quad \text{or} \quad v \le u$$

On the other hand, if we replace P in (2) by $\bar{P}_2$, we have

$$\bar{P}_2A\bar{Q}_1 \le v$$

and if we replace Q in (3) by $\bar{Q}_1$, we have

$$\bar{P}_2A\bar{Q}_1 \ge u \quad \text{or} \quad u \le \bar{P}_2A\bar{Q}_1$$

Thus

$$u \le \bar{P}_2A\bar{Q}_1 \le v \quad \text{or} \quad u \le v$$

Therefore, we have at the same time

$$u \le v \quad \text{and} \quad v \le u$$

which implies that

$$u = v$$

The value of the game is unique.

EXAMPLE   Assume that the game

$$A = \begin{bmatrix} 1 & 4 \\ 2 & -2 \end{bmatrix}$$

has one pair of optimum strategies

$$\bar{P} = [\tfrac{4}{7} \quad \tfrac{3}{7}], \quad \bar{Q} = \begin{bmatrix} \tfrac{6}{7} \\ \tfrac{1}{7} \end{bmatrix}$$

If $\bar{P}_1$ and $\bar{Q}_1$ are a second pair of optimum strategies, what is the value of $\bar{P}_1 A \bar{Q}_1$?

Solution   Since $\bar{P}_1 A \bar{Q}_1 = v$, the unique value of the game,

$$\bar{P}_1 A \bar{Q}_1 = \bar{P} A \bar{Q} = [\tfrac{4}{7} \quad \tfrac{3}{7}] \begin{bmatrix} 1 & 4 \\ 2 & -2 \end{bmatrix} \begin{bmatrix} \tfrac{6}{7} \\ \tfrac{1}{7} \end{bmatrix}$$

$$= [\tfrac{10}{7} \quad \tfrac{10}{7}] \begin{bmatrix} \tfrac{6}{7} \\ \tfrac{1}{7} \end{bmatrix}$$

$$= \tfrac{60}{49} + \tfrac{10}{49} = \tfrac{70}{49}$$

THEOREM

Given a game with payoff matrix A, value v, and optimum strategies $\bar{P}$ and $\bar{Q}$, then the game kA, where k is any positive constant, is such that its value is kv and $\bar{P}$ and $\bar{Q}$ are optimum strategies for it.

Proof: Since $\bar{P}$ is an optimum strategy for A,

$$\bar{P} A Q \geq v \quad \text{for all} \quad Q$$

$$k\bar{P} A Q \geq kv \quad \text{for all} \quad Q$$

$$\bar{P}(kA)Q \geq kv \quad \text{for all} \quad Q$$

Since every strategy for kA is a strategy for A and vice versa, we can conclude that $\bar{P}$ meets the requirements of an optimum of kA and kv is not bigger than the value of kA. From player C's point of view,

$$P(kA)\bar{Q} \leq kv \quad \text{for all} \quad P$$

and $\bar{Q}$ meets the requirements for an optimum strategy for kA. Also, kv is not smaller than the value of kA. Thus kv must be the value of kA.

EXAMPLE If optimum strategies for the game $A = \begin{bmatrix} 1 & 4 \\ 2 & -2 \end{bmatrix}$ are

$\bar{P} = [\frac{4}{7} \ \frac{3}{7}]$ and $\bar{Q} = \begin{bmatrix} \frac{6}{7} \\ \frac{1}{7} \end{bmatrix}$, with $v = \frac{70}{49} = \frac{10}{7}$, show that for the game

$A' = \begin{bmatrix} 49 & 196 \\ 98 & -98 \end{bmatrix}$, i.e., $49A$, $\bar{P}$ and $\bar{Q}$ are such that $\bar{P}A'\bar{Q} = 49 \cdot v = 49(\frac{70}{49})$

$= 70.$

Solution

$$\bar{P}A'\bar{Q} = [\frac{4}{7} \ \frac{3}{7}]\begin{bmatrix} 49 & 196 \\ 98 & -98 \end{bmatrix}\begin{bmatrix} \frac{6}{7} \\ \frac{1}{7} \end{bmatrix}$$

$$= [\frac{4}{7}\cdot 49 + \frac{3}{7}\cdot 98 \quad \frac{4}{7}\cdot 196 + \frac{3}{7}\cdot(-98)]\begin{bmatrix} \frac{6}{7} \\ \frac{1}{7} \end{bmatrix}$$

$$= [70 \quad 70]\begin{bmatrix} \frac{6}{7} \\ \frac{1}{7} \end{bmatrix}$$

$$= 60 + 10$$

$$= 70$$

THEOREM

Consider a given $m \times n$ matrix game A, with value v and optimum strategies $\bar{P}$ and $\bar{Q}$. Then for a positive number k, the $m \times n$ matrix game

$$\begin{bmatrix} a_{11} + k & a_{12} + k & \cdots & a_{1n} + k \\ \vdots & & & \\ a_{m1} + k & \cdots\cdots\cdots\cdots & a_{mn} + k \end{bmatrix}$$

has optimum strategies $\bar{P}$ and $\bar{Q}$ and value $v + k$.

Proof: We can define an $m \times n$ matrix K:

$$K = \begin{bmatrix} k & \cdots & k \\ \vdots & & \\ k & \cdots & k \end{bmatrix}$$

The game matrix described above becomes

$$A + K$$

Then since $A + K$ is an $m \times n$ matrix, any strategy for A is one for $A + K$, and vice versa. Now

$$\bar{P}(A + K)Q = \bar{P}AQ + \bar{P}KQ \geq v + \bar{P}KQ$$

Let us estimate $\bar{P}KQ$:

$$\bar{P}K = [\bar{p}_1 \quad \cdots \quad \bar{p}_m]\begin{bmatrix} k & \cdots & k \\ \vdots & & \\ k & \cdots & k \end{bmatrix}$$

$$= [k\bar{p}_1 + k\bar{p}_2 + \cdots + k\bar{p}_m \quad k\bar{p}_1 + \cdots + k\bar{p}_m \cdots k\bar{p}_1 + \cdots + k\bar{p}_m]$$

But

$$k\bar{p}_1 + \cdots + k\bar{p}_m = k(\bar{p}_1 + \cdots + \bar{p}_m)$$
$$= k(1) = k$$

Thus

$$\bar{P}K = [k \quad \cdots \quad k]$$

For any Q,

$$\bar{P}KQ = [k \quad \cdots \quad k]\begin{bmatrix} q_1 \\ \vdots \\ q_n \end{bmatrix}$$

$$= kq_1 + kq_2 + \cdots + kq_n$$
$$= k(q_1 + q_2 + \cdots + q_n)$$
$$= k$$

Therefore,

$$\bar{P}(A + K)Q \geq v + k \quad \text{for all possible} \quad Q$$

In a similar manner,

$$P(A + K)\bar{Q} \leq v + k \quad \text{for all possible} \quad P$$

Therefore, $\bar{P}$ and $\bar{Q}$ are optimum strategies for $A + K$, and $v + k$ is its value.

This latter theorem will be useful in applying linear programming methods to find optimum strategies.

EXAMPLE   Assume that the game

$$A = \begin{bmatrix} 1 & 4 \\ 2 & -2 \end{bmatrix}$$

has optimum strategies

$$\bar{P} = [\tfrac{4}{7} \quad \tfrac{3}{7}] \quad \text{and} \quad \bar{Q} = \begin{bmatrix} \tfrac{6}{7} \\ \tfrac{1}{7} \end{bmatrix}, \quad \text{with value } \tfrac{70}{49}$$

Show that for

$$A' = \begin{bmatrix} 1 & 4 \\ 2 & -2 \end{bmatrix} + \begin{bmatrix} 5 & 5 \\ 5 & 5 \end{bmatrix}$$

$$= \begin{bmatrix} 6 & 9 \\ 7 & 3 \end{bmatrix}$$

$$\bar{P}A'\bar{Q} = \tfrac{70}{49} + 5 = \tfrac{315}{49}$$

Solution

$$\bar{P}A'\bar{Q} = [\tfrac{4}{7} \quad \tfrac{3}{7}]\begin{bmatrix} 6 & 9 \\ 7 & 3 \end{bmatrix}\begin{bmatrix} \tfrac{6}{7} \\ \tfrac{1}{7} \end{bmatrix}$$

$$= [\tfrac{24}{7} + \tfrac{21}{7} \quad \tfrac{36}{7} + \tfrac{9}{7}]\begin{bmatrix} \tfrac{6}{7} \\ \tfrac{1}{7} \end{bmatrix}$$

$$= [\tfrac{45}{7} \quad \tfrac{45}{7}]\begin{bmatrix} \tfrac{6}{7} \\ \tfrac{1}{7} \end{bmatrix}$$

$$= \tfrac{270}{49} + \tfrac{45}{49}$$

$$= \tfrac{315}{49}$$

EXERCISES 7.3

For each of the matrix games given in Exercises 1–10, assume that $\bar{P}$ and $\bar{Q}$ are optimum strategies and use the fact that $\bar{P}A\bar{Q} = v$ to find the value of the game. Also verify by direct calculation that $\bar{P}A \geq V$ and $A\bar{Q} \leq V^T$.

1. $A \stackrel{\text{!}}{=} \begin{bmatrix} 1 & 2 \\ -1 & 0 \end{bmatrix}$, $\bar{P} = [1 \quad 0]$, $\bar{Q} = \begin{bmatrix} 1 \\ 0 \end{bmatrix}$

2. $A = \begin{bmatrix} 3 & -1 \\ 2 & -2 \end{bmatrix}$, $\bar{P} = [1 \quad 0]$, $\bar{Q} = \begin{bmatrix} 0 \\ 1 \end{bmatrix}$

3. $A = \begin{bmatrix} 1 & 1 & 2 \\ -2 & 0 & -3 \\ 1 & -1 & 2 \end{bmatrix}$, $\bar{P} = [1 \quad 0 \quad 0]$, $\bar{Q} = \begin{bmatrix} 1 \\ 0 \\ 0 \end{bmatrix}$

4. $A = \begin{bmatrix} 1 & 1 & 2 \\ -2 & 0 & -3 \\ 1 & -1 & 2 \end{bmatrix}$, $\bar{P} = [1 \quad 0 \quad 0]$, $\bar{Q} = \begin{bmatrix} 0 \\ 1 \\ 0 \end{bmatrix}$

5. $A = \begin{bmatrix} 0 & -2 & 1 & 3 \\ 4 & -3 & -4 & 1 \\ -5 & -4 & 3 & 1 \end{bmatrix}$, $\bar{P} = [1 \quad 0 \quad 0]$, $\bar{Q} = \begin{bmatrix} 0 \\ 1 \\ 0 \\ 0 \end{bmatrix}$

6. $A = \begin{bmatrix} 3 & 2 & 1 \\ -1 & 2 & -2 \\ 3 & -4 & -3 \\ -1 & 1 & -4 \end{bmatrix}$, $\bar{P} = [1 \quad 0 \quad 0 \quad 0]$, $\bar{Q} = \begin{bmatrix} 0 \\ 0 \\ 1 \end{bmatrix}$

7. $A = \begin{bmatrix} 1 & -1 \\ -1 & 1 \end{bmatrix}$, $\bar{P} = [\tfrac{1}{2} \quad \tfrac{1}{2}]$, $\bar{Q} = \begin{bmatrix} \tfrac{1}{2} \\ \tfrac{1}{2} \end{bmatrix}$

8. $A = \begin{bmatrix} 3 & 2 & 2 \\ 2 & 3 & 2 \\ 2 & 2 & 3 \end{bmatrix}$, $\bar{P} = [\frac{1}{3} \quad \frac{1}{3} \quad \frac{1}{3}]$, $\bar{Q} = \begin{bmatrix} \frac{1}{3} \\ \frac{1}{3} \\ \frac{1}{3} \end{bmatrix}$

9. $A = \begin{bmatrix} 3 & -1 & 0 \\ 2 & -2 & 4 \\ -3 & 4 & 5 \end{bmatrix}$, $\bar{P} = [\frac{7}{11} \quad 0 \quad \frac{4}{11}]$, $\bar{Q} = \begin{bmatrix} \frac{5}{11} \\ \frac{6}{11} \\ 0 \end{bmatrix}$

10. $A = \begin{bmatrix} 3 & 2 & 3 \\ 0 & 1 & 2 \\ 1 & 4 & 2 \end{bmatrix}$, $\bar{P} = [\frac{3}{4} \quad 0 \quad \frac{1}{4}]$, $\bar{Q} = \begin{bmatrix} \frac{1}{2} \\ \frac{1}{2} \\ 0 \end{bmatrix}$

11. Considering Exercises 3 and 4 above, are $\bar{P}$ and $\bar{Q}$ unique?

Use the previous exercises to find the optimum strategies and value of the game for the games in Exercises 12–15. (Hint: A and kA are related.)

12. $A = \begin{bmatrix} 3 & 6 \\ -3 & 0 \end{bmatrix}$
13. $A = \begin{bmatrix} 6 & -2 \\ 4 & -4 \end{bmatrix}$

14. $A = \begin{bmatrix} 2 & 2 & 3 \\ -1 & 1 & -2 \\ 2 & 0 & 3 \end{bmatrix}$
15. $A = \begin{bmatrix} 5 & 4 & 3 \\ 1 & 4 & 0 \\ 5 & -2 & -1 \\ 1 & 3 & -2 \end{bmatrix}$

In Exercises 16 and 17, verify by direct calculation that the new matrix game found by adding a positive amount k to the elements of a matrix with game value of v and using the same optimum strategies has game value v + k.

16. Form a new game by adding 3 to every element of the matrix in Exercise 8.

17. Form a new game by adding 2 to every element of the matrix of Exercise 6.

7.4 STRICTLY DETERMINED GAMES

Before we attempt to develop procedures for finding optimum strategies using linear programming methods, it will be instructive to consider two special cases, strictly determined games and 2 × 2 games.

DEFINITION

An $m \times n$ matrix game is **strictly determined** provided it has an entry which is a minimum in its row and a maximum in its column. The element is called a **saddle point**.

EXAMPLE Determine if the matrix

$$A = \begin{bmatrix} 0 & -2 & 1 & 3 \\ 4 & -3 & -4 & 1 \\ -5 & -4 & 3 & 1 \end{bmatrix}$$

is a strictly determined game.

Solution We can circle the minimum of each row and place a box around the maximum of each column:

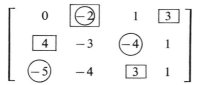

Notice that the -2 entry in row 1, column 2 is both a column maximum and a row minimum. It is a saddle point.

Why is a saddle point of interest? Consider the game in the above example. If R always chooses row 1, he or she is assured that his or her result will be greater than or equal to -2. For any other choice of a row, R may be less fortunate. Thus it is clear that if R chooses

$$\bar{P} = [1 \quad 0 \quad 0]$$

$$\bar{P}AQ \geq -2$$

no matter what choice C makes for Q. From C's point of view, he or she can achieve a similar result by letting

$$\bar{Q} = \begin{bmatrix} 0 \\ 1 \\ 0 \\ 0 \end{bmatrix}$$

In this case $PA\bar{Q} \leq -2$ for any possible choices for P.

It follows from the definition of optimum strategies that $\bar{P}$ and $\bar{Q}$ as given are optimum and that -2 is the value of the game.

EXAMPLE Does

$$A = \begin{bmatrix} 3 & 1 & -2 \\ -4 & 3 & 0 \\ -1 & 4 & 2 \end{bmatrix}$$

define a strictly determined game?

Solution

$$A = \begin{bmatrix} \boxed{3} & 1 & \widehat{-2} \\ \widehat{-4} & 3 & 0 \\ \widehat{-1} & \boxed{4} & \boxed{2} \end{bmatrix}$$

Since no row minimum is also a column maximum, there is no saddle point, and the game is not strictly determined.

Any game with a saddle point is then strictly determined both in a formal, defined sense and because the strategies of the players are fixed in that player R will choose the row of the saddle point and C will choose the column of the saddle point. This is formally stated in the following theorem.

THEOREM

Let a_{ij} be a saddle point of an $m \times n$ matrix game A. An optimum strategy for R is to always pick the ith row and for C is to always to pick the jth column. Further, a_{ij} is the value of the game.

EXAMPLE Find, if possible, the optimum strategies for

$$A = \begin{bmatrix} 1 & 1 & 2 \\ -2 & 0 & -3 \\ 1 & -1 & 2 \end{bmatrix}$$

Solution Circling row minimums and boxing column maximums of A, we have

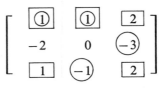

It appears that there are two saddle points, the 1 in row 1, column 1, and the 1 in row 1, column 2. Notice that none of the results and theorems above imply or require a single saddle point. We may take $\bar{P} = [1 \quad 0 \quad 0]$ and

$$\bar{Q}_1 = \begin{bmatrix} 1 \\ 0 \\ 0 \end{bmatrix} \quad \text{or} \quad \bar{Q}_2 = \begin{bmatrix} 0 \\ 1 \\ 0 \end{bmatrix}$$

as optimum strategies for C.

The following theorem points out that the value of a saddle point is unique, even though there may be several.

THEOREM

If for an $m \times n$ matrix game A, a_{11} and a_{ij} are both saddle points, then

$$a_{11} = a_{ij} = a_{1j} = a_{i1}$$

Proof: Consider A with the two saddle points circled:

$$A = \begin{bmatrix} \boxed{a_{11}} & \cdots & a_{1j} & \cdots & a_{1n} \\ \vdots & & & & \\ a_{i1} & \cdots & \boxed{a_{ij}} & \cdots & a_{in} \\ \cdot & & & & \\ a_{m1} & \cdots & a_{mj} & \cdots & a_{mn} \end{bmatrix}$$

Since a_{11} is a saddle point, it is a row minimum value. Thus

$$a_{11} \leq a_{1j}$$

Since a_{ij} is a saddle point, it is a column maximum. Thus

$$a_{ij} \geq a_{1j}$$

which is equivalent to

$$a_{1j} \leq a_{ij}$$

Thus

$$a_{11} \leq a_{1j} \leq a_{ij}$$

which implies

$$a_{11} \leq a_{ij}$$

On the other hand, we can also change our view as follows. Since a_{11} is a saddle point, it is a column maximum. Thus

$$a_{11} \geq a_{i1} \quad \text{or equivalently,} \quad a_{i1} \leq a_{11}$$

Since a_{ij} is a saddle point, it is a row minimum:

$$a_{ij} \leq a_{i1}$$

Thus

$$a_{ij} \leq a_{i1} \quad \text{and} \quad a_{i1} \leq a_{11}$$

which can be stated

$$a_{ij} \leq a_{i1} \leq a_{11}$$

So

$$a_{ij} \leq a_{11}$$

Summarizing both results,

$$a_{ij} \leq a_{11} \quad \text{and} \quad a_{11} \leq a_{ij}$$

we can conclude that

$$a_{ij} = a_{11}$$

Also, since

$$a_{ij} \leq a_{i1} \leq a_{11} \quad \text{and} \quad a_{11} \leq a_{1j} \leq a_{ij}$$

we can conclude that

$$a_{ij} = a_{i1} = a_{11} = a_{1j}$$

EXERCISES 7.4

Determine which of the games in Exercises 1–12 are strictly determined and find optimum strategies for those which are.

1. $A = \begin{bmatrix} 1 & 2 \\ -1 & 0 \end{bmatrix}$ 　　　　　　　　　　　**2.** $A = \begin{bmatrix} 5 & -5 \\ -2 & 4 \end{bmatrix}$

3. $A = \begin{bmatrix} 1 & -2 \\ -1 & 0 \end{bmatrix}$ 　　　　　　　　　**4.** $A = \begin{bmatrix} 1 & -2 \\ -3 & 4 \end{bmatrix}$

5. $A = \begin{bmatrix} 1 & -3 & 0 \\ -4 & -7 & 6 \\ 8 & -7 & -3 \end{bmatrix}$ 　　　　　**6.** $A = \begin{bmatrix} 9 & 1 & 4 \\ 10 & 11 & 7 \\ 3 & 6 & 4 \end{bmatrix}$

7. $A = \begin{bmatrix} -5 & -15 & -7 & -2 \\ -6 & -9 & -7 & -1 \end{bmatrix}$ 　　**8.** $A = \begin{bmatrix} 5 & 2 & 3 & 2 \\ 4 & -1 & 7 & 1 \\ 6 & 0 & -2 & -1 \end{bmatrix}$

9. $A = \begin{bmatrix} 1 & 1 & 1 & 1 \\ 0 & -1 & 1 & -2 \\ -1 & -2 & 0 & -3 \end{bmatrix}$ 　　**10.** $A = \begin{bmatrix} -2 & -1 & 4 \\ -3 & 5 & 2 \\ -2 & 4 & 3 \\ -6 & 6 & 1 \end{bmatrix}$

11. $A = \begin{bmatrix} 1 & 4 & 6 \\ 2 & -2 & 1 \end{bmatrix}$

12. $A = \begin{bmatrix} 1 & 2 \\ 3 & 4 \end{bmatrix}$

13. Suppose that $A = \begin{bmatrix} a & b \\ c & d \end{bmatrix}$ is any 2×2 matrix game. If a, b, c, and d are single digits chosen at random, what are the conditions on a, b, c, and d which would result in A being strictly determined?

14. A razor blade company plans to run ads on one of two television networks. The effect of their choice will be modified by the type of ads carried out by its major competitor. The economic data is given in the following table (the figures are for additional sales in hundreds of thousands of dollars):

Company G	Company S		
	Newspapers	Radio	TV
Network 1	5	3	4
Network 2	2	1	-5

Is this a strictly determined game, and if so, what are the optimum strategies?

15. A senator plans to support one of two tax bills. A senator of the opposite party also plans to support one of the two bills. Each can gain support for conflicting bills of their own according to the following table:

Senator T	Senator C	
	Support bill 1	Support bill 2
Support bill 1	1 extra vote	5 extra votes
Support bill 2	0 extra votes	-2 extra votes

How should each senator act?

7.5 2 × 2 GAMES

In this section we will attack the problem of determining optimum strategies for 2×2 games and games that can be reduced to equivalent 2×2 games. Rather than using subscript notation for the game payoffs, it will be more convenient to refer to them as a, b, c, and d. That is, we will consider an abstract 2×2 game as

$$A = \begin{bmatrix} a & b \\ c & d \end{bmatrix}$$

It will be useful to state two simple theorems about 2×2 games. We have omitted the proofs, but the reader can provide them by considering specific examples of what happens.

THEOREM

In any 2×2 matrix game, if the two entries in either row or in either column are equal, the game is strictly determined.

THEOREM

$A = \begin{bmatrix} a & b \\ c & d \end{bmatrix}$ is a *non–strictly determined* game if and only if one of the following conditions holds:

    1. $a < b, a < c$ and $d < b, d < c$

    2. $b < a, b < d$ and $c < a, c < d$

The above theorem plays a key role in proving the theorem below.

THEOREM

Let

$$A = \begin{bmatrix} a & b \\ c & d \end{bmatrix}$$

be a non–strictly determined game. Then

$$\overline{P} = [\overline{p}_1 \quad \overline{p}_2]$$

with

$$\overline{p}_1 = \frac{d - c}{a + d - b - c} \quad \text{and} \quad \overline{p}_2 = \frac{a - b}{a + d - b - c}$$

$$\overline{Q} = \begin{bmatrix} \overline{q}_1 \\ \overline{q}_2 \end{bmatrix}$$

with

$$\bar{q}_1 = \frac{d - b}{a + d - b - c} \quad \text{and} \quad \bar{q}_2 = \frac{a - c}{a + d - b - c}$$

are optimum strategies, and the value of the game is

$$v = \frac{ad - bc}{a + d - b - c}$$

Proof: We will not discuss where these formulas come from but will consider the relatively simple matter of showing them to be valid.

From the previous theorem we can see that

$$a + d - b - c \neq 0$$

Let Q be any strategy for C. Since

$$Q = \begin{bmatrix} q_1 \\ q_2 \end{bmatrix} \quad \text{and} \quad q_1 + q_2 = 1$$

we can write every Q in the form

$$Q = \begin{bmatrix} q \\ 1 - q \end{bmatrix}, \quad \text{where} \quad 0 \leq q \leq 1$$

Consider

$$\bar{P}A = \begin{bmatrix} \dfrac{d - c}{a + d - b - c} & \dfrac{a - b}{a + d - b - c} \end{bmatrix} \begin{bmatrix} a & b \\ c & d \end{bmatrix}$$

$$= \begin{bmatrix} \dfrac{a(d - c)}{a + d - b - c} + \dfrac{c(a - b)}{a + d - b - c} & \dfrac{b(d - c)}{a + d - b - c} + \dfrac{d(a - b)}{a + d - b - c} \end{bmatrix}$$

$$= \begin{bmatrix} \dfrac{ad - ac + ac - cb}{a + d - b - c} & \dfrac{bd - bc + ad - bd}{a + d - b - c} \end{bmatrix}$$

$$= \begin{bmatrix} \dfrac{ad - cb}{a + d - b - c} & \dfrac{ad - cb}{a + d - b - c} \end{bmatrix}$$

Therefore,

$$\bar{P}AQ = \begin{bmatrix} \dfrac{ad - bc}{a + d - b - c} & \dfrac{ad - bc}{a + d - b - c} \end{bmatrix} \begin{bmatrix} q \\ 1 - q \end{bmatrix}$$

$$= q \left(\frac{ad - bc}{a + d - b - c} \right) + (1 - q) \left(\frac{ad - bc}{a + d - b - c} \right)$$

$$= \frac{ad - bc}{a + d - b - c} = v$$

In a like manner, for every possible P,

$$PA\bar{Q} = \frac{ad - bc}{a + d - b - c} \quad \text{for} \quad \bar{Q} \quad \text{as above}$$

$\bar{P}$, $\bar{Q}$, and v as defined above meet the requirements of optimum strategies and the value of the game.

EXAMPLE   Find optimum strategies for the game and the value of the game: $\begin{bmatrix} 1 & 4 \\ 2 & -2 \end{bmatrix}$.

Solution   Circling the row minimums and column maximums, we have

$$\begin{bmatrix} \boxed{①} & \boxed{4} \\ \boxed{2} & \boxed{-2} \end{bmatrix}$$

There is no saddle point; thus the game is not strictly determined. We can apply the formulas with $a = 1$, $b = 4$, $c = 2$, and $d = -2$:

$$\bar{p}_1 = \frac{d - c}{a + d - b - c} = \frac{-2 - 2}{1 - 2 - 4 - 2} = \frac{-4}{-7} = \frac{4}{7}$$

$$\bar{p}_2 = \frac{a - b}{a + d - b - c} = \frac{1 - 4}{-7} = \frac{-3}{-7} = \frac{3}{7}$$

$$\bar{q}_1 = \frac{d - b}{a + d - b - c} = \frac{-2 - 4}{-7} = \frac{6}{7}$$

$$\bar{q}_2 = \frac{a - c}{a + d - b - c} = \frac{1 - 2}{-7} = \frac{1}{7}$$

Thus $\bar{P} = \begin{bmatrix} \frac{4}{7} & \frac{3}{7} \end{bmatrix}$ and $\bar{Q} = \begin{bmatrix} \frac{6}{7} \\ \frac{1}{7} \end{bmatrix}$ are optimum strategies.

The value of the game is given by

$$\begin{aligned} v &= \frac{ad - bc}{a + d - b - c} \\ &= \frac{1(-2) - 4(2)}{1 + (-2) - 4 - 2} \\ &= \frac{-2 - 8}{-7} \\ &= \tfrac{10}{7} \end{aligned}$$

Many seemingly more complex matrix games can be reduced to 2×2 forms through the concept of recessive rows and recessive columns. We shall begin with a definition and an example.

To **solve a matrix game** means to find optimum strategies for the game
and to find the value of the game.

EXAMPLE   Solve the game specified by

$$A = \begin{bmatrix} 1 & 2 & 5 & 6 & 3 \\ 2 & 3 & 7 & 8 & 4 \\ 1 & -1 & 4 & 5 & 3 \\ 6 & 1 & 2 & 8 & 5 \end{bmatrix}$$

Solution   Naturally there is something special about this game. One
notices that term by term the elements of row 2 of A are larger than those of
row 1. Upon a moment's reflection, any rational player R would realize that
there is *absolutely no reason* to select row 1. He or she will *always* do better
by selecting row 2 over row 1. From a game theory point of view, an optimum strategy for R *must* assign a zero probability to row 1. Thus one would
only need to analyze a reduced matrix

$$A_1 = \begin{bmatrix} 2 & 3 & 7 & 8 & 4 \\ 1 & -1 & 4 & 5 & 3 \\ 6 & 1 & 2 & 8 & 5 \end{bmatrix}$$

and recall how its rows relate to A.

If one examines A_1 from player C's point of view, one notices a similar
result. Column 3 of A_1 is term by term smaller and thus better (from C's
view) than column 4. Any rational strategy for C would assign zero probability to the selection of column 4. Thus the game analysis need only be
applied to

$$A_2 = \begin{bmatrix} 2 & 3 & 7 & 4 \\ 1 & -1 & 4 & 3 \\ 6 & 1 & 2 & 5 \end{bmatrix}$$

This is A_1 with column 4 deleted. Looking at A_2, one notes that row 1 is
term by term larger than row 2; thus one can assign a zero probability to
row 2 and analyze

$$A_3 = \begin{bmatrix} 2 & 3 & 7 & 4 \\ 6 & 1 & 2 & 5 \end{bmatrix}$$

In A_3, column 2 is term by term smaller than column 3, so one moves to

$$A_4 = \begin{bmatrix} 2 & 3 & 4 \\ 6 & 1 & 5 \end{bmatrix}$$

In A_4, column 2 is term by term smaller than column 3; thus one is finally led to

$$A_5 = \begin{bmatrix} 2 & 3 \\ 6 & 1 \end{bmatrix}$$

We can view the position of A_5 elements in A by circling them in A:

$$A = \begin{bmatrix} 1 & 2 & 5 & 6 & 3 \\ 2 & 3 & 7 & 8 & 4 \\ 1 & -1 & 4 & 5 & 3 \\ 6 & 1 & 2 & 8 & 5 \end{bmatrix}$$

A_5 can be analyzed as a 2×2 game. First, one sees that A_5 does not have a saddle point. Thus it is not strictly determined. We can apply the formulas

$$\bar{P} = [\bar{p}_1 \quad \bar{p}_2] \quad \text{and} \quad \bar{Q} = \begin{bmatrix} \bar{q}_1 \\ \bar{q}_2 \end{bmatrix}$$

with $a = 2$, $b = 3$, $c = 6$, and $d = 1$:

$$\bar{p}_1 = \frac{d-c}{a+d-b-c} = \frac{1-6}{2+1-3-6} = \frac{-5}{-6} = \frac{5}{6}$$

$$\bar{p}_2 = \frac{a-b}{a+d-b-c} = \frac{2-3}{-6} = \frac{-1}{-6} = \frac{1}{6}$$

$$\bar{q}_1 = \frac{d-b}{a+d-b-c} = \frac{1-3}{-6} = \frac{-2}{-6} = \frac{1}{3}$$

$$\bar{q}_2 = \frac{a-c}{a+d-b-c} = \frac{2-6}{-6} = \frac{-4}{-6} = \frac{2}{3}$$

$$v = \frac{ad-bc}{a+d-b-c} = \frac{(2)(1)-3(6)}{-6} = \frac{2-18}{-6} = \frac{-16}{-6} = \frac{8}{3}$$

$$\bar{P} = [\tfrac{5}{6} \quad \tfrac{1}{6}] \quad \text{and} \quad \bar{Q} = \begin{bmatrix} \tfrac{1}{3} \\ \tfrac{2}{3} \end{bmatrix}$$

For A,

$$\bar{P} = [0 \quad \tfrac{5}{6} \quad 0 \quad \tfrac{1}{6}]$$

and

$$\bar{Q} = \begin{bmatrix} \tfrac{1}{3} \\ \tfrac{2}{3} \\ 0 \\ 0 \\ 0 \end{bmatrix}$$

The example leads to the following definitions and an obvious theorem.

DEFINITION

Let

$$A = \begin{bmatrix} a_{11} & \cdots & a_{1n} \\ \vdots & & \\ a_{m1} & \cdots & a_{mn} \end{bmatrix}$$

be an $m \times n$ matrix game. Then if every element of the ith row is term by term *greater than or equal to* the corresponding elements of the kth row, the ith row **dominates** the kth row, and the kth row is said to be **recessive** to the ith row.

DEFINITION

Let A be an $m \times n$ matrix game as in the previous definition. If every element of the jth column is term by term *smaller than or equal to* the corresponding elements of the lth column, then the jth column **dominates** the lth column and the lth column is **recessive** to the jth column.

These definitions can be written in symbolic form as:

The ith row dominates the kth row provided

$$a_{i1} \geq a_{k1}, a_{i2} \geq a_{k2}, \ldots, a_{in} \geq a_{kn}$$

The jth column dominates the lth row provided

$$a_{1j} \leq a_{1l}, a_{2j} \leq a_{2l}, \ldots, a_{mj} \leq a_{ml}$$

THEOREM

Any recessive row or column may be dropped from a matrix game without affecting the solution of the game.

EXAMPLE Solve the game

$$\begin{bmatrix} 5 & -1 & 1 & 3 \\ 4 & 1 & 2 & 4 \\ 6 & -1 & 3 & 4 \end{bmatrix}$$

Solution   We can test this game directly to see if it is strictly determined. Circling row minimums and boxing column maximums, we have

$$\begin{bmatrix} 5 & \boxed{-1} & 1 & 3 \\ 4 & \boxed{1} & 2 & \boxed{4} \\ \boxed{6} & \boxed{-1} & \boxed{3} & \boxed{4} \end{bmatrix}$$

This is a strictly determined game with value 1 and optimum strategies

$$\bar{P} = [0 \quad 1 \quad 0] \quad \text{and} \quad \bar{Q} = \begin{bmatrix} 0 \\ 1 \\ 0 \\ 0 \end{bmatrix}$$

We might also try reducing this matrix using recessive rows and columns. We may "strike" these out. The matrix can be reduced as

$$\begin{bmatrix} 5 & -1 & 1 & 3 \\ 4 & 1 & 2 & 4 \\ 6 & -1 & 3 & 4 \end{bmatrix}$$

Solving,

$$\begin{bmatrix} \boxed{1} & 2 \\ \boxed{-1} & \boxed{3} \end{bmatrix}$$

we observe it is still strictly determined, yielding the same solution as before.

EXERCISES 7.5

Solve the matrix games in Exercises 1–13.

1. $\begin{bmatrix} -1 & 0 \\ -7 & 5 \end{bmatrix}$

2. $\begin{bmatrix} 6 & -5 \\ -5 & 6 \end{bmatrix}$

3. $\begin{bmatrix} 3 & -5 \\ 0 & 4 \end{bmatrix}$

4. $\begin{bmatrix} 9 & 5 \\ 1 & -2 \end{bmatrix}$

5. $\begin{bmatrix} -9 & 0 & 1 \\ -2 & 3 & 4 \end{bmatrix}$

6. $\begin{bmatrix} 0 & -5 \\ -1 & -6 \\ -7 & 8 \end{bmatrix}$

7. $\begin{bmatrix} -7 & -2 \\ -6 & -1 \\ -3 & -7 \end{bmatrix}$

8. $\begin{bmatrix} 6 & 3 & 4 \\ 0 & 5 & 8 \end{bmatrix}$

9.
$$\begin{bmatrix} -1 & 7 & 13 \\ 9 & -3 & -2 \\ 8 & -4 & -3 \end{bmatrix}$$

10.
$$\begin{bmatrix} 2 & 9 & 8 \\ 1 & 7 & 6 \\ 8 & 0 & -3 \end{bmatrix}$$

11.
$$\begin{bmatrix} 2 & 3 & -3 & 2 & 3 \\ 4 & 5 & 1 & 2 & 4 \end{bmatrix}$$

12.
$$\begin{bmatrix} 1 & 0 & 6 & 8 & 5 \\ -8 & -1 & -2 & -9 & 8 \\ -3 & -4 & 9 & 2 & -7 \\ 9 & -3 & 4 & -3 & 3 \end{bmatrix}$$

13.
$$\begin{bmatrix} 4 & -9 & -4 & 1 \\ 7 & 0 & -2 & -7 \\ 2 & 5 & -1 & 3 \\ -2 & 1 & -1 & -2 \end{bmatrix}$$

14. An auto dealer plans to take on one of two new lines of cars. A local government agency plans to buy a stock of one of the two lines. From the dealer's point of view, he will make extra profit as given in the following table (in tens of thousands of dollars):

Dealer Stocks	Agency Picks Line 1	Line 2
Line 1	7	2
Line 2	5	6

Assuming that the government is an active opponent, how should the dealer behave?

15. A dealer in farm futures plans to invest in one of three crops, while a foreign government plans to buy large amounts of one of the crops or delay such a purchase until next year. From the dealer's point of view, his profit is given by the following table (in hundreds of thousands of dollars):

Dealer Invests	Government Buys Crop 1	Crop 2	Crop 3	None
Crop 1	8	2	5	4
Crop 2	9	3	7	5
Crop 3	4	9	0	-2

Assuming that the government is an active opponent, how should the dealer proceed?

7.6 LINEAR PROGRAMMING AND GAME THEORY

In this section we will examine how to use the methods of linear programming to solve games that are not strictly determined or reducible to equivalent 2×2 games.

EXAMPLE Solve the game with $A = \begin{bmatrix} 8 & 7 & 4 \\ 2 & 3 & 5 \end{bmatrix}$.

Solution There are no saddle points, nor are there any recessive rows or columns. Since the game has only positive entries, the value of the game v will be positive. Assume that we have an optimum strategy for player C:

$$\bar{Q} = \begin{bmatrix} \bar{q}_1 \\ \bar{q}_2 \\ \bar{q}_3 \end{bmatrix}$$

We know that

$$A\bar{Q} \leq V^T, \quad \text{where} \quad V^T = \begin{bmatrix} v \\ \vdots \\ v \end{bmatrix}$$

or component by component

$$8\bar{q}_1 + 7\bar{q}_2 + 4\bar{q}_3 \leq v$$
$$2\bar{q}_1 + 3\bar{q}_2 + 5\bar{q}_3 \leq v$$

We can divide each of these equations by the positive number v:

$$\frac{8\bar{q}_1}{v} + \frac{7\bar{q}_2}{v} + \frac{4\bar{q}_3}{v} \leq 1$$

$$\frac{2\bar{q}_1}{v} + \frac{3\bar{q}_2}{v} + \frac{5\bar{q}_3}{v} \leq 1$$

Let

$$x_1 = \frac{\bar{q}_1}{v}, \quad x_2 = \frac{\bar{q}_2}{v}, \quad \text{and} \quad x_3 = \frac{\bar{q}_3}{v}$$

The equations become

$$8x_1 + 7x_2 + 4x_3 \leq 1$$
$$2x_1 + 3x_2 + 5x_3 \leq 1$$

Since

$$\bar{q}_1 + \bar{q}_2 + \bar{q}_3 = 1$$

$$x_1 + x_2 + x_3 = \frac{1}{v}$$

The game theory problem as C sees it is to find a minimum value of v, which implies finding a maximum value for $1/v$. In short, we want to maximize the linear expression $x_1 + x_2 + x_3$ provided

$$8x_1 + 7x_2 + 4x_3 \leq 1$$
$$2x_1 + 3x_2 + 5x_3 \leq 1$$

THIS IS A BASIC PROBLEM OF LINEAR PROGRAMMING.

We can apply the simplex method to its solution. The simplex tableau is

$$\begin{bmatrix} 8 & 7 & 4 & 1 & 0 & 1 \\ 2 & 3 & 5 & 0 & 1 & 1 \\ \hline -1 & -1 & -1 & 0 & 0 & 0 \end{bmatrix}$$

Selecting at random the left-hand -1 as the most negative indicator, we would then have 8 as the pivot element. Dividing the first row by 8 and tagging that row with a 1, we have

$$\begin{array}{c} 1 \\ \\ \\ \end{array} \begin{bmatrix} \textcircled{1} & \frac{7}{8} & \frac{1}{2} & \frac{1}{8} & 0 & \frac{1}{8} \\ 2 & 3 & 5 & 0 & 1 & 1 \\ \hline -1 & -1 & -1 & 0 & 0 & 0 \end{bmatrix}$$

Zeroing the remaining members of column 1,

$$\begin{array}{c} 1 \\ \\ \\ \end{array} \begin{bmatrix} 1 & \frac{7}{8} & \frac{1}{2} & \frac{1}{8} & 0 & \frac{1}{8} \\ 0 & \frac{5}{4} & 4 & -\frac{1}{4} & 1 & \frac{3}{4} \\ \hline 0 & -\frac{1}{8} & -\frac{1}{2} & \frac{1}{8} & 0 & \frac{1}{8} \end{bmatrix}$$

Selecting the most negative indicator and a new pivot, we have

$$\begin{array}{c} 1 \\ 3 \\ \\ \end{array} \begin{bmatrix} 1 & \frac{7}{8} & \frac{1}{2} & \frac{1}{8} & 0 & \frac{1}{8} \\ 0 & \frac{5}{4} & \textcircled{4} & -\frac{1}{4} & 1 & \frac{3}{4} \\ \hline 0 & -\frac{1}{8} & -\frac{1}{2} & \frac{1}{8} & 0 & \frac{1}{8} \end{bmatrix}$$

Dividing row 2 through by 4,

$$\begin{array}{c} 1 \\ 3 \\ \\ \end{array} \begin{bmatrix} 1 & \frac{7}{8} & \frac{1}{2} & \frac{1}{8} & 0 & \frac{1}{8} \\ 0 & \frac{5}{16} & \textcircled{1} & -\frac{1}{16} & \frac{1}{4} & \frac{3}{16} \\ \hline 0 & -\frac{1}{8} & -\frac{1}{2} & \frac{1}{8} & 0 & \frac{1}{8} \end{bmatrix}$$

Zeroing column 3, we have

$$\begin{array}{c} 1 \\ 3 \\ \\ \end{array} \begin{bmatrix} 1 & \frac{23}{32} & 0 & \frac{5}{32} & -\frac{1}{8} & \frac{1}{32} \\ 0 & \frac{5}{16} & 1 & -\frac{1}{16} & \frac{1}{4} & \frac{3}{16} \\ \hline 0 & \frac{1}{32} & 0 & \frac{3}{32} & \frac{1}{8} & \frac{7}{32} \end{bmatrix}$$

This is the terminal tableau; thus the maximum value of

$$x_1 + x_2 + x_3 = \frac{1}{v} \quad \text{is} \quad \frac{7}{32}$$

Thus

$$v = \frac{32}{7}$$

with

$$x_1 = \frac{1}{32}$$
$$x_2 = 0$$
$$x_3 = \frac{3}{16}$$

Since $x_1 = \dfrac{\bar{q}_1}{v}$, $\bar{q}_1 = x_1 \cdot v = \dfrac{1}{32} \cdot \dfrac{32}{7} = \dfrac{1}{7}$

$x_2 = \dfrac{\bar{q}_2}{v}$, $\bar{q}_2 = x_2 \cdot v = 0 \cdot \dfrac{32}{7} = 0$

$x_3 = \dfrac{\bar{q}_3}{v}$, $\bar{q}_3 = x_3 \cdot v = \dfrac{3}{16} \cdot \dfrac{32}{7} = \dfrac{6}{7}$

$$\bar{Q} = \begin{bmatrix} \frac{1}{7} \\ 0 \\ \frac{6}{7} \end{bmatrix}$$

is an optimum strategy.

We can perform a similar analysis to find an expression for $\bar{P}$, one of player R's optimum strategies. Assume that we have such a strategy:

$$\bar{P} = [\bar{p}_1 \quad \bar{p}_2]$$

Since $\bar{P}A \geq V$ with $V = [v \quad \cdots \quad v]$, we have

$$8\bar{p}_1 + 2\bar{p}_2 \geq v$$
$$7\bar{p}_1 + 3\bar{p}_2 \geq v$$
$$4\bar{p}_1 + 5\bar{p}_2 \geq v$$

or dividing both sides by v, which is possible,

$$\frac{8\bar{p}_1}{v} + \frac{2\bar{p}_2}{v} \geq 1$$

$$\frac{7\bar{p}_1}{v} + \frac{3\bar{p}_2}{v} \geq 1$$

$$\frac{4\bar{p}_1}{v} + \frac{5\bar{p}_2}{v} \geq 1$$

We also know that

$$\bar{p}_1 + \bar{p}_2 = 1$$

Let

$$y_1 = \frac{\bar{p}_1}{v} \quad \text{and} \quad y_2 = \frac{\bar{p}_2}{v}$$

Then

$$y_1 + y_2 = \frac{1}{v}$$

and

$$8y_1 + 2y_2 \geq 1$$
$$7y_1 + 3y_2 \geq 1$$
$$4y_1 + 5y_2 \geq 1$$

From R's point of view, v is to be made as large as possible, which would make $1/v$ as small as possible, a minimum. Thus we can solve for $\bar{p}_1$ and $\bar{p}_2$ *provided* we can find y_1 and y_2 such that $y_1 + y_2$ is a minimum under the conditions

$$8y_1 + 2y_2 \geq 1$$
$$7y_1 + 3y_2 \geq 1$$
$$4y_1 + 5y_2 \geq 1$$

This is a minimum linear programming problem. We solve it using the concept of a dual. The dual of this problem is: Maximize $x_1 + x_2 + x_3$ provided

$$8x_1 + 7x_2 + 4x_3 \leq 1$$
$$2x_1 + 3x_2 + 5x_3 \leq 1$$

This is *exactly* the problem we solved above. We already know that the solutions to the dual are in the slack variable columns of the terminal tableau. Specifically, $y_1 = \frac{3}{32}$ and $y_2 = \frac{1}{8}$ with $v = \frac{32}{7}$ as before. Thus

$$\bar{p}_1 = y_1 \cdot v = \frac{3}{32} \cdot \frac{32}{7} = \frac{3}{7}$$
$$\bar{p}_2 = y_2 \cdot v = \frac{1}{8} \cdot \frac{32}{7} = \frac{4}{7}$$

The solution to our game is

$$\bar{P} = \begin{bmatrix} \frac{3}{7} & \frac{4}{7} \end{bmatrix} \quad \text{and} \quad \bar{Q} = \begin{bmatrix} \frac{1}{7} \\ 0 \\ \frac{6}{7} \end{bmatrix}$$

with $v = \frac{32}{7}$ the value of the game.

If one studies the example, it is clear that the method used should work on any game matrix *provided* the value of the game is *positive*. This is not an important restriction. We can get around this restriction by using a previously proven theorem. Recall that we showed that any game matrix A and any other matrix formed by adding a positive constant k to A, element by element, will have common strategies, and their game values will differ by k.

EXAMPLE   Solve the game $A = \begin{bmatrix} -2 & -7 \\ -5 & -4 \end{bmatrix}$.

Solution   We could solve this as a 2×2 matrix game. Rather, let us apply the linear programming procedure. Since we cannot use the linear programming methods unless the terms of our matrix are positive, we will arbitrarily add a suitable positive amount k to each term of A to produce a matrix with all positive terms. Let us use $k = 8$. We then have

$$A_k = \begin{bmatrix} 6 & 1 \\ 3 & 4 \end{bmatrix}$$

Our simplex tableau is

$$\begin{bmatrix} 6 & 1 & 1 & 0 & 1 \\ 3 & 4 & 0 & 1 & 1 \\ \hline -1 & -1 & 0 & 0 & 0 \end{bmatrix}$$

At random we will select the first -1 as the most negative indicator, which makes 6 the pivot:

$$\begin{array}{c} 1 \\ \\ \\ \end{array} \begin{bmatrix} ① & \frac{1}{6} & \frac{1}{6} & 0 & \frac{1}{6} \\ 3 & 4 & 0 & 1 & 1 \\ \hline -1 & -1 & 0 & 0 & 0 \end{bmatrix}$$

Zeroing column 1,

$$\begin{array}{c} 1 \\ \\ \\ \end{array} \begin{bmatrix} 1 & \frac{1}{6} & \frac{1}{6} & 0 & \frac{1}{6} \\ 0 & ⑦⁄₂ & -\frac{1}{2} & 1 & \frac{1}{2} \\ \hline 0 & -\frac{5}{6} & \frac{1}{6} & 0 & \frac{1}{6} \end{bmatrix}$$

Selecting a new pivot,

$$\begin{array}{c} 1 \\ 2 \\ \\ \end{array} \begin{bmatrix} 1 & \frac{1}{6} & \frac{1}{6} & 0 & \frac{1}{6} \\ 0 & ① & -\frac{1}{7} & \frac{2}{7} & \frac{1}{7} \\ \hline 0 & -\frac{5}{6} & \frac{1}{6} & 0 & \frac{1}{6} \end{bmatrix}$$

$$\begin{array}{c} 1 \\ 2 \\ \\ \end{array} \begin{bmatrix} 1 & 0 & \frac{4}{21} & -\frac{1}{21} & \frac{1}{7} \\ 0 & 1 & -\frac{1}{7} & \frac{2}{7} & \frac{1}{7} \\ \hline 0 & 0 & \frac{1}{21} & \frac{5}{21} & \frac{2}{7} \end{bmatrix}$$

This is our terminal tableau. The solutions are
$$x_1 = \tfrac{1}{7}, \; x_2 = \tfrac{1}{7}, \; y_1 = \tfrac{1}{21}, \quad \text{and} \quad y_2 = \tfrac{5}{21}$$
with
$$\frac{1}{v} = \frac{2}{7} \quad \text{or} \quad v = \frac{7}{2}$$

Then this result can be converted to the game solution
$$\bar{q}_1 = x_1 \cdot v = \tfrac{1}{7} \cdot \tfrac{7}{2} = \tfrac{1}{2}$$
$$\bar{q}_2 = x_2 \cdot v = \tfrac{1}{7} \cdot \tfrac{7}{2} = \tfrac{1}{2}$$
$$\bar{p}_1 = y_1 \cdot v = \tfrac{1}{21} \cdot \tfrac{7}{2} = \tfrac{1}{6}$$
$$\bar{p}_2 = y_2 \cdot v = \tfrac{5}{21} \cdot \tfrac{7}{2} = \tfrac{5}{6}$$
or
$$\bar{P} = \begin{bmatrix} \tfrac{1}{6} & \tfrac{5}{6} \end{bmatrix} \quad \text{and} \quad \bar{Q} = \begin{bmatrix} \tfrac{1}{2} \\ \tfrac{1}{2} \end{bmatrix}$$

with the value of $A_k = \tfrac{7}{2}$. This implies that the value of A, the original matrix game, is
$$\tfrac{7}{2} - k = \tfrac{7}{2} - 8 = -\tfrac{9}{2}$$

We can now outline this procedure in the abstract.

THEOREM

Let A be an $m \times n$ matrix game with elements a_{ij}, $i = 1$ to $i = m$ and $j = 1$ to $j = n$. Further assume that

1. $a_{ij} > 0$ for all i and j.

2. A is not a strictly determined game.

3. A does not contain any recessive rows or recessive columns.

Let

$$X = \begin{bmatrix} x_1 \\ x_2 \\ \vdots \\ x_n \end{bmatrix}$$

and let

$$\phi = \begin{bmatrix} 1 \\ \vdots \\ 1 \end{bmatrix}$$

be an n-dimensional column vector with 1 as each component. Then the solutions to the linear programming problem of finding the maximum value of

$$x_1 + x_2 + \cdots + x_n$$

provided
$$AX \le \phi$$

are related to the solution to the game theory problem in the following manner:

A. The value of the game v is the reciprocal of the maximum value of the linear program.

B. An optimum strategy for C

$$\bar{Q} = \begin{bmatrix} \bar{q}_1 \\ \vdots \\ \bar{q}_n \end{bmatrix}$$

is given by

$$\bar{q}_j = \frac{x_j}{v}, \quad j = 1 \quad \text{to} \quad j = n$$

C. An optimum strategy for R

$$\bar{P} = [\bar{p}_1 \quad \cdots \quad \bar{p}_m]$$

is given by

$$\bar{p}_i = \frac{y_i}{v}, \quad i = 1 \quad \text{to} \quad i = m$$

where $y_1, y_2, \ldots, y_m$ are the components of the solution of the dual linear programming problem.

EXAMPLE   A distributor of television sets plans to stock one of three lines of sets, as does a very large competitor. The results, from the first distributor's point of view, are given in the following table (in tens of thousands of dollars):

Distributor	Competitor		
	Line 1	Line 2	Line 3
Line 1	0	1	−1
Line 2	−2	−1	0
Line 3	−1	−2	2

How should the distributor act?

Solution   The game matrix is

$$A = \begin{bmatrix} 0 & 1 & -1 \\ -2 & -1 & 0 \\ -1 & -2 & 2 \end{bmatrix}$$

First, we must adjust the elements of A to make them positive. This can be done by adding 3 to each member of A. The result is

$$A_1 = \begin{bmatrix} 3 & 4 & 2 \\ 1 & 2 & 3 \\ 2 & 1 & 5 \end{bmatrix}$$

The reader can verify that neither A nor A_1 is strictly determined or contains recessive rows or columns. The linear programming tableau is

$$\begin{bmatrix} 3 & 4 & 2 & 1 & 0 & 0 & 1 \\ 1 & 2 & 3 & 0 & 1 & 0 & 1 \\ 2 & 1 & 5 & 0 & 0 & 1 & 1 \\ \hline -1 & -1 & -1 & 0 & 0 & 0 & 0 \end{bmatrix}$$

Let us choose at random the -1 in column 1 as the most negative indicator. This makes the 3 in column 1, row 1 the pivot. We can proceed in the usual manner. We have

$$1 \begin{bmatrix} \boxed{1} & \frac{4}{3} & \frac{2}{3} & \frac{1}{3} & 0 & 0 & \frac{1}{3} \\ 1 & 2 & 3 & 0 & 1 & 0 & 1 \\ 2 & 1 & 5 & 0 & 0 & 1 & 1 \\ \hline -1 & -1 & -1 & 0 & 0 & 0 & 0 \end{bmatrix}$$

Zeroing the first column and selecting a new pivot using the most negative indicator,

$$\begin{array}{c} 1 \\ \\ 3 \\ \\ \end{array} \begin{bmatrix} 1 & \frac{4}{3} & \frac{2}{3} & \frac{1}{3} & 0 & 0 & \frac{1}{3} \\ 0 & \frac{2}{3} & \frac{7}{3} & -\frac{1}{3} & 1 & 0 & \frac{2}{3} \\ 0 & -\frac{5}{3} & \boxed{\frac{11}{3}} & -\frac{2}{3} & 0 & 1 & \frac{1}{3} \\ \hline 0 & \frac{1}{3} & -\frac{1}{3} & \frac{1}{3} & 0 & 0 & \frac{1}{3} \end{bmatrix}$$

which becomes

$$\begin{array}{c} 1 \\ \\ 3 \\ \\ \end{array} \begin{bmatrix} 1 & \frac{4}{3} & \frac{2}{3} & \frac{1}{3} & 0 & 0 & \frac{1}{3} \\ 0 & \frac{2}{3} & \frac{7}{3} & -\frac{1}{3} & 1 & 0 & \frac{2}{3} \\ 0 & -\frac{5}{11} & \boxed{1} & -\frac{2}{11} & 0 & \frac{3}{11} & \frac{1}{11} \\ \hline 0 & \frac{1}{3} & -\frac{1}{3} & \frac{1}{3} & 0 & 0 & \frac{1}{3} \end{bmatrix}$$

Zeroing column 3,

$$\begin{array}{c} 1 \\ \\ \\ 3 \\ \\ \end{array} \begin{bmatrix} 1 & \frac{18}{11} & 0 & \frac{5}{11} & 0 & -\frac{2}{11} & \frac{3}{11} \\ 0 & \frac{19}{11} & 0 & \frac{1}{11} & 1 & -\frac{7}{11} & \frac{5}{11} \\ 0 & -\frac{5}{11} & 1 & -\frac{2}{11} & 0 & \frac{3}{11} & \frac{1}{11} \\ \hline 0 & \frac{2}{11} & 0 & \frac{3}{11} & 0 & \frac{1}{11} & \frac{4}{11} \end{bmatrix}$$

This is the terminal tableau. The solution of the linear program is $\frac{4}{11}$. Thus $A_1 = \frac{11}{4}$. We have

$$X = \begin{bmatrix} \frac{3}{11} \\ 0 \\ \frac{1}{11} \end{bmatrix} \quad \text{or} \quad x_1 = \frac{3}{11}, \, x_2 = 0, \quad \text{and} \quad x_3 = \frac{1}{11}$$

The solution to the dual is

$$Y = [\frac{3}{11} \quad 0 \quad \frac{1}{11}]$$

or

$$y_1 = \frac{3}{11}, \, y_2 = 0, \quad \text{and} \quad y_3 = \frac{1}{11}$$

Therefore, the value of the game A is

$$\frac{11}{4} - 3 = -\frac{1}{4}$$

$$\bar{P} = [\bar{p}_1 \quad \bar{p}_2 \quad \bar{p}_3]$$

with

$$\bar{p}_1 = y_1 \cdot v = \frac{3}{11} \cdot \frac{11}{4} = \frac{3}{4}$$

$$\bar{p}_2 = y_2 \cdot v = 0 \cdot \frac{11}{4} = 0$$

$$\bar{p}_3 = y_3 \cdot v = \frac{1}{11} \cdot \frac{11}{4} = \frac{1}{4}$$

$$\bar{Q} = \begin{bmatrix} \bar{q}_1 \\ \bar{q}_2 \\ \bar{q}_3 \end{bmatrix}$$

with

$$\bar{q}_1 = x_1 \cdot v = \frac{3}{11} \cdot \frac{11}{4} = \frac{3}{4}$$

$$\bar{q}_2 = x_2 \cdot v = 0 \cdot \frac{11}{4} = 0$$

$$\bar{q}_3 = x_3 \cdot v = \frac{1}{11} \cdot \frac{11}{4} = \frac{1}{4}$$

In summary, the value of A is $v = -\frac{1}{4}$ with one set of optimum strategies

$$\bar{P} = [\frac{3}{4} \quad 0 \quad \frac{1}{4}] \quad \text{and} \quad \bar{Q} = \begin{bmatrix} \frac{3}{4} \\ 0 \\ \frac{1}{4} \end{bmatrix}$$

We will end our discussion of game theory here with many unanswered questions. For example, how can we find other optimum strategies, as we know they are not unique? What do we do when the games are not zero sum, i.e., when the game requires two matrices for definition? For many of the questions we might ask there are no answers: game theory is too new. One thing is clear, however; it is a powerful tool of analysis.

EXERCISES 7.6

1. Check the 2 × 2 matrix game example in this section using the formulas given for the solution of such games in the previous section.

2. Is it possible the optimum strategies found in Exercise 1 do not match those for the same game found in this section?

3. In the text we stated that for a given matrix game A, if $a_{ij} > 0$ for all possible i and j, the value of the game, v, is also positive; argue this.

Solve the games in Exercises 4–16 by the linear programming method unless strictly determined. Note that some may contain recessive rows or columns.

4. $\begin{bmatrix} 2 & 4 \\ 3 & 1 \end{bmatrix}$

5. $\begin{bmatrix} 0 & 5 \\ 1 & -1 \end{bmatrix}$

6. $\begin{bmatrix} -5 & 3 \\ -1 & -3 \end{bmatrix}$

7. $\begin{bmatrix} -3 & -5 & -2 \\ -4 & -3 & -4 \end{bmatrix}$

8. $\begin{bmatrix} 4 & 2 & 0 \\ 3 & 0 & 1 \end{bmatrix}$

9. $\begin{bmatrix} -3 & -1 \\ 2 & -2 \\ 1 & 3 \end{bmatrix}$

10. $\begin{bmatrix} 0 & 3 \\ -2 & 1 \\ 3 & -4 \end{bmatrix}$

11. $\begin{bmatrix} -2 & 3 & -3 \\ 2 & 0 & -3 \\ 3 & -2 & -4 \end{bmatrix}$

12. $\begin{bmatrix} -2 & 0 & 2 \\ -5 & -2 & 1 \\ -1 & -6 & 0 \end{bmatrix}$

13. $\begin{bmatrix} -3 & 4 & -1 \\ 3 & 3 & -4 \\ -5 & 3 & -1 \end{bmatrix}$

14. $\begin{bmatrix} -4 & 4 & -5 & 3 \\ -3 & -3 & -1 & 2 \\ -2 & -4 & 4 & 5 \end{bmatrix}$

15. $\begin{bmatrix} 1 & -4 & 0 & 0 \\ 1 & 0 & 4 & 2 \\ -4 & -2 & 0 & 1 \end{bmatrix}$

16. $\begin{bmatrix} 4 & 7 & 8 & 9 & 6 \\ 1 & 2 & 9 & 4 & 0 \end{bmatrix}$

17. A gasoline maker must invest in one of three methods of making lead-free gasoline. The government must select one of three types of anti-pollution devices. The profit the gas company will make with each process depends on the device chosen. Their profit (in millions of dollars) is given by the following table:

	Device 1	Device 2	Device 3
Method 1	1	−4	−6
Method 2	2	7	−5
Method 3	−6	9	1

Assume that the government is an active opponent and determine the optimum strategies for the gas maker.

18. Assuming that the government is an active opponent of the farmer in Exercises 7.2.13 and 7.2.14, what strategy should the farmer use?

REVIEW EXERCISES

For each zero-sum game in Exercises 1 and 2, determine the maximum amount R can win and the maximum amount C can win.

1. $\begin{bmatrix} 3 & 7 & -5 & 4 \\ 2 & -6 & 7 & 1 \\ 8 & -3 & 9 & -2 \end{bmatrix}$

2. $\begin{bmatrix} 4 & 3 & -5 \\ -2 & 8 & 3 \\ -5 & 4 & 4 \\ 2 & 1 & -6 \end{bmatrix}$

3. Construct payoff matrices suitable for a *positive* sum game.

4. Construct payoff matrices suitable for a *negative* sum game.

Solve the following strictly determined zero-sum games in Exercises 5–8.

5. $\begin{bmatrix} 4 & 3 \\ -1 & 2 \end{bmatrix}$

6. $\begin{bmatrix} 5 & 0 & 8 \\ 3 & -4 & 2 \end{bmatrix}$

7. $\begin{bmatrix} 4 & -2 & 7 \\ 4 & 1 & 6 \\ -2 & -5 & 5 \end{bmatrix}$

8. $\begin{bmatrix} -2 & 7 & -2 & 4 \\ 0 & 1 & 2 & 3 \\ -3 & 8 & 5 & 1 \end{bmatrix}$

Solve the 2 × 2 zero-sum games in Exercises 9–12.

9. $\begin{bmatrix} 2 & 1 \\ 4 & -3 \end{bmatrix}$

10. $\begin{bmatrix} 7 & -5 \\ -4 & 6 \end{bmatrix}$

11. $\begin{bmatrix} -1 & 0 \\ 0 & -1 \end{bmatrix}$

12. $\begin{bmatrix} 1 & 1 \\ 0 & 4 \end{bmatrix}$

Solve the zero-sum games in Exercises 13–16 by any method.

13. $\begin{bmatrix} 6 & 0 & 4 \\ 0 & -4 & 2 \\ 2 & 4 & 4 \end{bmatrix}$ **14.** $\begin{bmatrix} -1 & 4 & 3 \\ -7 & -8 & 5 \\ 6 & -2 & 4 \end{bmatrix}$

15. $\begin{bmatrix} 1 & -4 & 2 \\ 3 & -2 & 1 \\ 7 & 6 & 4 \end{bmatrix}$ **16.** $\begin{bmatrix} 2 & 3 & 4 & 3 \\ 6 & 2 & -3 & 7 \\ -8 & 4 & 2 & -3 \end{bmatrix}$

17. In the political problem described in the first section of this chapter, the parties were to "play" a vote game with the following matrix:

$$\begin{bmatrix} 0 & 40,000 & 100,000 \\ -40,000 & 0 & 60,000 \\ -100,000 & -60,000 & 0 \end{bmatrix}$$

What strategies should the two parties select?

18. In Exercises 7.2.15 and 7.2.16 a pro quarterback was involved with a game with the following payoff (in yards):

	Blitz	Regular
Run	5	1
Short Pass	15	3
Long Pass	0	20

Based on this information, what strategy should the quarterback adopt?

CHAPTER

This chapter includes a number of optional topics which may be incorporated with other chapters or taught as a unit.

A.I NUMBER BASES

The numerals that are commonly used—for instance, those which number the pages of this book—are called base ten or *decimal* numerals because they count by grouping by tens. The word *twenty* is a modification of "two tens." *Thirty* is a modification of "three tens," and so on. One hundred is ten tens. One thousand is ten hundreds, etc. A base of ten is satisfactory for most, but not all, of our number needs. In one important case—in the construction of computers—it is a total failure. In order to work efficiently, the internal parts of a computer must compute in base two numerals, also called *binary* numbers. The word *number* is commonly misused in this manner.

Numerals in bases other than ten use the same essential characteristics as base ten. They are positional, and the values of the positions are powers of the base. *Positional* means that the value a digit has depends on its position in a numeral. The 3 in 4,739 indicates three tens by virtue of its position in the numeral. If the digits of 4,739 are rearranged, a new number is indicated: 4,739 and 7,349 are different, and although both contain a 3, the 3 means different values in each number. The positional values in base ten are powers of ten. To show some of these values we write 4,739 in expanded notation

$$4,739 = 4 \cdot 1,000 + 7 \cdot 100 + 3 \cdot 10 + 9 \cdot 1$$
$$= 4 \cdot 10^3 + 7 \cdot 10^2 + 3 \cdot 10^1 + 9 \cdot 10^0$$

Then, starting from the rightmost digit, the first digit indicates a number of units, or 10^0. The second digit indicates a number of 10^1, the third digit a number of 10^2, and so on.

NOTATION

A written subscript will be used to show the number base that is being used when it is not ten.

EXAMPLES

$(34)_{\text{seven}}$ indicates a base of seven

$(10,101)_{\text{two}}$ indicates a base of two

A number written in a base other than ten will follow the same pattern as base ten.

EXAMPLE   Change $(2,243)_{\text{five}}$ to base ten notation.

Solution   The digits will indicate numbers of the powers of five starting on the right with 5^0, or 1:

$$(2,243)_{\text{five}} = 2 \cdot 5^3 + 2 \cdot 5^2 + 4 \cdot 5^1 + 3 \cdot 5^0$$
$$= 2 \cdot 125 + 2 \cdot 25 + 4 \cdot 5 + 3 \cdot 1$$
$$= 250 + 50 + 20 + 3$$
$$= 323$$

$(2,243)_{\text{five}}$ in base ten notation is 323.

EXAMPLE   Write $(403)_{\text{seven}}$ in base ten notation.

Solution $(403)_{\text{seven}} = 4 \cdot 7^2 + 0 \cdot 7^1 + 3 \cdot 7^0$
$$= 4 \cdot 49 + 0 + 3$$
$$= 199$$

$(403)_{\text{seven}}$ is 199 in base ten notation.

EXAMPLE   Write $(10,110,101)_{\text{two}}$ in base ten notation.

Solution   The digits show the presence or absence of powers of two.

$(10,110,101)_{\text{two}} = 1 \cdot 2^7 + 0 \cdot 2^6 + 1 \cdot 2^5 + 1 \cdot 2^4 + 0 \cdot 2^3 + 1 \cdot 2^2 + 0 \cdot 2^1 + 1 \cdot 2^0$
$$= 128 + 0 + 32 + 16 + 0 + 4 + 0 + 1$$
$$= 181$$

In the binary system there are two digits, 0 and 1. In the base five numeral system the digits are 0, 1, 2, 3, and 4. Table A.1.1 shows the numerals for the numbers from zero through twenty-five in bases ten, two, and five.

TABLE A.1.1 THE NUMERALS FROM ZERO TO TWENTY-FIVE IN BASES TEN, TWO, AND FIVE

Base Ten	Base Two	Base Five
0	0	0
1	1	1
2	10	2
3	11	3
4	100	4
5	101	10
6	110	11
7	111	12
8	1000	13
9	1001	14
10	1010	20
11	1011	21
12	1100	22
13	1101	23
14	1110	24
15	1111	30
16	10000	31
17	10001	32
18	10010	33
19	10011	34
20	10100	40
21	10101	41
22	10110	42
23	10111	43
24	11000	44
25	11001	100

A "point" in a number will indicate negative powers of the base, as it does in base ten. The number 45.72 in expanded notation is

$$45.72 = 4 \cdot 10^1 + 5 \cdot 10^0 + 7 \cdot 10^{-1} + 2 \cdot 10^{-2}$$

A point divides a number. The powers of the base increase moving from the decimal point to the left and decrease moving from the decimal point to the right.

EXAMPLE   Write $(2.14)_{\text{five}}$ in base ten notation.

Solution $(2.14)_{\text{five}} = 2 \cdot 5^0 + 1 \cdot 5^{-1} + 4 \cdot 5^{-2}$

$$= 2 + \tfrac{1}{5} + \tfrac{4}{25}$$
$$= 2 + 0.20 + 0.16$$
$$= 2.36$$

$(2.14)_{\text{five}}$ is 2.36 in base ten notation.

EXAMPLE   Write the binary numeral $(110.111)_{\text{two}}$ in base ten notation.

Solution

$(110.111)_{\text{two}} = 1 \cdot 2^2 + 1 \cdot 2^1 + 0 \cdot 2^0 + 1 \cdot 2^{-1} + 1 \cdot 2^{-2} + 1 \cdot 2^{-3}$

$$= 4 + 2 + 0 + \tfrac{1}{2} + \tfrac{1}{4} + \tfrac{1}{8}$$
$$= 6 + 0.5 + 0.25 + 0.125$$
$$= 6.875$$

EXAMPLE   Write the decimal number 486 in base five notation.

Solution   The place values of base five are $5^0 = 1$, $5^1 = 5$, $5^2 = 25$, $5^3 = 125$, $5^4 = 625, \ldots$. The largest place value of base five that is not larger than 486 is $5^3 = 125$. There are three 125s in 486 with a remainder of $486 - 375 = 111$. There are four 25s in 111 with a remainder of 11. The 11 contains two 5s with a remainder of 1. Therefore,

$$486 = (3,421)_{\text{five}}$$

Any base ten number can be written in the notation of any other base by dividing the base ten number repeatedly by the base to which it is to be translated until a quotient of zero is achieved. The remainders of each division form the digits for the translation. This technique could have been used to translate 486 in the above example to base five:

$$
\begin{array}{r|l}
0 & 3 \\
5)\overline{3} & 4 \\
5)\overline{19} & 2 \\
5)\overline{97} & 1 \\
5)\overline{486} &
\end{array}
$$

The remainders taken in the order shown by the arrow form the base five translation of 486: $486 = (3,421)_{\text{five}}$. This agrees with the results of the last example.

EXAMPLE   Translate the decimal number 95 into base three notation.

Solution   Using the repeated division technique,

$$
\begin{array}{r}
0 \quad 1 \\
3\overline{)1} \quad 0 \\
3\overline{)3} \quad 1 \\
3\overline{)10} \quad 1 \\
3\overline{)31} \quad 2 \\
3\overline{)95}
\end{array}
$$

$$95 = (10{,}112)_{\text{three}}$$

Checking this result gives

$$
\begin{aligned}
(10{,}112)_{\text{three}} &= 1\cdot 3^4 + 0\cdot 3^3 + 1\cdot 3^2 + 1\cdot 3^1 + 2\cdot 3^0 \\
&= 81 + 0 + 9 + 3 + 2 \\
&= 95
\end{aligned}
$$

The results check.

Bases greater than ten require special symbols for some of the digits. In base twelve two such symbols are required, one for ten and one for eleven. The digits for base twelve numerals will be 0, 1, 2, 3, 4, 5, 6, 7, 8, 9, A, and B, where A represents ten and B represents eleven. Such symbols are necessary because $(10)_{\text{twelve}} = 1\cdot 12^1 + 0\cdot 12^0 = 12$ and $(11)_{\text{twelve}} = 1\cdot 12^1 + 1\cdot 12^0 = 13$.

EXAMPLE   Translate $(3AB)_{\text{twelve}}$ into base ten notation.

Solution   The expanded notation in base twelve follows the same pattern as in any other base:

$$
\begin{aligned}
(3AB)_{\text{twelve}} &= 3\cdot 12^2 + 10\cdot 12^1 + 11\cdot 12^0 \\
&= 3\cdot 144 + 120 + 11 \\
&= 432 + 120 + 11 \\
&= 563
\end{aligned}
$$

EXERCISES A.I

Write the number in each of Exercises 1–12 in base ten notation.

1. $(33)_{\text{five}}$ **2.** $(424)_{\text{five}}$ **3.** $(3{,}131)_{\text{five}}$

4. $(202)_{\text{three}}$ **5.** $(1{,}022)_{\text{three}}$ **6.** $(10{,}201)_{\text{three}}$

7. $(110{,}110)_{\text{two}}$ **8.** $(1{,}101{,}100)_{\text{two}}$ **9.** $(111.1)_{\text{two}}$

10. $(111.10101)_{\text{two}}$ **11.** $(432.1)_{\text{five}}$ **12.** $(222.22)_{\text{five}}$

In Exercises 13–20, change the decimal numeral to the indicated base notation.

13. $381 = ($ $)_{\text{five}}$ **14.** $721 = ($ $)_{\text{five}}$

15. $721 = ($ $)_{\text{seven}}$ **16.** $721 = ($ $)_{\text{nine}}$

17. $4{,}385 = ($ $)_{\text{five}}$ **18.** $70 = ($ $)_{\text{two}}$

19. $100 = ($ $)_{\text{two}}$ **20.** $7{,}391 = ($ $)_{\text{five}}$

In Exercises 21–26, let $A = 10$ and $B = 11$ and translate the base twelve numbers to base ten.

21. $(148)_{\text{twelve}}$ **22.** $(3A)_{\text{twelve}}$ **23.** $(3A1)_{\text{twelve}}$

24. $(4AB)_{\text{twelve}}$ **25.** $(10AB)_{\text{twelve}}$ **26.** $(1B0A)_{\text{twelve}}$

27. Add $75 + 41$ by translating the addends to binary numerals and add in that base. Check by translating the binary answer to base ten.

28. Repeat Exercise 27 for $81 + 57$.

29. Multiply 17×10 by translating 17 and 10 to base two notation and performing the multiplication in that base. Check by translating the answer back to base ten.

30. Repeat Exercise 29 for 25×25.

31. Repeat Exercise 29 in base five.

32. Repeat Exercise 30 in base five.

A.2 MATHEMATICAL INDUCTION

It is often necessary to prove that some statement is true for all positive integers. For instance, it seems that

$$1 + 2 + 3 + 4 + \cdots + n = \frac{n(n+1)}{2} \quad \text{for} \quad n \quad \text{a positive integer} \qquad (1)$$

since for $n = 1$, $\qquad\qquad 1 = \dfrac{1(1+1)}{2} = 1$

for $n = 2$, $\qquad\qquad 1 + 2 = \dfrac{2(2+1)}{2} = 3$

for $n = 3$, $\qquad\qquad 1 + 2 + 3 = \dfrac{3(3+1)}{2} = \dfrac{3 \cdot 4}{2} = 6$

for $n = 4$, $\qquad 1 + 2 + 3 + 4 = \dfrac{4(4+1)}{2} = \dfrac{4 \cdot 5}{2} = 10$

Showing that (1) is true for $n = 1, 2, 3,$ and 4 does not prove that it is true for all positive integers. We could continue to demonstrate the truth of (1) for other specific values of n, but proof for specific values does not prove (1) for all values, and clearly, we cannot substitute every positive integer into (1). Another method of proof is needed, and it is known as *mathematical induction.*

The principle of mathematical induction is based on an axiom of set theory that states if a set S contains the number 1 and if it is also true that if any other positive integer k is in the set, then its successor, $k + 1$, is in S, and S contains all the natural numbers. In other words, if S contains 1 and also the successor of every natural number, then S contains 2, 3, 4, . . ., which is the same as saying S contains all the natural numbers. Note that we have used the terms *positive integer* and *natural number* interchangeably.

THE PRINCIPLE OF MATHEMATICAL INDUCTION

If for a given statement P:

    1. P is true for $n = 1$.

    2. If P is true for $n = k$, then P is true for $n = k + 1$.

Then P is true for all positive integers n.

EXAMPLE   Use the principle of mathematical induction to prove that

$$1 + 2 + 3 + \cdots + n = \frac{n(n + 1)}{2}$$

for every positive integer n.

Solution   Mathematical induction requires that two things be shown: that the proposition is true when $n = 1$, and that assuming it is true for $n = k$ proves that it is true for $k + 1$. In the discussion of (1) above it was shown to be true for $n = 1$. To establish the second condition necessary in a mathematical induction proof, we *assume* that the following is true:

$$1 + 2 + 3 + \cdots + k = \frac{k(k + 1)}{2}$$

We must show the statement is true for $n = k + 1$—that is,

$$1 + 2 + 3 + \cdots + k + (k + 1) = \frac{(k + 1)((k + 1) + 1)}{2}$$

$$= \frac{(k + 1)(k + 2)}{2}$$

To prove this, start with the assumed equation and add $k + 1$ to both sides:

$$1 + 2 + 3 + \cdots + k + (k + 1) = \frac{k(k + 1)}{2} + (k + 1)$$

$$= \frac{k(k + 1)}{2} + \frac{2(k + 1)}{2}$$

$$= \frac{k(k + 1) + 2(k + 1)}{2}$$

$$= \frac{k^2 + k + 2k + 2}{2}$$

$$= \frac{k^2 + 3k + 2}{2}$$

$$= \frac{(k + 1)(k + 2)}{2}$$

Thus the conditions of mathematical induction are satisfied and

$$1 + 2 + 3 + \cdots + n = \frac{n(n + 1)}{2} \quad \text{for every positive integer } n$$

There are many analogies used to describe mathematical induction. One of these asks you to visualize the positive integers as a ladder of numbers. You demonstrate that you can get onto the ladder by showing a statement true for $n = 1$. You then demonstrate that from any step you can always advance to the next one by showing that if the statement is true for $n = k$, then it is true for $n = k + 1$. Thus if you can get on the ladder and always advance upward, you will touch all the steps.

EXAMPLE   Show that $2n - 1$ is the nth odd integer.

Solution   We use mathematical induction. Part 1 of the principle requires that substituting $n = 1$ into $2n - 1$ will yield the first odd integer:

$$2(1) - 1 = 2 - 1$$
$$= 1$$

Since 1 is the first odd integer, part 1 of the principle of mathematical induction is satisfied. For part 2 it is necessary to assume that $2k - 1$ is the kth odd integer and to show that $2(k + 1) - 1$ is the $(k + 1)$th odd integer. If $2k - 1$ is the kth odd integer, as assumed, then the $(k + 1)$th odd integer is two greater than $2k - 1$. That is, the $(k + 1)$th odd integer is $2k - 1 + 2 = 2k + 1$. But $2(k + 1) - 1 = 2k + 2 - 1 = 2k + 1$, which is the $(k + 1)$th odd integer. Thus if $2k - 1$ is the kth odd integer, then $2(k + 1) - 1$ is the

$(k + 1)$th odd integer. By mathematical induction, then, $2n - 1$ is the nth odd integer for all positive integers n.

EXAMPLE   Prove that $1 + 3 + 5 + \cdots + (2n - 1) = n^2$ for every positive integer n.

Solution   The statement is that the sum of the first n odd numbers is n^2. The proof uses mathematical induction. It is obviously true for $n = 1$ since $1 = 1^2$. To comply with the second requirement of mathematical induction, we assume that

$$1 + 3 + 5 + \cdots + (2k - 1) = k^2$$

and must show that

$$1 + 3 + 5 + \cdots + (2(k + 1) - 1) = (k + 1)^2$$

which simplifies to

$$1 + 3 + 5 + \cdots + (2k + 1) = (k + 1)^2$$

By hypothesis, the sum of the first k odd integers is k^2. That is,

$$1 + 3 + 5 + \cdots + (2k - 1) = k^2$$

The $(k + 1)$th odd integer is $2k + 1$, and when it is added to both sides of the above:

$$1 + 3 + 5 + \cdots + (2k - 1) + (2k + 1) = k^2 + (2k + 1)$$
$$= k^2 + 2k + 1$$
$$= (k + 1)^2$$

This establishes the second requirement in the principle of mathematical induction. Therefore, the sum of the first n odd integers is n^2 for all positive integers n.

EXERCISES A.2

Prove that the statements in Exercises 1–10 are true for all positive integers n.

1. $2n$ is the nth even positive integer.

2. $2 + 4 + 6 + \cdots + (2n) = n(n + 1)$

3. $1 + 4 + 9 + \cdots + n^2 = \dfrac{n(n + 1)(2n + 1)}{6}$

4. $1 + 4 + 7 + \cdots + (3n - 2) = \dfrac{n(3n - 1)}{2}$

5. $1 + 5 + 9 + \cdots + (4n - 3) = n(2n - 1)$

6. $1 + 6 + 11 + \cdots + (5n - 4) = \dfrac{n(5n - 3)}{2}$

7. $\dfrac{1}{2} + \dfrac{1}{4} + \cdots + \dfrac{1}{2^n} = 1 - \dfrac{1}{2^n}$

8. $\dfrac{1}{1 \cdot 2} + \dfrac{1}{2 \cdot 3} + \dfrac{1}{3 \cdot 4} + \cdots + \dfrac{1}{n(n + 1)} = \dfrac{n}{n + 1}$

9. $\dfrac{1}{1 \cdot 3} + \dfrac{1}{3 \cdot 5} + \dfrac{1}{5 \cdot 7} + \cdots + \dfrac{1}{(2n - 1)(2n + 1)} = \dfrac{n}{2n + 1}$

10. $a + (a + d) + (a + 2d) + \cdots + [a + (n - 1)d] = \dfrac{n}{2}[2a + (n - 1)d]$

if a and d are constants.

A.3 SUMMATION NOTATION

A useful notation, mentioned previously in this text, is summation or sigma notation to indicate the sum of a finite or infinite number of terms in a sequence. The Greek letter $\sum$ (sigma) is used for this purpose, hence the name *sigma notation*.

Let $f(i)$ be an expression involving the letter i. Then

$$\sum_{i=m}^{n} f(i) = f(m) + f(m + 1) + f(m + 2) + \cdots + f(n - 1) + f(n)$$

where $m \le n$, and m and n are integers.

EXAMPLE Let $f(i) = i$; expand

$$\sum_{i=1}^{5} f(i)$$

and find this sum.

Solution To *expand* means to write out each term. Thus

$$\sum_{i=1}^{5} f(i) = \sum_{i=1}^{5} i = 1 + 2 + 3 + 4 + 5 = 15$$

EXAMPLE Let $f(i) = (i + 2)/i$. Expand

$$\sum_{i=3}^{6} f(i)$$

and find this sum.

Solution

$$\sum_{i=3}^{6} f(i) = \sum_{i=3}^{6} \frac{i+2}{i} = \frac{3+2}{3} + \frac{4+2}{4} + \frac{5+2}{5} + \frac{6+2}{6}$$

$$= \tfrac{5}{3} + \tfrac{6}{4} + \tfrac{7}{5} + \tfrac{8}{6}$$

$$= 5\tfrac{9}{10}$$

The symbol $\sum$ is called the **summation symbol**; the letter i is called the **index of summation**. The set of integers which i can assume is called the **range of summation**.

The letter i is called a *dummy variable* and need not be i; often k, j, n, or m are used, and any symbol can be used for this purpose. For example,

$$\sum_{i=2}^{6} f(i) = \sum_{k=2}^{6} f(k) = \sum_{n=2}^{6} f(n), \quad \text{and so on}$$

There are several useful theorems associated with sigma notation, and they are stated here without proof.

Theorems on Sigma Notation

1. $\displaystyle\sum_{i=1}^{n} k = nk \qquad$ if k is a constant.

   Examples: $\displaystyle\sum_{i=1}^{5} 3 = 5(3) = 15$

   $$\sum_{i=1}^{50} \tfrac{1}{3} = 50(\tfrac{1}{3}) = 16\tfrac{2}{3}$$

2. $\displaystyle\sum_{i=1}^{n} \{f(i) \pm g(i)\} = \sum_{i=1}^{n} f(i) \pm \sum_{i=1}^{n} g(i)$

   Examples: $\displaystyle\sum_{i=1}^{3} \{i^2 + 2i\} = \sum_{i=1}^{3} i^2 + \sum_{i=1}^{3} 2i$

   $$= (1^2 + 2^2 + 3^2) + (2\cdot 1 + 2\cdot 2 + 2\cdot 3)$$

   $$\sum_{i=1}^{4} \left\{\frac{1}{i} - 6\right\} = \sum_{i=1}^{4} \frac{1}{i} - \sum_{i=1}^{4} 6$$

   $$= (\tfrac{1}{1} + \tfrac{1}{2} + \tfrac{1}{3} + \tfrac{1}{4}) - 4(6)$$

3. $\displaystyle\sum_{i=1}^{n} kf(i) = k \sum_{i=1}^{n} f(i) \qquad$ where k is a constant.

   Examples: $\displaystyle\sum_{i=1}^{5} 3i = 3 \sum_{i=1}^{5} i = 3(1 + 2 + 3 + 4 + 5)$

   $$\sum_{i=1}^{4} \frac{7}{i+1} = 7 \sum_{i=1}^{4} \frac{1}{i+1} = 7\left(\frac{1}{2} + \frac{1}{3} + \frac{1}{4} + \frac{1}{5}\right)$$

4. $\displaystyle\sum_{i=1}^{n+1} f(i) = \sum_{i=1}^{n} f(i) + f(n+1)$

   Example: $\displaystyle\sum_{i=1}^{6} i = \sum_{i=1}^{5} i + 6$

   $= (1 + 2 + 3 + 4 + 5) + 6$

It must be pointed out again that the range of summation need not start with 1; it can begin with any integer and can continue sequentially as far as indicated by the summation.

EXAMPLE

$$\sum_{k=3}^{5} (k+1) = \sum_{k=3}^{5} k + \sum_{k=3}^{5} 1$$

$$= (3 + 4 + 5) + 3(1)$$

$$= 15$$

Since 1 is a constant,

$$\sum_{k=3}^{5} 1 = \sum_{k=1}^{3} 1 = 3(1)$$

because there are three terms in the expansion.

EXAMPLE

$$\sum_{j=0}^{4} \frac{1}{j+2} = \frac{1}{0+2} + \frac{1}{1+2} + \frac{1}{2+2} + \frac{1}{3+2} + \frac{1}{4+2}$$

$$= \tfrac{1}{2} + \tfrac{1}{3} + \tfrac{1}{4} + \tfrac{1}{5} + \tfrac{1}{6}$$

EXAMPLE

$$\sum_{n=-2}^{2} (n+2) = (-2+2) + (-1+2) + (0+2) + (1+2) + (2+2)$$

$$= 0 + 1 + 2 + 3 + 4$$

Sigma notation is a very compact notation, and it is relatively simple to expand an expression in this notation. Of greater difficulty is the problem of replacing an expanded expression with its compact sigma form, and often some trial and error is necessary.

EXAMPLE Write $\tfrac{1}{3} + \tfrac{1}{4} + \tfrac{1}{5} + \tfrac{1}{6}$ in sigma notation.

Solution The denominators are a sequence of natural numbers from 3 to 6. Thus we can write

$$\sum_{i=3}^{6} \frac{1}{i}$$

We could also have written

$$\sum_{i=1}^{4} \frac{1}{i+2}$$

There are, of course, many more ways that this expression could be written.

EXAMPLE   Write $5 + 10 + 15 + 20 + \cdots$ in sigma notation.

Solution   The three dots, $\ldots$, indicate that the progression goes on indefinitely; this is indicated in the summation notation by the symbol ∞.
If $f(i) = 5i$, we test to see that this generates the desired terms:

$$f(1) = 5(1) = 5; \quad f(2) = 5(2) = 10;$$
$$f(3) = 5(3) = 15; \quad f(4) = 5(4) = 20$$

so

$$5 + 10 + 15 + 20 + \cdots = \sum_{i=1}^{\infty} 5i$$

EXAMPLE   Write $1 - \frac{1}{2} + \frac{1}{3} - \frac{1}{4} + \frac{1}{5}$ in sigma notation.

Solution   Since $1 = \frac{1}{1}$, it is relatively obvious that $f(i) = 1/i$. The signs which alternate from $+$ to $-$ are a little more complicated. Recognize that (-1) raised to an odd power is negative, whereas (-1) raised to an even power is positive. Since we want the even terms (the second and fourth terms) to be negative, we must write $(-1)^{i+1}$. Thus

$$\text{when } i = 1, (-1)^{1+1} = (-1)^2 = 1$$
$$\text{when } i = 2, (-1)^{2+1} = (-1)^3 = -1$$
$$\text{when } i = 3, (-1)^{3+1} = (-1)^4 = 1$$
$$\text{when } i = 4, (-1)^{4+1} = (-1)^5 = -1$$
$$\text{when } i = 5, (-1)^{5+1} = (-1)^6 = 1$$

Thus

$$1 - \frac{1}{2} + \frac{1}{3} - \frac{1}{4} + \frac{1}{5} = \sum_{i=1}^{5} (-1)^{i+1}\left(\frac{1}{i}\right)$$

Sigma notation is frequently used in statistics. The following example shows one such use for finding the arithmetic mean of a set of numerical data. Some other applications are to be found in the exercises.

EXAMPLE   Suppose we are told that for a given day the arrests for traffic violations at six intersections in Capitol City were 13, 24, 4, 18, 1, 6. The average number of arrests at any intersection is called the *arithmetic mean* of the data and is designated by $\bar{x}$. The formula for finding the arithmetic mean of n numerical values is

$$\bar{x} = \frac{1}{n} \sum_{i=1}^{n} x_i$$

where x_i is a value in the set of data. Find $\bar{x}$ for the traffic arrests cited above.

Solution $x_1 = 13$, $x_2 = 24$, $x_3 = 4$, $x_4 = 18$, $x_5 = 1$, $x_6 = 6$, and $n = 6$. Thus

$$\bar{x} = \frac{1}{n} \sum_{i=1}^{n} x_i$$

implies

$$\bar{x} = \tfrac{1}{6} \sum_{i=1}^{6} x_i = \tfrac{1}{6}(13 + 24 + 4 + 18 + 1 + 6)$$

$$\bar{x} = \tfrac{1}{6}(66) = 11$$

EXERCISES A.3

Write each of Exercises 1–10 in expanded notation.

1. $\displaystyle\sum_{i=1}^{7} (3i - 2)$

2. $\displaystyle\sum_{i=3}^{6} (i^2 + i)$

3. $\displaystyle\sum_{k=0}^{15} \frac{1}{k + 2}$

4. $\displaystyle\sum_{k=1}^{10} (-1)^{k+1}k$

5. $\displaystyle\sum_{n=0}^{\infty} (-1)^n n^2$

6. $\displaystyle\sum_{n=77}^{80} \frac{n + 1}{n - 1}$

7. $\displaystyle\sum_{i=1}^{6} (i^{-1} + i)$

8. $\displaystyle\sum_{i=1}^{3} (i^{-2} + 3)$

9. $\displaystyle\sum_{k=1}^{\infty} \frac{1}{3k - 1}$

10. $\displaystyle\sum_{k=1}^{6} 15(10)^{-k}$

Write each of Exercises 11–21 in sigma notation.

11. $2 + 4 + 6 + 8 + 10 + 12 + 14$
12. $1 + 3 + 5 + 7 + 9 + 11 + 13$
13. $1 + 4 + 9 + 16 + 25$
14. $2 - 2^2 + 2^3 - 2^4 + 2^5 - 2^6 + 2^7 - 2^8$
15. $(1 \times 2) + (2 \times 3) + (3 \times 4) + (4 \times 5)$

16. $\frac{1}{2} + \frac{1}{4} + \frac{1}{8} + \frac{1}{16} + \frac{1}{32} + \frac{1}{64} + \frac{1}{128}$

17. $-1 + 2 - 3 + 4 - 5 + 6 - 7 + 8 - 9 + 10 - 11 + 12$

18. $7 + \frac{7}{3} + \frac{7}{9} + \frac{7}{27} + \frac{7}{81}$

19. $1 + \frac{1}{5} + \frac{1}{25} + \frac{1}{125} + \frac{1}{625} + \frac{1}{3125}$

20. $\frac{3}{2} + \frac{4}{3} + \frac{5}{4} + \frac{6}{5}$

21. $\dfrac{1}{3\cdot5} + \dfrac{2}{4\cdot6} + \dfrac{3}{5\cdot7} + \dfrac{4}{6\cdot8} + \dfrac{5}{7\cdot9} + \cdots$

Use the theorems on summation to simplify each of Exercises 22–26, and find the sum.

22. $\displaystyle\sum_{k=1}^{6} (2k + 3)$ **23.** $\displaystyle\sum_{k=0}^{4} (k^2 + 2k + 1)$

24. $\displaystyle\sum_{n=1}^{5} \frac{4}{n+1}$ **25.** $\displaystyle\sum_{n=1}^{3} n(n + 2)$

26. $\displaystyle\sum_{j=3}^{n+1} 5j$

Use mathematical induction to verify each of Exercises 27–29.

27. $\displaystyle\sum_{i=1}^{n} i = \frac{n(n+1)}{2}$

28. $\displaystyle\sum_{i=1}^{n} i^2 = \frac{n(n+1)(2n+1)}{6}$

29. $\displaystyle\sum_{i=1}^{n} i^3 = \left[\frac{n(n+1)}{2}\right]^2$

30. A salesperson at a department store receives a basic hourly wage plus 6% commission on all sales which average more than $30 per hour worked. For five consecutive pay periods an employee's gross pay amounted to $320, $185, $216, $380, $275. Find the arithmetic mean of this data.

31. If x_i is a variable which can take on all values in a set of data, $\bar{x}$ is the arithmetic mean of this data, and n is the number of elements in the data set, then the *standard deviation* of x, denoted by σ (sigma), may be defined as

$$\sigma = \sqrt{\frac{1}{n} \sum_{i=1}^{n} (x_i - \bar{x})^2}$$

Find the standard deviation of the following: $x_1 = 5$, $x_2 = 8$, $x_3 = 2$, $x_4 = 3$, $x_5 = 2$.

32. The intelligence quotients of 20 pupils in the fifth grade of John F. Kennedy Elementary School were found to be as follows: 110, 112, 130, 108, 95, 92, 115, 132, 106, 111, 114, 98, 125, 118, 102, 114, 126, 110, 122, 102.

  a. Find the arithmetic mean, $\bar{x}$, of this set of data.

  b. Find the standard deviation, σ.

33. The *variance* of a set of numerical data is defined as the square of the standard deviation, σ^2. The table below describes the price of several small appliances and the "frequency" of occurrence (how many appliances at each price).

  a. Fill in the values $(\bar{x} - x_i)$.

  b. Compute and complete the entries for $(\bar{x} - x_i)^2$.

  c. Multiply $(\bar{x} - x_i)^2$ by its corresponding frequency entry to obtain $f_i(\bar{x} - x_i)^2$.

  d. Compute the variance,

$$\sigma^2 = \frac{1}{n} \sum_{i=1}^{n} f_i(\bar{x} - x_i)^2$$

The first line of the table is completed as an illustration.

Price of Appliance x_i	Frequency (f_i)	$(\bar{x} - x_i)$	$(\bar{x} - x_i)^2$	$f_i(\bar{x} - x_i)^2$
$20	12	14	196	2,352
$24	18			
$30	22			
$38	6			
$42	16			
$50	5			
$\bar{x} = \$34$				

34. For Exercise 33, find

$$\sum_{i=1}^{6} f_i(\bar{x} - x_i)$$

A.4 REVIEW OF LINEAR SYSTEMS OF EQUATIONS

DEFINITION

A **linear equation in n variables**, where n is any natural number, is an equation of the form

$$a_1 x_1 + a_2 x_2 + a_3 x_3 + \cdots + a_n x_n = k$$

where $x_1, x_2, x_3, \ldots, x_n$ are n variables and $a_1, a_2, a_3, \ldots, a_n$, and k are constants, and not all a's are zero.

EXAMPLES

1. $3x_1 + 2x_2 = 5$ is an equation in two variables. This equation could also have been written

$$3x + 2y = 5$$

   where $x = x_1, y = x_2$.

2. $6x_1 - 2x_2 + x_3 + 4x_4 - 7x_5 = 0$ is an equation in five variables.

A solution of a linear equation is an ordered n-tuple $(p_1, p_2, p_3, \ldots, p_n)$ which makes the linear equation true when x_1 is replaced by p_1, x_2 by p_2, x_3 by $p_3, \ldots, x_n$ by p_n.

For example, the ordered pair $(1, 1)$ is a solution for $3x_1 + 2x_2 = 5$. There are infinitely many ordered pairs which are solutions for $3x_1 + 2x_2 = 5$. Some of these are $(0, \frac{5}{2})$, $(\frac{5}{3}, 0)$, $(-1, 4)$, $(10, -\frac{25}{2})$.

DEFINITIONS

The **solution set of a linear equation** in n variables is the set consisting of all possible solutions of the equation.

A **linear system of equations** is a set of two or more linear equations.

EXAMPLES

1. $3x_1 + 2x_2 = 5$
   $x_1 - x_2 = 2$

   is a system of two linear equations in two variables.

2. $x + 3y + z = 2$
   $5x - y + 4z = 6$
   $x + 2y - z = 0$

   is a system of three linear equations in three variables.

DEFINITION

The **solution set of a linear system** is the intersection of the solution sets of each equation in the system.

For example, symbolically, the solution set of the system of equations

$$a_1x_1 + b_1x_2 = k_1$$
$$a_2x_1 + b_2x_2 = k_2$$

is the set

$$\{(x_1, x_2)\mid a_1x_1 + b_1x_2 = k_1\} \cap \{(x_1, x_2)\mid a_2x_1 + b_2x_2 = k_2\}$$

where a_1 and a_2 are constants, not both zero, and b_1 and b_2 are constants, not both zero, and k_1 and k_2 are constants.

We will start the discussion by finding the solution set of linear systems of two equations in two variables. There are three possibilities for the solution set:

1. There is exactly one ordered pair in the solution set. Graphically, we know that a linear equation in two variables graphs as a straight line, so the two lines whose equations make up the system intersect and have exactly one point in common.

2. The solution set is empty. The graph of the equations is a pair of parallel lines, and parallel lines do not intersect.

3. The solution set is infinite, and all the solutions of one equation also satisfy the second equation. Graphically, the two equations represent the same line.

There are several ways of finding the solution of this type of system.

1. Graph the two lines and determine the solution set from the graph. The method is suitable for two equations in two variables (although it has shortcomings even then in terms of reading nonintegral solutions), but it does not readily extend to the solution of n equations in n variables.

2. *The addition method.* Consider the system

$$a_1x_1 + b_1x_2 = k_1$$
$$a_2x_1 + b_2x_2 = k_2$$

Multiply the first equation by a_2 and the second equation by $(-a_1)$:

$$a_2a_1x_1 + a_2b_1x_2 = a_2k_1$$
$$-a_1a_2x_1 - a_1b_2x_2 = -a_1k_2$$

By adding these two equations, we can eliminate the x_1 term and arrive at one equation in one variable:

$$(a_2b_1 - a_1b_2)x_2 = a_2k_1 - a_1k_2$$

Solving for x_2,

$$x_2 = \frac{a_2k_1 - a_1k_2}{a_2b_1 - a_1b_2}$$

provided $a_2b_1 - a_1b_2 \neq 0$.

In order to find x_1, we can either repeat the process as outlined, but eliminate x_2 instead of x_1, or we can substitute the value

$$\frac{a_2k_1 - a_1k_2}{a_2b_1 - a_1b_2}$$

for x_2 in either equation and obtain the x_1 value of the ordered pair, $(x_1, x_2) = (p_1, p_2)$, of the solution.

EXAMPLE Use the addition method to find the solution set of the system

$$3x_1 + 2x_2 = 5$$
$$x_1 - 3x_2 = -13$$

Solution By multiplying the second equation by -3 and adding the two equations, we obtain one equation in one unknown:

$$3x_1 + 2x_2 = 5$$
$$\underline{-3x_1 + 9x_2 = 39}$$
$$11x_2 = 44$$
$$x_2 = 4$$

Substituting $x_2 = 4$ in the second equation gives

$$x_1 - 3(4) = -13$$
$$x_1 = -1$$

Thus $\{(-1, 4)\}$ is the solution set for this system. To check,

$$3(-1) + 2(4) = -3 + 8 = 5$$
$$(-1) - 3(4) = -1 - 12 = -13$$

3. *The substitution method.* In this method we solve for one variable in either equation, and substitute this value in the other equation. Then solve the resulting equation for the one variable.

EXAMPLE   Use the substitution method to solve the system

$$2x_1 - x_2 = 5$$
$$x_1 + 4x_2 = 3$$

Solution   Select x_1 in the second equation and solve for x_1 in terms of x_2:

$$x_1 = 3 - 4x_2$$

Substituting this value for x_1 in the first equation yields

$$2(3 - 4x_2) - x_2 = 5$$
$$6 - 8x_2 - x_2 = 5$$
$$-9x_2 = -1$$
$$x_2 = \tfrac{1}{9}$$

Substituting $x_2 = \tfrac{1}{9}$ in the equation,

$$x_1 = 3 - 4x_2$$
$$x_1 = 3 - 4(\tfrac{1}{9})$$
$$x_1 = 2\tfrac{5}{9}$$

Thus the solution set is $\{(2\tfrac{5}{9}, \tfrac{1}{9})\}$.
Checking,

$$2(2\tfrac{5}{9}) - \tfrac{1}{9} = 5\tfrac{1}{9} - \tfrac{1}{9} = 5$$
$$2\tfrac{5}{9} + 4(\tfrac{1}{9}) = 2\tfrac{5}{9} + \tfrac{4}{9} = 3$$

Other methods of solution are by matrices, as shown in a previous chapter, and by determinants, which will be discussed in the next section.

The elements in the nonempty solution set of three equations in three variables are ordered triples.

For example, the solution set for the equation $3x_1 + x_2 - x_3 = 4$ contains the ordered triple $(0, 5, 1)$ because $3(0) + 5 - 1 = 4$. Since the solution set of a system is the intersection of the solution sets for all equations in the system, any element in the solution set of a linear system of three equations in three variables must be an ordered triple.

To solve such a system it is desirable to eliminate a variable in order to reduce the system to two equations in two unknowns, and then to proceed as outlined for the two-equation, two-variable system.

EXAMPLE   Find the solution set for the system

$$2x_1 + x_2 - x_3 = 3 \tag{1}$$
$$4x_1 + 2x_2 + 3x_3 = 1 \tag{2}$$
$$6x_1 + 3x_2 - 2x_3 = 2 \tag{3}$$

Solution Any variable may be selected for elimination; let us select x_3. Multiply equation (1) by 3 and add to equation (2):

$$6x_1 + 3x_2 - 3x_3 = 9$$
$$\underline{4x_1 + 2x_2 + 3x_3 = 1}$$
$$10x_1 + 5x_2 \qquad = 10 \qquad\qquad (4)$$

Now multiply equation (1) by -2 and add to equations (3):

$$-4x_1 - 2x_2 + 2x_3 = -6$$
$$\underline{6x_1 + 3x_2 - 2x_3 = 2}$$
$$2x_1 + x_2 \qquad = -4 \qquad\qquad (5)$$

Now solve the system represented by equations (4) and (5):

$$10x_1 + 5x_2 = 10 \qquad\qquad (4)$$
$$2x_1 + x_2 = -4 \qquad\qquad (5)$$

Multiply equation (5) by -5 and add to equation (4), thus eliminating x_2:

$$10x_1 + 5x_2 = 10$$
$$\underline{-10x_1 - 5x_2 = 20}$$
$$0 = 30$$

This cannot happen. $0 \neq 30$, and the system has an empty solution set, $\varnothing$.

EXAMPLE Solve the system

$$x_1 - x_2 + 2x_3 = 2 \qquad\qquad (1)$$
$$x_1 + x_2 + x_3 = 3 \qquad\qquad (2)$$
$$3x_1 - x_2 + 2x_3 = 4 \qquad\qquad (3)$$

Solution Select x_2 as the first variable to be eliminated. Add equations (1) and (2), and add equations (2) and (3):

$$2x_1 + 3x_3 = 5 \qquad\qquad (4)$$
$$4x_1 + 3x_3 = 7 \qquad\qquad (5)$$

Now multiply equation (4) by -1 and add to equation (5) to eliminate x_3:

$$-2x_1 - 3x_3 = -5$$
$$\underline{4x_1 + 3x_3 = 7}$$
$$2x_1 \qquad = 2$$
$$x_1 \qquad = 1$$

Substituting $x_1 = 1$ in equation (4) gives

$$2(1) + 3x_3 = 5$$
$$\underline{3x_3 = 3}$$
$$x_3 = 1$$

Now taking $x_1 = 1$ and $x_3 = 1$ and substituting in equation (2) gives

$$1 + x_2 + 1 = 3$$
$$x_2 = 1$$

Thus the ordered triple $(x_1, x_2, x_3) = (1, 1, 1)$ is the solution to the system, and $\{(1, 1, 1)\}$ is the solution set.

The solution set for a linear system of n equations in n variables contains either

1. Exactly one ordered n-tuple
2. No elements—the set is empty
3. An infinite number of n-tuples

Systems with exactly one element in the solution set and an empty solution set have been given as examples. Below is an example of a system of three equations in three variables with infinitely many elements in its solution set.

EXAMPLE   Find the solution set for the system

$$x_1 - x_2 + 2x_3 = 0 \tag{1}$$
$$2x_1 + x_2 - x_3 = 0 \tag{2}$$
$$x_1 + 5x_2 - 8x_3 = 0 \tag{3}$$

Solution   Select x_2 as the first variable to be eliminated. Add equations (1) and (2), resulting in equation (4); then multiply equation (1) by 5 and add equations (1) and (3):

$$3x_1 + x_3 = 0 \tag{4}$$
$$6x_1 + 2x_3 = 0 \tag{5}$$

Divide (5) by 2:

$$3x_1 + x_3 = 0 \tag{4}$$
$$3x_1 + x_3 = 0 \tag{5}$$

Thus equations (4) and (5) have the same solution set.
Let

$$x_1 = c, \quad \text{any constant}$$

Then

$$x_3 = -3c \quad \text{and} \quad x_2 = -5c$$

The solution set for this system is

$$\{(x_1, x_2, x_3)|\; x_1 = c, x_2 = -5c, x_3 = -3c, c \text{ is any constant}\}$$

Among the elements in the solution set are the ordered triples $(1, -5, -3)$, $(0, 0, 0)$, $(-1, 5, 3)$. The reader should verify that these ordered triples do indeed satisfy the system.

Solving four equations in four variables involves a similar procedure to the one outlined for the three-equation, three-variable system, and in theory this method of solution can be extended to n equations in n variables. In practice, this method is very tedious when $n \geq 4$, and a matrix solution aided by a computer is more practical.

EXERCISES A.4

Find the solution set for each of the linear systems in Exercises 1–12.

1. $x_1 + 3x_2 = 4$
$\quad x_1 - x_2 = 0$

2. $x + y = 2$
$\quad x + 2y = 8$

3. $\quad x_1 - x_2 = 3$
$\quad -2x_1 + 2x_2 = -6$

4. $\quad 6x + 8y = 3$
$\quad -3x - 4y = 5$

5. $x_1 + x_2 + x_3 = 1$
$\quad 2x_1 - x_2 + 3x_3 = 2$
$\quad 2x_1 - x_2 - x_3 = 2$

6. $x + y - z = 2$
$\quad 4x - y + 3z = 1$
$\quad 2x - 3y + z = 4$

7. $x_1 - 2x_2 + x_3 = 2$
$\quad 3x_1 - x_2 + 2x_3 = 4$
$\quad x_1 + x_2 - x_3 = 1$

8. $3x_1 - x_2 - x_3 = -2$
$\quad 5x_1 + 2x_2 + x_3 = 22$
$\quad 2x_1 + 4x_2 - 5x_3 = 0$

9. $3x_1 - x_2 + 2x_3 = 7$
$\quad 2x_1 - 2x_2 + x_3 = 4$
$\quad x_1 + x_2 + x_3 = 5$

10. $2x_1 - x_2 = 2$
$\quad x_1 + x_3 = 4$
$\quad -x_2 + x_3 = -6$

11. $x_1 + 2x_2 - 3x_3 + 2x_4 = 6$
$\quad 2x_1 + 3x_2 + 4x_3 + x_4 = -5$
$\quad 2x_1 + x_2 + x_3 - x_4 = -4$
$\quad 3x_1 - x_2 + 2x_3 - x_4 = 0$

12. $4x_1 - 3x_2 - x_3 - 2x_4 = 1$
$\quad 3x_1 + 2x_2 - 6x_3 + x_4 = 2$
$\quad 2x_1 - 2x_2 + 3x_3 + 2x_4 = 2$
$\quad x_1 + 3x_2 - 2x_3 + 5x_4 = 0$

A system of equations which has more variables than equations is said to be **underdetermined.**

$$2x_1 + 3x_2 + x_3 = 5$$
$$x_1 - x_2 - 2x_3 = 0$$

is such a system. *An underdetermined system has either an infinite number of solutions or no solution. The procedure for solution is the same as for n equations in n variables. Solve each of the underdetermined systems in Exercises 13 and 14.*

13. $2x_1 + 3x_2 + x_3 = 5$
  $x_1 - x_2 - 2x_3 = 0$

14. $3x_1 + x_2 - 2x_3 = 1$
  $5x_1 - x_2 + 3x_3 = 4$

A system of equations which has more equations than variables is said to be **overdetermined**. *An example of an overdetermined system is*

$$x_1 - 2x_2 = 5$$
$$2x_1 + 3x_2 = 7$$
$$x_1 + 5x_2 = 2$$

To solve an overdetermined system, select n equations in n variables and solve as above. Then check the unused equation(s) to see if the solutions satisfy these equations also. If they do, then the system has a solution. If they do not, then the solution set is empty. Solve each of the overdetermined systems in Exercises 15 and 16.

15. $x_1 - 2x_2 = 5$
  $2x_1 + 3x_2 = 7$
  $x_1 + 5x_2 = 2$

16. $x_1 + x_2 + x_3 = 6$
  $5x_1 + 2x_2 - 2x_3 = 12$
  $2x_1 - x_2 + x_3 = 3$
  $3x_1 - x_2 - x_3 = 2$

A.5 DETERMINANTS

A real number can be assigned to every square matrix whose elements are real numbers. The real number is called the *determinant* of the matrix.

DEFINITION

A determinant of the second order, denoted by the symbol $\begin{vmatrix} a_1 & b_1 \\ a_2 & b_2 \end{vmatrix}$, where a_1, b_1, a_2, and b_2 are real numbers, is the real number, $a_1b_2 - a_2b_1$—that is,

$$\begin{vmatrix} a_1 & b_1 \\ a_2 & b_2 \end{vmatrix} = a_1b_2 - a_2b_1$$

EXAMPLE   Evaluate the determinant $\begin{vmatrix} 3 & 1 \\ 2 & 4 \end{vmatrix}$.

Solution $\begin{vmatrix} 3 & 1 \\ 2 & 4 \end{vmatrix} = 3(4) - 2(1) = 12 - 2 = 10$

EXAMPLE   Evaluate the determinant $\begin{vmatrix} 2 & -1 \\ -3 & -5 \end{vmatrix}$.

Solution $\begin{vmatrix} 2 & -1 \\ -3 & -5 \end{vmatrix} = 2(-5) - (-1)(-3) = -10 - 3 = -13$

DEFINITION

A determinant of the third order, denoted by the symbol

$$\begin{vmatrix} a_1 & b_1 & c_1 \\ a_2 & b_2 & c_2 \\ a_3 & b_3 & c_3 \end{vmatrix}$$

where the elements are real numbers, is the real number

$$a_1 \begin{vmatrix} b_2 & c_2 \\ b_3 & c_3 \end{vmatrix} - a_2 \begin{vmatrix} b_1 & c_1 \\ b_3 & c_3 \end{vmatrix} + a_3 \begin{vmatrix} b_1 & c_1 \\ b_2 & c_2 \end{vmatrix}$$

that is,

$$\begin{vmatrix} a_1 & b_1 & c_1 \\ a_2 & b_2 & c_2 \\ a_3 & b_3 & c_3 \end{vmatrix} = a_1 \begin{vmatrix} b_2 & c_2 \\ b_3 & c_3 \end{vmatrix} - a_2 \begin{vmatrix} b_1 & c_1 \\ b_3 & c_3 \end{vmatrix} + a_3 \begin{vmatrix} b_1 & c_1 \\ b_2 & c_2 \end{vmatrix}$$

Thus it may be observed that the association of a real number, the determinant, to a square matrix M defines a function $d(M)$. The domain is the set of square matrices, and the range is the set of real numbers. The function is the set of ordered pairs $\{(M, d(M))\}$.

The **order** of a determinant is the number of rows (or columns) in the square array of the matrix associated with the determinant.

When the determinant is being indicated, the array of numbers is enclosed by *vertical lines*, in contrast to the use of brackets used to designate the matrix.

The *expansion* of a determinant is the indicated calculation in the definition of the determinant.

The **minor** of an element of a determinant is the determinant obtained by deleting the row and column in which the element lies.

For example, for a determinant of order three,

$$A_1 = \begin{vmatrix} b_2 & c_2 \\ b_3 & c_3 \end{vmatrix} \quad \text{is the minor of } a_1$$

$$A_2 = \begin{vmatrix} b_1 & c_1 \\ b_3 & c_3 \end{vmatrix} \quad \text{is the minor of } a_2$$

$$A_3 = \begin{vmatrix} b_1 & c_1 \\ b_2 & c_2 \end{vmatrix} \quad \text{is the minor of } a_3$$

$$B_2 = \begin{vmatrix} a_1 & c_1 \\ a_3 & c_3 \end{vmatrix} \quad \text{is the minor of } b_2 \text{, and so on}$$

Thus

$$D = \begin{vmatrix} a_1 & b_1 & c_1 \\ a_2 & b_2 & c_2 \\ a_3 & b_3 & c_3 \end{vmatrix} = a_1 A_1 - a_2 A_2 + a_3 A_3$$

is called an expansion of D by the minors of the elements of the first column.

THEOREM I

Any row or column may be used to expand a determinant of the third order—that is

$$\begin{aligned} D &= & a_1 A_1 - a_2 A_2 + a_3 A_3 \\ &= & -b_1 B_1 + b_2 B_2 - b_3 B_3 \\ &= & c_1 C_1 - c_2 C_2 + c_3 C_3 \\ &= & a_1 A_1 - b_1 B_1 + c_1 C_1 \\ &= & -a_2 A_2 + b_2 B_2 - c_2 C_2 \\ &= & a_3 A_3 - b_3 B_3 + c_3 C_3 \end{aligned}$$

Whether a product in the expansion is to be multiplied by $+1$ or -1 is determined by the following array, called the "checkerboard of signs":

$$\begin{vmatrix} + & - & + \\ - & + & - \\ + & - & + \end{vmatrix}$$

The proof given below shows that $a_1A_1 - a_2A_2 + a_3A_3 = -b_1B_1 + b_2B_2 - b_3B_3$. The other cases are proved similarly and are left for the student to verify.

$$a_1A_1 - a_2A_2 + a_3A_3 = a_1\begin{vmatrix} b_2 & c_2 \\ b_3 & c_3 \end{vmatrix} - a_2\begin{vmatrix} b_1 & c_1 \\ b_3 & c_3 \end{vmatrix} + a_3\begin{vmatrix} b_1 & c_1 \\ b_2 & c_2 \end{vmatrix}$$

$$= a_1(b_2c_3 - b_3c_2) - a_2(b_1c_3 - b_3c_1) + a_3(b_1c_2 - b_2c_1)$$

$$= a_1b_2c_3 + a_2b_3c_1 + a_3b_1c_2 - a_1b_3c_2 - a_2b_1c_3$$
$$\quad - a_3b_2c_1$$

$$= -b_1(a_2c_3 - a_3c_2) + b_2(a_1c_3 - a_3c_1)$$
$$\quad - b_3(a_1c_2 - a_2c_1)$$

$$= -b_1\begin{vmatrix} a_2 & c_2 \\ a_3 & c_3 \end{vmatrix} + b_2\begin{vmatrix} a_1 & c_1 \\ a_3 & c_3 \end{vmatrix} - b_3\begin{vmatrix} a_1 & c_1 \\ a_2 & c_2 \end{vmatrix}$$

$$= -b_1B_1 + b_2B_2 - b_3B_3$$

EXAMPLE Expand

$$\begin{vmatrix} 2 & 3 & -1 \\ 1 & -2 & 2 \\ -4 & -1 & 5 \end{vmatrix}$$

by row 1.

Solution

$$D = 2\begin{vmatrix} -2 & 2 \\ -1 & 5 \end{vmatrix} - 3\begin{vmatrix} 1 & 2 \\ -4 & 5 \end{vmatrix} + (-1)\begin{vmatrix} 1 & -2 \\ -4 & -1 \end{vmatrix}$$

$$= 2(-10 - (-2)) - 3(5 - (-8)) - (-1 - 8)$$

$$= 2(-8) - 3(13) - (-9)$$

$$= -16 - 39 + 9 = -46$$

EXAMPLE Expand

$$\begin{vmatrix} 3 & 2 & -5 \\ 2 & 0 & 1 \\ 1 & 0 & 4 \end{vmatrix}$$

by column 2.

Solution

$$D = -2\begin{vmatrix} 2 & 1 \\ 1 & 4 \end{vmatrix} + 0\begin{vmatrix} 3 & -5 \\ 1 & 4 \end{vmatrix} - 0\begin{vmatrix} 3 & -5 \\ 2 & 1 \end{vmatrix}$$

$$= -2(8 - 1) + 0 - 0$$

$$= -14$$

The two preceding examples illustrate that the determinant is much easier to evaluate when there are zeros in a row or column. Thus it is desirable to establish some theorems indicating the transformations that can be performed on determinants in order to obtain an equal determinant with zeros as some of its elements.

THEOREM 2

Two determinants of the same order are equal if one is obtained from the other by **interchanging the rows and columns.** In symbols,

$$\begin{vmatrix} a_1 & b_1 \\ a_2 & b_2 \end{vmatrix} = \begin{vmatrix} a_1 & a_2 \\ b_1 & b_2 \end{vmatrix} \quad \text{and} \quad \begin{vmatrix} a_1 & b_1 & c_1 \\ a_2 & b_2 & c_2 \\ a_3 & b_3 & c_3 \end{vmatrix} = \begin{vmatrix} a_1 & a_2 & a_3 \\ b_1 & b_2 & b_3 \\ c_1 & c_2 & c_3 \end{vmatrix}$$

This is proved by applying the definition, regrouping the terms, and applying the definition again.

THEOREM 3

If **two rows (or two columns) of a determinant are interchanged,** then the resulting determinant is the negative of the original one. For example,

$$\text{let } D = \begin{vmatrix} a_1 & b_1 & c_1 \\ a_2 & b_2 & c_2 \\ a_3 & b_3 & c_3 \end{vmatrix} \quad \text{and} \quad \text{let } E = \begin{vmatrix} c_1 & b_1 & a_1 \\ c_2 & b_2 & a_2 \\ c_3 & b_3 & a_3 \end{vmatrix}$$

Then $E = -D$.

This is proved by expanding E, regrouping the terms, and applying the definition.

THEOREM 4

If **each element of a row (or column) is multiplied by a constant,** then the determinant is multiplied by a constant. For example,

$$\begin{vmatrix} ka_1 & kb_1 & kc_1 \\ a_2 & b_2 & c_2 \\ a_3 & b_3 & c_3 \end{vmatrix} = k \begin{vmatrix} a_1 & b_1 & c_1 \\ a_2 & b_2 & c_2 \\ a_3 & b_3 & c_3 \end{vmatrix}$$

and

$$\begin{vmatrix} 6 & 9 & -12 \\ 2 & 1 & 5 \\ 3 & -1 & 2 \end{vmatrix} = 3 \begin{vmatrix} 2 & 3 & -4 \\ 2 & 1 & 5 \\ 3 & -1 & 2 \end{vmatrix}$$

THEOREM 5

If the **corresponding elements of two rows (or columns) are equal**, then the value of the determinant is 0. For example,

$$\begin{vmatrix} 1 & 4 & 1 \\ 3 & 1 & 3 \\ -2 & 5 & -2 \end{vmatrix} = 0 \quad \text{and} \quad \begin{vmatrix} 2 & 1 & 5 \\ 2 & 1 & 5 \\ 3 & 7 & 9 \end{vmatrix} = 0$$

THEOREM 6

If each element of a row (or column) is multiplied by a constant and then added to the corresponding element of another row (or column), then the resulting determinant is equal to the original determinant. For example,

$$\begin{vmatrix} a_1 + ka_3 & b_1 + kb_3 & c_1 + kc_3 \\ a_2 & b_2 & c_2 \\ a_3 & b_3 & c_3 \end{vmatrix} = \begin{vmatrix} a_1 & b_1 & c_1 \\ a_2 & b_2 & c_2 \\ a_3 & b_3 & c_3 \end{vmatrix}$$

and

$$\begin{vmatrix} a_1 & b_1 + kc_1 & c_1 \\ a_2 & b_2 + kc_2 & c_2 \\ a_3 & b_3 + kc_3 & c_3 \end{vmatrix} = \begin{vmatrix} a_1 & b_1 & c_1 \\ a_2 & b_2 & c_2 \\ a_3 & b_3 & c_3 \end{vmatrix}$$

This theorem may be proved by expanding the determinant by the altered row (or column) and expressing the expansion as the sum of two determinants, one of which is zero.

EXAMPLE   Find a determinant equal to

$$\begin{vmatrix} -2 & 3 & 1 \\ 1 & 2 & 3 \\ 2 & 3 & 3 \end{vmatrix}$$

having zeros everywhere in column 3 except the first row. Expand the resulting determinant.

Solution

$$\begin{vmatrix} -2 & 3 & 1 \\ 1 & 2 & 3 \\ 2 & 3 & 3 \end{vmatrix} = \begin{vmatrix} -2 & 3 & 1 \\ 7 & -7 & 0 \\ 2 & 3 & 3 \end{vmatrix} \quad \text{(Multiply row 1 by } -3 \text{ and add to row 2)}$$

$$= \begin{vmatrix} -2 & 3 & 1 \\ 7 & -7 & 0 \\ 8 & -6 & 0 \end{vmatrix} \quad \text{(Multiply row 1 by } -3 \text{ and add to row 3)}$$

$$= 1 \begin{vmatrix} 7 & -7 \\ 8 & -6 \end{vmatrix} = 7(-6) - 8(-7) = -42 + 56 = 14$$

EXAMPLE   Expand the determinant having zeros everywhere in row 2 except column 2 and equal to

$$\begin{vmatrix} 2 & -1 & 4 \\ 5 & 1 & -2 \\ 3 & -3 & -4 \end{vmatrix}$$

Solution

$$\begin{vmatrix} 2 & -1 & 4 \\ 5 & 1 & -2 \\ 3 & -3 & -4 \end{vmatrix} = \begin{vmatrix} 7 & -1 & 4 \\ 0 & 1 & -2 \\ 18 & -3 & -4 \end{vmatrix}$$ (Multiply column 2 by -5 and add to column 1)

$$= \begin{vmatrix} 7 & -1 & 2 \\ 0 & 1 & 0 \\ 18 & -3 & -10 \end{vmatrix}$$ (Multiply column 2 by 2 and add to column 3)

$$= 1 \begin{vmatrix} 7 & 2 \\ 18 & -10 \end{vmatrix}$$

$$= 7(-10) - 18(2) = -70 - 36 = -106$$

Determinants may be used to solve linear systems of equations. Two cases of systems are shown: two equations in two variables and three equations in three variables. The method of solution may be extended to n equations in n variables, but again, the arithmetic practicality for $n > 4$ is questionable.

Case 1—Two equations, two variables: In solving the system

$$a_1 x + b_1 y = k_1$$
$$a_2 x + b_2 y = k_2$$

by the addition method, the following equations were obtained:

$$(a_1 b_2 - a_2 b_1)x = b_2 k_1 - b_1 k_2$$
$$(a_1 b_2 - a_2 b_1)y = a_1 k_2 - a_2 k_1$$

Now letting

$$D = \begin{vmatrix} a_1 & b_1 \\ a_2 & b_2 \end{vmatrix}, \quad X = \begin{vmatrix} k_1 & b_1 \\ k_2 & b_2 \end{vmatrix}, \quad \text{and} \quad Y = \begin{vmatrix} a_1 & k_1 \\ a_2 & k_2 \end{vmatrix}$$

these equations become $Dx = X$, $Dy = Y$.

The determinant, D, is called the *determinant of the coefficients*.

If $D \neq 0$, then there is a unique solution, $(X/D, Y/D)$ (the case of two intersecting lines).

If $D = 0$ and $X = Y = 0$, then the solution set is infinite (the case of coincident lines).

If $D = 0$ and either $X \neq 0$ or $Y \neq 0$, then the solution set is empty (the case of parallel lines).

Case 2—Three equations, three variables: If in the system

$$a_1x + b_1y + c_1z = k_1$$
$$a_2x + b_2y + c_2z = k_2$$
$$a_3x + b_3y + c_3z = k_3$$

the first equation is multiplied by A_1, the minor of a_1 for the determinant of the coefficients, and the second equation is multiplied by $-A_2$, and the third equation is multiplied by A_3, and the resulting three equations are added, then

$$(a_1A_1 - a_2A_2 + a_3A_3)x + (b_1A_1 - b_2A_2 + b_3A_3)y$$
$$+ (c_1A_1 - c_2A_2 + c_3A_3)z = k_1A_1 - k_2A_2 + k_3A_3$$

The coefficients of the variables and the constant term can be recognized as the expansions of determinants as follows:

$$\begin{vmatrix} a_1 & b_1 & c_1 \\ a_2 & b_2 & c_2 \\ a_3 & b_3 & c_3 \end{vmatrix}x + \begin{vmatrix} b_1 & b_1 & c_1 \\ b_2 & b_2 & c_2 \\ b_3 & b_3 & c_3 \end{vmatrix}y + \begin{vmatrix} c_1 & b_1 & c_1 \\ c_2 & b_2 & c_2 \\ c_3 & b_3 & c_3 \end{vmatrix}z = \begin{vmatrix} k_1 & b_1 & c_1 \\ k_2 & b_2 & c_2 \\ k_3 & b_3 & c_3 \end{vmatrix}$$

Thus
$$\begin{vmatrix} a_1 & b_1 & c_1 \\ a_2 & b_2 & c_2 \\ a_3 & b_3 & c_3 \end{vmatrix}x + 0 \cdot y + 0 \cdot z = \begin{vmatrix} k_1 & b_1 & c_1 \\ k_2 & b_2 & c_2 \\ k_3 & b_3 & c_3 \end{vmatrix}$$

since the determinants which are the coefficients of y and z each contain two identical columns and thus equal 0.

Similar equations may be obtained where x and y or x and z are eliminated. Finally,

$$\begin{vmatrix} a_1 & b_1 & c_1 \\ a_2 & b_2 & c_2 \\ a_3 & b_3 & c_3 \end{vmatrix}x = \begin{vmatrix} k_1 & b_1 & c_1 \\ k_2 & b_2 & c_2 \\ k_3 & b_3 & c_3 \end{vmatrix}$$

$$\begin{vmatrix} a_1 & b_1 & c_1 \\ a_2 & b_2 & c_2 \\ a_3 & b_3 & c_3 \end{vmatrix}y = \begin{vmatrix} a_1 & k_1 & c_1 \\ a_2 & k_2 & c_2 \\ a_3 & k_3 & c_3 \end{vmatrix}$$

$$\begin{vmatrix} a_1 & b_1 & c_1 \\ a_2 & b_2 & c_2 \\ a_3 & b_3 & c_3 \end{vmatrix}z = \begin{vmatrix} a_1 & b_1 & k_1 \\ a_2 & b_2 & k_2 \\ a_3 & b_3 & k_3 \end{vmatrix}$$

The determinant of the coefficients, D, may be readily obtained from the equations when they are expressed in the form stated earlier. The determinants on the right of each equation above can be obtained from the determinant of the coefficients by replacing the column containing the coefficients of the variable in the equation by the constant terms of the system.

Now designating the determinants on the right of each equation by X, Y, and Z, respectively, these equations can be expressed as follows:

$$Dx = X$$
$$Dy = Y$$
$$Dz = Z$$

If $D \neq 0$, then there is exactly one solution: $(X/D,\ Y/D,\ Z/D)$.

If $D = 0$ and $X = Y = Z = 0$, then the solution set is infinite.

If $D = 0$ and either $X \neq 0$ or $Y \neq 0$ or $Z \neq 0$, then the solution set is empty.

This solution of linear systems using determinants is known as *Cramer's rule* in honor of the Swiss mathematician Gabriel Cramer (1704–52).

EXAMPLE   Solve by using determinants and check:

$$x + y + z = 4$$
$$2x - y - 2z = -1$$
$$x - 2y - z = 1$$

Solution

$$D = \begin{vmatrix} 1 & 1 & 1 \\ 2 & -1 & -2 \\ 1 & -2 & -1 \end{vmatrix} = \begin{vmatrix} 1 & 1 & 2 \\ 2 & -1 & 0 \\ 1 & -2 & 0 \end{vmatrix} = 2\begin{vmatrix} 2 & -1 \\ 1 & -2 \end{vmatrix} = 2(-4 + 1) = -6$$

$$X = \begin{vmatrix} 4 & 1 & 1 \\ -1 & -1 & -2 \\ 1 & -2 & -1 \end{vmatrix} = \begin{vmatrix} 4 & -3 & -7 \\ -1 & 0 & 0 \\ 1 & -3 & -3 \end{vmatrix} = \begin{vmatrix} -3 & -7 \\ -3 & -3 \end{vmatrix} = \begin{vmatrix} 3 & 7 \\ 3 & 3 \end{vmatrix} = 9 - 21$$
$$= -12$$

$$Y = \begin{vmatrix} 1 & 4 & 1 \\ 2 & -1 & -2 \\ 1 & 1 & -1 \end{vmatrix} = \begin{vmatrix} 1 & 4 & 1 \\ 0 & -3 & 0 \\ 1 & 1 & -1 \end{vmatrix} = -3\begin{vmatrix} 1 & 1 \\ 1 & -1 \end{vmatrix} = -3(-1 - 1)$$
$$= 6$$

$$Z = \begin{vmatrix} 1 & 1 & 4 \\ 2 & -1 & -1 \\ 1 & -2 & 1 \end{vmatrix} = \begin{vmatrix} 1 & 1 & 4 \\ 2 & -1 & -1 \\ 3 & -3 & 0 \end{vmatrix} = \begin{vmatrix} 1 & 2 & 4 \\ 2 & 1 & -1 \\ 3 & 0 & 0 \end{vmatrix} = 3\begin{vmatrix} 2 & 4 \\ 1 & -1 \end{vmatrix}$$
$$= -18$$

Thus

$$x = \frac{X}{D} = \frac{-12}{-6} = 2$$

$$y = \frac{Y}{D} = \frac{6}{-6} = -1$$

$$z = \frac{Z}{D} = \frac{-18}{-6} = 3$$

The solution set is $(2, -1, 3)$.

Check: $x + y + z = 2 - 1 + 3 = 4$

$2x - y - 2z = 2(2) - (-1) - 2(3) = 4 + 1 - 6 = -1$

$x - 2y - z = 2 - 2(-1) - 3 = 2 + 2 - 3 = 1$

EXAMPLE   Solve by using determinants:

$$2x - 4y + 2z = 3$$
$$x + y - z = 2$$
$$3x - 6y + 3z = 2$$

Solution

$$D = \begin{vmatrix} 2 & -4 & 2 \\ 1 & 1 & -1 \\ 3 & -6 & 3 \end{vmatrix} = \begin{vmatrix} 4 & -2 & 2 \\ 0 & 0 & -1 \\ 6 & -3 & 3 \end{vmatrix} = -(-1)\begin{vmatrix} 4 & -2 \\ 6 & -3 \end{vmatrix} = -12 + 12 = 0$$

$$X = \begin{vmatrix} 3 & -4 & 2 \\ 2 & 1 & -1 \\ 2 & -6 & 3 \end{vmatrix} = \begin{vmatrix} 7 & -2 & 2 \\ 0 & 0 & -1 \\ 8 & -3 & 3 \end{vmatrix} = -(-1)\begin{vmatrix} 7 & -2 \\ 8 & -3 \end{vmatrix} = -21 + 16$$
$$= -5$$

Since $D = 0$ and $X \neq 0$, the solution set is the empty set, $\varnothing$.

EXAMPLE   Solve by using determinants and check:

$$x + 2y - 3z = 4$$
$$2x - y + z = 1$$
$$3x + y - 2z = 5$$

Solution

$$D = \begin{vmatrix} 1 & 2 & -3 \\ 2 & -1 & 1 \\ 3 & 1 & -2 \end{vmatrix} = \begin{vmatrix} 1 & 0 & 0 \\ 2 & -5 & 7 \\ 3 & -5 & 7 \end{vmatrix} = 0$$

$$X = \begin{vmatrix} 4 & 2 & -3 \\ 1 & -1 & 1 \\ 5 & 1 & -2 \end{vmatrix} = \begin{vmatrix} 6 & 2 & -1 \\ 0 & -1 & 0 \\ 6 & 1 & -1 \end{vmatrix} = -1\begin{vmatrix} 6 & -1 \\ 6 & -1 \end{vmatrix} = 0$$

$$Y = \begin{vmatrix} 1 & 4 & -3 \\ 2 & 1 & 1 \\ 3 & 5 & -2 \end{vmatrix} = \begin{vmatrix} 1 & 1 & -3 \\ 2 & 2 & 1 \\ 3 & 3 & -2 \end{vmatrix} = 0$$

$$Z = \begin{vmatrix} 1 & 2 & 4 \\ 2 & -1 & 1 \\ 3 & 1 & 5 \end{vmatrix} = \begin{vmatrix} 1 & 2 & 4 \\ 3 & 1 & 5 \\ 3 & 1 & 5 \end{vmatrix} = 0$$

Thus the solution set is infinite. Now try to solve two equations for which the determinant of the coefficients of two of the variables, say x and y, is *not* zero. Using the first and second equations,

$$x + 2y = 4 + 3z$$
$$2x - y = 1 - z$$

$$D = \begin{vmatrix} 1 & 2 \\ 2 & -1 \end{vmatrix} = -1 - 4 = -5, \quad X = \begin{vmatrix} 4 + 3z & 2 \\ 1 - z & -1 \end{vmatrix} = -6 - z$$

$$Y = \begin{vmatrix} 1 & 4 + 3z \\ 2 & 1 - z \end{vmatrix} = -7 - 7z$$

Thus $x = \dfrac{X}{D} = \dfrac{-6 - z}{-5} = \dfrac{6 + z}{5}$ and $y = \dfrac{Y}{D} = \dfrac{-7 - 7z}{-5} = \dfrac{7 + 7z}{5}$

The solution set is $\left\{ \left(\dfrac{6 + z}{5}, \dfrac{7 + 7z}{5}, z \right) \middle| z \text{ is any real number} \right\}.$

Check: $x + 2y - 3z = \dfrac{6 + z}{5} + 2\left(\dfrac{7 + 7z}{5} \right) - 3z$

$$= \frac{6 + 14 + z + 14z - 15z}{5} = 4$$

$$2x - y + z = 2\left(\frac{6 + z}{5} \right) - \frac{7 + 7z}{5} + z$$

$$= \frac{12 - 7 + 2z - 7z + 5z}{5} = 1$$

$$3x + y - 2z = 3\left(\frac{6 + z}{5} \right) + \frac{7 + 7z}{5} - 2z$$

$$= \frac{18 + 7 + 3z + 7z - 10z}{5} = 5$$

EXERCISES A.5

Expand each of the determinants in Exercises 1–4.

1. $\begin{vmatrix} 3 & 5 \\ 4 & 2 \end{vmatrix}$

2. $\begin{vmatrix} 2 & 1 \\ -3 & -5 \end{vmatrix}$

3. $\begin{vmatrix} 1 & 2 & 3 \\ 3 & 2 & 1 \\ -1 & 0 & 2 \end{vmatrix}$

4. $\begin{vmatrix} 7 & -3 & 5 \\ 1 & 2 & -3 \\ 2 & -1 & 0 \end{vmatrix}$

In Exercises 5–8, find a determinant equal to the given one and satisfying the stated conditions; expand the determinant.

5. $\begin{vmatrix} -2 & 3 & 1 \\ 1 & 2 & 3 \\ 2 & 3 & 3 \end{vmatrix}$; zeros everywhere in row 2 except column 1

6. $\begin{vmatrix} 1 & 1 & 1 \\ 1 & 4 & 9 \\ 1 & 8 & 27 \end{vmatrix}$; zeros everywhere in column 1 except row 1

7. $\begin{vmatrix} -1 & 4 & -4 \\ 2 & 3 & 2 \\ 1 & -1 & 2 \end{vmatrix}$; zeros everywhere in column 3 except row 3

8. $\begin{vmatrix} 5 & -2 & 3 \\ -12 & 3 & 9 \\ 4 & 1 & -2 \end{vmatrix}$; zeros everywhere in row 2 except column 2

Find an equal determinant having zeros everywhere below the main diagonal in Exercises 9 and 10.

9. $\begin{vmatrix} 1 & 2 & 3 \\ 1 & 4 & 9 \\ 1 & 8 & 27 \end{vmatrix}$ **10.** $\begin{vmatrix} 2 & 3 & 4 \\ 1 & -5 & 6 \\ 4 & -7 & -1 \end{vmatrix}$

Exercises 11 and 12 are based on the following data:

The definition and theorems for third-order determinants can be generalized for fourth-order or higher-order determinants. Thus

$$\begin{vmatrix} a_1 & b_1 & c_1 & d_1 \\ a_2 & b_2 & c_2 & d_2 \\ a_3 & b_3 & c_3 & d_3 \\ a_4 & b_4 & c_4 & d_4 \end{vmatrix} = a_1A_1 - a_2A_2 + a_3A_3 - a_4A_4$$

11. Find an equal determinant having zeros everywhere in column 1 except row 1, then expand the determinant:

$$\begin{vmatrix} 1 & 3 & 2 & -2 \\ 2 & 5 & -1 & 1 \\ -3 & 2 & 1 & 6 \\ -1 & 4 & 5 & 2 \end{vmatrix}$$

12. Find an equal determinant having zeros everywhere below the main diagonal, then expand the determinant:

$$\begin{vmatrix} 1 & 1 & 1 & 1 \\ 1 & 2 & 3 & 4 \\ 1 & 4 & 9 & 16 \\ 1 & 8 & 27 & 64 \end{vmatrix}$$

Solve each of the systems in Exercises 13–22 by using determinants.

13. $3x + 2y = 6$
$x - 4y = 1$

14. $2x_1 - x_2 = 0$
$5x_1 - 3x_2 = 2$

15. $x - 2y - 3z = -20$
$2x + 4y - 5z = 11$
$3x + 7y - 4z = 33$

16. $4x + y - 3z = 0$
$x - y + z = -7$
$3x + 2y - z = -5$

17. $8x + 2y = 5$
$4y - 3z = 0$
$4x + 6z = -1$

18. $x + y = z$
$2x + 2z = 3 - 2y$
$5x + 2y = z + 2$

19. $3x + 2y - 4 = 0$
$4y - z + 2 = 0$

20. $6x + 3y - 3z = 2$
$2x + 2y - 2z = 5$
$3x - 3y + 3z = 7$

If

$$D = \begin{vmatrix} a_1 & b_1 & c_1 & d_1 \\ a_2 & b_2 & c_2 & d_2 \\ a_3 & b_3 & c_3 & d_3 \\ a_4 & b_4 & c_4 & d_4 \end{vmatrix}, \quad X = \begin{vmatrix} k_1 & b_1 & c_1 & d_1 \\ k_2 & b_2 & c_2 & d_2 \\ k_3 & b_3 & c_3 & d_3 \\ k_4 & b_4 & c_4 & d_4 \end{vmatrix},$$

$$Y = \begin{vmatrix} a_1 & k_1 & c_1 & d_1 \\ a_2 & k_2 & c_2 & d_2 \\ a_3 & k_3 & c_3 & d_3 \\ a_4 & k_4 & c_4 & d_4 \end{vmatrix}, \quad Z = \begin{vmatrix} a_1 & b_1 & k_1 & d_1 \\ a_2 & b_2 & k_2 & d_2 \\ a_3 & b_3 & k_3 & d_3 \\ a_4 & b_4 & k_4 & d_4 \end{vmatrix},$$

and

$$T = \begin{vmatrix} a_1 & b_1 & c_1 & k_1 \\ a_2 & b_2 & c_2 & k_2 \\ a_3 & b_3 & c_3 & k_3 \\ a_4 & b_4 & c_4 & k_4 \end{vmatrix},$$

then the system

$$a_1x + b_1y + c_1z + d_1t = k_1$$
$$a_2x + b_2y + c_2z + d_2t = k_2$$
$$a_3x + b_3y + c_3z + d_3t = k_3$$
$$a_4x + b_4y + c_4z + d_4t = k_4$$

is equivalent to $Dx = X$, $Dy = Y$, $Dz = Z$, $Dt = T$. *Solve the systems in Exercises 21 and 22.*

21.
$$2x + y - z + t = 1$$
$$x - y + 2z - t = 2$$
$$-x - y + z - t = -1$$
$$3x + y - z + 2t = 0$$

22.
$$x \quad\quad - 2z \quad\quad = 4$$
$$y + z + t = 6$$
$$x \quad\quad\quad + 3t = 6$$
$$x + y + z + t = 0$$

A.6 COMPUTERS AND BASIC

Since throughout this text we have referred to the expanded role of finite mathematics due to the widespread availability of computers, it is incumbent upon us to discuss their nature and use. Historically, of course, computers are extensions of the calculating machine. One of the main advantages of a computer over a calculator is that a computer can carry out a sequence of operations automatically, and the user can adjust the sequence to fit an unlimited number of problems and their associated calculations. As new machines were developed, it soon became clear that the machines were far more than simple extensions of a calculator. Rather, they were devices that could manipulate symbols according to fixed rules—rules that were not "natural" rules but rather the arbitrary choice of the machine's designer. This is exactly what we do when we "think" about a problem in the special set of symbols we call the English language. The language that most machines use involves numerals rather than such English symbols as a, b, c, etc., but the machines can handle English by using simple codes. For example, a computer can store letters with a simple cipher such as $a = 01$, $b = 02$, etc.

One method that can help the beginner understand the function of modern computers is to view them as automatic machines that understand numerical codes as instructions to perform different operations. For example, one machine might understand a 17-digit instruction like

$$01 \quad 00623 \quad 00745 \quad 00811$$

as follows:

> 01 means add the number stored in memory cell 00623 to the number in memory cell 00745 and store the result in memory cell 00811.

The above example is not an actual instruction, but rather it illustrates the general type of instruction one might have used on an early computer. The programmer would make up a list of such instructions. This list would then be loaded into the memory of the computer, and someone would "push the button." The computer would then execute the first instruction in the list and proceed automatically down the list, executing the instructions in sequence. Hopefully the result was the appropriate calculation and thus the problem's solution.

To use a machine one had to learn the specific codes or *machine language* the computer in question used. In the late 1950s work was done on "teaching" computers to understand English so that the computer rather than the human user could translate the problem from English to machine language. A universal translator of programs was beyond the capability of the machines, but a happy solution to this problem was the design of *programming languages*. These are special languages whose role is to bridge the gap between the language the human user thinks in and the numerical codes of the machine. The human half of the team translates his solution into the programming language. This set of instructions is loaded into the computer. The computer uses a translator program to convert the programming language instructions into machine language.

Most modern computers can translate several programming languages. Each language is designed to meet the needs of the user in a specific area of application. For example, there is COBOL, which stands for *com*mon *b*usiness *o*riented *l*anguage and which is designed to be used in accounting and allied fields. There is FORTRAN, short for *for*mula *tran*slation, a language designed for the scientist and engineer. We can illustrate the nature of such languages and how they are applied by studying a specific language. The one we will consider is called BASIC because it is simple to learn and use. We will consider a version designed for use on a time-sharing basis—that is, one used on terminals connected to a central computer.

In such an application one computer is used by a number of terminals at one time; hence time is shared.

A program written in BASIC consists of a list of statements, each identified with a statement number. The computer "translator" understands that the instructions are to be executed in the numerical order of the program statements.

EXAMPLE Consider the following program designed to evaluate a 2×2 determinant. Recall that for $\begin{vmatrix} A & B \\ C & D \end{vmatrix}$, the value of the determinant is given by $AD - BC$. A BASIC program which could be used to evaluate this expression might read

```
10     INPUT A, B, C, D
20     LET E = A * D − B * C
30     PRINT E
```

In the example the computer would execute the three statements of the program in numerical order. That is, it would first execute statement 10, then statement 20, then statement 30. The exact meaning of these statements is covered below. If statement numbers are missing, the computer skips until it finds the next highest statement number. The statements of the BASIC program need not be listed in numerical order, but they are always carried out that way.

In the remainder of this section we will consider some actual instructions one can use in BASIC. In the next section we will examine and analyze several sample programs. From an operational point of view, BASIC instructions fall into three classes:

1. Arithmetic statements: statements which instruct the computer to carry out some calculation.

2. Control statements: statements which change or modify the sequence in which the program's instructions are to be carried out.

3. Input/output statements: statements that cause the computer to read into its memory numerical or other data and to assign values thus read to symbolic values within the program, or which print out the results of calculations.

To illustrate this language we will consider a short, and by no means exhaustive, list of BASIC instructions.

I. Arithmetic Statements

Arithmetic statements involve the use of constants and variables. Constants are inserted into a program by using their actual value. Variables are given symbolic names. These names can be any single capital letter or any single capital letter followed by a single digit.

Variables and constants are linked together with an equals sign to form arithmetic statements. The following symbols indicate arithmetic operations:

+ addition
− subtraction
* multiplication
/ division
↑ exponentiation (raising to a power)
( ) grouping

The general form of an arithmetic statement is:

$$\text{statement} \ \# \ \text{LET variable} = \text{expression}$$

EXAMPLE

$$50 \quad \text{LET} \quad X = (A * B - A * C)/(A + B) \uparrow 3$$

is understood to say that the variable X is to be given the value

$$\frac{(A \cdot B - A \cdot C)}{(A + B)^3}$$

The indicated statement number is 50. Prior to this statement, values for A, B, and C must have been either input data or calculated in previous statements.

EXAMPLE Write a BASIC statement that calculates a value for C if $C = \frac{5}{9}(F - 32)$. Assume that F has been calculated, and call the statement number 75.

Solution 75   LET $C = (5/9) * (F - 32)$

II. Control Statements

Control statements determine or modify the sequence in which the list of instructions are executed. There are many control statements in BASIC, and we will consider only a few.

1. The GO TO statement.
   The general form of the GO TO statement is

statement # GO TO statement #

   The computer understands the GO TO statement as an instruction to jump to the indicated statement and carry out its instructions and continue from that point.

EXAMPLE

$$76 \quad \text{GO TO} \quad 37$$

instructs the computer to jump to statement number 37 for its next instruction.

2. The IF $\cdots$ THEN statement.
   The IF $\cdots$ THEN statement causes the computer to jump to a new statement provided a given condition is met. The general form of the IF $\cdots$ THEN statement is

statement # IF condition THEN statement #

The conditions take the form of mathematical sentences using the following:

$=$	equals
$>$	greater than
$<$	less than
$> \, =$	greater than or equal to
$< \, =$	less than or equal to
$\neq$	not equal to

EXAMPLE   The BASIC statement

$$110 \quad \text{IF} \quad A < \, = (B + C) \quad \text{THEN} \quad 60$$

is understood to mean that the computer is to take its next instruction from statement number 60 *provided* the value of the variable A is less than or equal to the sum of the values of B and C. If this condition is not met, the computer executes the instruction numerically after 110.

3. The MAX and MIN operators.
   The MAX operator selects the larger of two expressions; the MIN selects the smaller. The general form is

variable or constant	MAX	variable or constant

variable or constant	MIN	variable or constant

EXAMPLE

$$115 \quad \text{LET} \quad B = (7 \text{ MAX } X)$$

instructs the computer to give B the value of X or the value of 7, whichever is larger.

EXAMPLE

$$120 \quad \text{IF} \quad (A \text{ MIN } Z) > (C \text{ MAX } D) \quad \text{THEN} \quad 75$$

instructs the computer to transfer to statement number 75 *provided* the smaller of the two values A or Z is greater than the larger of the two values C or D.

4. The END statement.
   The END statement tells the computer that the statement is the end of the program.

III. Input/Output Statements

1. The INPUT statement.
The general form of the INPUT statement is

> statement # INPUT   variable, variable, . . ., variable

The computer executes the INPUT statement by turning on its
terminal and printing a question mark (?). The user then types in a
numerical value. This value is assigned to the first variable in the
list following the word INPUT. The computer then types a second
question mark. The user responds with a second numerical value,
which is assigned to the second variable in the list. This process is
continued until a numerical value has been given to each variable
in the list.

EXAMPLE

$$80 \quad \text{INPUT} \quad A1, B, I$$

instructs the computer to—in sequence—read in from the terminal the values
for the variable A1, then for the variable B, then for the variable I.

2. The PRINT statement.
The general form of the PRINT statement is

> statement # PRINT   expression or list of variables

EXAMPLE

$$85 \quad \text{PRINT} \quad A, B6, C$$

instructs the computer to print out on the terminal the calculated values for
the variables A, B6, C.

EXAMPLE

$$90 \quad \text{PRINT} \quad A; \quad \text{``CUBED IS''}; \quad A \uparrow 3$$

would cause the computer to print the value for A, followed by the phrase
"cubed is," followed by the value of A^3—i.e., $A \uparrow 3$ in BASIC. Suppose that
in a prior statement A has been given a value of 4. This instruction would
cause the computer to type

$$4 \quad \text{CUBED IS} \quad 64$$

EXAMPLE

75   PRINT   A = (D   MAX   C2)

would cause the computer to print out the value of A, which would be equal
to the larger of the values of the two variables D or C2.

This list of BASIC statements is far from complete, but it is sufficient to
develop some sample programs.

EXERCISES A.6

Define each of the terms in Exercises 1–5.

1. Machine language   2. Programming language
3. Statement number   4. Constant
5. Variable

In Exercises 6 and 7, identify if the symbol given is a legal BASIC variable.

6. a. A   7. a. 2X
   b. A3   b. XY
   c. 3X   c. X + Z
   d. XB   d. A2

*Name the arithmetic operation or phrase indicated by each of the symbols in
Exercises 8–11.*

8. < =   9. ↑
10. *   11. ≠

*In Exercises 12–15, translate each of the expressions written using the BASIC
arithmetic symbols into the normal algebraic notation.*

12. (A + B) ↑ 2   13. A/3 + B/2 − C/4
14. (A + B) * (C + D)   15. A ↑ 2 + B ↑ 2 + 2 * A * B

*Explain how each of the BASIC statements in Exercises 16–23 would be executed
by the computer if it appeared as part of a program.*

16. 70   LET   X = (B − C) * (B + C)
17. 80   LET   A = (C   MAX   8)
18. 65   LET   X = (A ↑ 2 − B ↑ 2)

19. 80   GO TO   75
20. 75   IF   A < B   THEN   110
21. 90   IF   X ≠ (A + B)   THEN   75
22. 115   IF   (A MAX B) > 10   THEN   80
23. 520   PRINT   A = (6.2 ∗ B)

In Exercises 24–30, write a BASIC *statement that will produce the indicated result.*

24. Evaluate A by the formula

$$A = \frac{(X^2 - Y^2)}{Z^3}$$

25. Set the variable B equal to the expression

$$\frac{X1 + X2 + X3}{3}$$

26. Shift the control of the computer to statement 75 provided that the value of a variable Z is not zero.

27. Shift the control of the computer to statement 600.

28. If the value of a variable Z is less than 10, then the computer is to execute statement 115.

29. Assign values from the computer terminal to the variables X, Y, Z, and A3.

30. Print out values for the variable K where K is equal to the expression $(2.714)^x$.

A.7   BASIC PROGRAMS

In this section we will consider the use of BASIC in carrying out various calculations.

Problem I

Let us consider how we might use the computer to calculate the probability table that for any given number of persons in a room, two or more will have the same birthday. From the material we studied on probability, we first found the probability that no pair of persons has the same birthday. This number is then subtracted from 1 to yield the probability that there are matching birthdays. If *n* is the number of persons, with *n* greater than or equal to 2, we might view the calculations in table form.

n	Probability of No Match	Probability of Match
2	$\dfrac{364}{365}$	$1 - \dfrac{364}{365}$
3	$\dfrac{364}{365} \times \dfrac{363}{365}$	$1 - \dfrac{364}{365} \times \dfrac{363}{365}$
4	$\dfrac{364}{365} \times \dfrac{363}{365} \times \dfrac{362}{365}$	$1 - \dfrac{364}{365} \times \dfrac{363}{365} \times \dfrac{362}{365}$
$\vdots$		
k	$\dfrac{364}{365} \times \dfrac{363}{365} \times \cdots \times \dfrac{(365-k+1)}{365}$	$1 - \dfrac{364}{365} \times \dfrac{363}{365} \times \cdots \times \dfrac{(365-k+1)}{365}$
$\vdots$		
365	$\dfrac{364}{365} \times \cdots \times \dfrac{3}{365} \times \dfrac{2}{365} \times \dfrac{1}{365}$	$1 - \dfrac{364}{365} \times \cdots \times \dfrac{1}{365}$

The object is to design a BASIC program that will print out this table. Consider the program in Figure A.7.1.

To analyze the effect of this program, let us consider the effect of each statement.

$$10 \quad \text{LET} \quad X = 2$$

This instruction initially assigns the value 2 to the variable X.

$$20 \quad \text{LET} \quad Z = 1$$

This instruction initially assigns the value 1 to the variable Z.

$$30 \quad \text{LET} \quad W = 364$$

This instruction initially assigns the value 364 to the variable W.

$$40 \quad \text{LET} \quad Z = (Z * W)/365$$

This step is somewhat unusual. First it calculates the expression ZW/365. Initially this will be $\frac{364}{365}$. It then reassigns this value to the variable Z,

```
10   LET  X=2
20   LET  Z=1
30   LET  W=364
40   LET  Z=(Z*W)/365
50   LET  Y=1-Z
60   PRINT  X,Z,Y
70   IF  X >= 365 THEN 110
80   LET  X=X+1
90   LET  W=W-1
100    GOTO 40
110    PRINT "END OF CALCULATION"
120    END
```

FIGURE A.7.1 The "birthday paradox" in BASIC.

replacing its old value. Thus on the first "pass" through the program $Z = \frac{364}{365}$.

$$50 \quad \text{LET} \quad Y = 1 - Z$$

This statement calculates the expression $1 - Z$ and assigns that value to the variable Y.

$$60 \quad \text{PRINT} \quad X, Z, Y$$

Initially this instruction prints the values of X, which is 2; Z, which is $\frac{364}{365} \approx$ .997260; and Y, which would be $1 - \frac{364}{365} \approx .0027397$. Thus we would have printed

$$2 \qquad .997260 \qquad .0027397$$

$$70 \quad \text{IF} \quad X > = 365 \quad \text{THEN} \quad 110$$

If the value of X is greater than or equal to 365, the computer is instructed to jump to statement 110 for its next instruction. If X is less than 365, the computer continues in sequence.

$$80 \quad \text{LET} \quad X = X + 1$$

This instruction in effect increases the value of X by 1. On the first pass the value of X will become $2 + 1$, or 3.

$$90 \quad \text{LET} \quad W = W - 1$$

This decreases the value of W by 1. On the first pass through the program W will move from its initial value, 364, to a new value, 363.

$$100 \quad \text{GO TO} \quad 40$$

The computer is told to take its next instruction from statement 40.

Let us now assume that we have returned to statement 40 on a second pass through the program. The computer is told

$$40 \quad Z = (Z * W)/365$$

On this second pass the value of Z is $\frac{364}{365}$; that of W is 363. Thus the expression $(Z * W)/365 = (\frac{364}{365} \times 363)/365 \approx .991796$. This value is assigned to Z.

We then return again to statement 50.

$$50 \quad \text{LET} \quad Y = 1 - Z$$

On this pass the value of Y would become

$$1 - Z = 1 - \frac{364}{365} \times \frac{363}{365} \approx .0082041$$

The computer would then execute

$$60 \quad \text{PRINT} \quad X, Z, Y$$

This would result in a second line of print, and the teletypeout would look like

2	0.997260	.0027397	(printed on the first pass)
3	0.991796	.0082041	(printed on the second pass)

We then move to

$$70 \quad \text{IF} \quad X > = 365 \quad \text{THEN} \quad 110$$

Since $X = 3$, the computer moves to

$$80 \quad \text{LET} \quad X = X + 1$$

and

$$90 \quad \text{LET} \quad W = W - 1$$

which assigns 4 to the variable X and 362 to W. The computer then returns to statement 40 as it executes

$$100 \quad \text{GO TO} \quad 40$$

The computer would continue executing statements 40 through 100 until the value of X reaches 365. Let us examine what happens then. Assume that the value of X is 365 and the computer has just executed statement 60. It moves to statement 70:

$$70 \quad \text{IF} \quad X > = 365 \quad \text{THEN} \quad 110$$

Since X is equal to 365, the computer jumps to statement 110 for its next instruction:

$$110 \quad \text{PRINT} \quad \text{"END OF CALCULATION"}$$

This causes the computer to print

END OF CALCULATION

at the end of the table. The output of the terminal looks like

2	0.997260	.0027397
3	0.991796	.0082041
4	0.986354	.0163559
⋮		
365	0.0	1.

END OF CALCULATION

This is the table we desired.

Problem 2

Let us examine a program designed to "solve" a 2×2 matrix game. Assume that we have such a 2×2 game with game matrix

$$\begin{bmatrix} A & B \\ C & D \end{bmatrix}$$

Recall that the game is either strictly determined or optimum strategies can be calculated from the formulas:

$$\bar{P} = [\bar{p}_1 \quad \bar{p}_2] \quad \text{and} \quad \bar{Q} = \begin{bmatrix} \bar{q}_1 \\ \bar{q}_2 \end{bmatrix}$$

with

$$\bar{p}_1 = \frac{D - C}{A + D - B - C}$$

$$\bar{p}_2 = \frac{A - B}{A + D - B - C}$$

$$\bar{q}_1 = \frac{D - B}{A + D - B - C}$$

$$\bar{q}_2 = \frac{A - C}{A + D - B - C}$$

with the value of the game

$$v = \frac{AD - BC}{A + D - B - C}$$

Our program must first check for a saddle point. If it exists, the program should cause the computer to indicate that the game is strictly determined and give the value of the game. If there is no saddle point, then the program should instruct the computer that optimum strategies and game value are to be calculated by the above formulas.

Consider the following BASIC program, Figure A.7.2, designed to "solve" a 2×2 zero-sum matrix game.

We can analyze this program in block segments designed to perform a specific portion of the total object. For example, the job of statements 5 and 10 is clear. They tell the program that the user will enter the values of the game matrix and these are to be assigned to the appropriate variables. The job of statements 15, 17, 19, 21, and 87 through 101 is to search for a saddle point. If in statement 15 the value of A is less than or equal to B, then A is a row minimum and thus a possible saddle point. In this case the computer jumps to statement 87 to see if A is also a column maximum. If it is, it is a saddle point, and the computer is instructed to jump to statement 89, then to 105 to inform the user that the game is strictly determined and to print out the game value. If A is not a column maximum in statement 87, statement 88 sends the computer to statement 17 to determine if B is a row minimum. If B is a row minimum, then in statement 91 a check is made to determine if B is also a column maximum and hence a saddle point. If neither A nor B are saddle points, statements 19, 21, 95, and 99 perform a similar check for C and D. If any A, B, C, or D are saddle points, the appropriate statement from 89, 93, 97, or 101 assigns the saddle point value to the variable X, which is then printed out together with appropriate comments by statements 105, 106, and 107. Statement 108 instructs the computer to return to statement 5 and a new set of game coefficients.

```
                            5    PRINT "ENTER VALUES FOR A, B, C, AND D"
                            10   INPUT A, B, C, D
Checking for         ⎧      15   IF A <= B THEN 87
row minimums         ⎨      17   IF B <= A THEN 91
                     ⎪      19   IF C <= D THEN 95
                     ⎩      21   IF D <= C THEN 99
                     ⎧      24   LET P1=(D-C)/(A+D-B-C)
                     ⎪      25   LET P2=(A-B)/(A+D-B-C)
                     ⎪      26   LET Q1=(D-B)/(A+D-B-C)
                     ⎪      27   LET Q2=(A-C)/(A+D-B-C)
                     ⎪      28   LET V=(A*D-B*C)/(A+D-B-C)
                     ⎪      30   PRINT "A NON STRICTLY DETERMINED GAME"
                     ⎪      31   PRINT "THE VALUE OF THE GAME IS"
Computing and        ⎨      32   PRINT V
printing strategies  ⎪      33   PRINT "THE OPTIMUM STRATEGIES FOR R ARE"
                     ⎪      34   PRINT "P1=";P1
                     ⎪      35   PRINT "P2=";P2
                     ⎪      36   PRINT "THE OPTIMUM STRATEGIES FOR C ARE"
                     ⎪      37   PRINT "Q1=";Q1
                     ⎪      38   PRINT "Q2=";Q2
                     ⎩      39   GOTO 5
                     ⎧      87   IF A >= C THEN 89
                     ⎪      88   GOTO 17
                     ⎪      89   LET X=A
                     ⎪      90   GOTO 105
                     ⎪      91   IF B >= D THEN 93
                     ⎪      92   GOTO 19
Checking for column  ⎨      93   LET X=B
maximums             ⎪      94   GOTO 105
                     ⎪      95   IF C >= A THEN 97
                     ⎪      96   GOTO 21
                     ⎪      97   LET X=C
                     ⎪      98   GOTO 105
                     ⎪      99   IF D >= B THEN 101
                     ⎪      100  GOTO 24
                     ⎩      101  LET X=D
                            105  PRINT "STRICTLY DETERMINED GAME"
                            106  PRINT " THE GAME VALUE IS"
                            107  PRINT X
                            108  GOTO 5
                            109  END
```

FIGURE A.7.2   A 2 × 2 matrix game in BASIC.

If the game is not strictly determined, the computer will find itself at statement 24. Statements 24 through 38 calculate optimum strategies for the game and the game value and print these with appropriate labels. Statement 39 performs the same role as statement 108, returning the program to statement 5 for new game coefficients.

The two sample problems considered here illustrate how a few simple instructions can be put together to "solve" relatively complex problems. We have not examined the rationale involved with the construction of these programs; we have only attempted to show how they reach the goal or solution of the problem once they have been written. In many respects a program is like a recipe. Most of us can follow it, but its initial creation is something of an art form. Students will find that, like driving a car, the key is actually writing a program oneself. Programming is a relatively easy task once some experience has been gained, and it is relatively hard to follow looking over someone else's shoulder, as was done here.

EXERCISES A.7

In Exercises 1–9, consider the sample program which tabulates the probability that two persons in a room share the same birthday.

1. Suppose that statement 80 was renumbered as statement 55. How would this change the values in the table?

2. If statement 80 is renumbered as statement 55, how could statement 10 be adjusted to keep the printout unchanged?

3. Write a BASIC statement that could be inserted in this program so that headings are supplied to the table printout.

4. Modify statement 60 so that the probability that two persons *do not* share a birthday is not printed.

5. If the statement numbers of statements 10, 20, and 30 are renumbered as 20, 10, and 30 or 20, 30, and 10, what effect would this have on the program?

6. How many times will the computer execute statement 60 during its processing of this program?

7. Modify statement 70 so that the condition on the ending of calculation is not that all 365 days have been calculated but rather that the number of people is such that there is a 50% or greater chance that two share a birthday.

8. Write a BASIC program that will accept as input a specific number of people and calculate only the probability that with the given number of people two or more share a birthday.

9. Write a BASIC program to calculate the probability that for any given number of persons in a room two or more were born on the same day of the week.

In Exercises 10–14, consider the program designed to solve a 2 × 2 matrix game.

10. What would be the effect on the final outcome if statements 24–28 were renumbered in the following manner: statement 24 becomes statement 28; the statement numbers for statements 25–28 are reduced by one.

11. Since $\bar{P}_1 + \bar{P}_2 = 1$ and $\bar{Q}_1 + \bar{Q}_2 = 1$, modify statements 25 and 27 to calculate $\bar{Q}_2$ and $\bar{P}_2$ without the direct use of the given formulas.

12. The program uses statements 105–107 to print out the value of a strictly determined game. How could this printout be carried out without using a dummy variable such as X?

13. Modify the program so that it indicates which coefficient—A, B, C, or D—was the saddle point in a strictly determined game.

14. Modify the program so that it prints out and labels optimum strategies for strictly determined games.

15. Construct a BASIC program which will find the arithmetic mean for ten numbers.

16. A statistics text indicates that for a set of n numbers $x_1, x_2, \ldots, x_n$, the variance S^2 of the set can be found using the formula

$$S^2 = \frac{n \sum_{i=1}^{n} x_i^2 - \left(\sum_{i=1}^{n} x_i \right)^2}{n(n-1)}$$

Set up a program to calculate this for a specific set of ten numbers.

ANSWERS
TO ODD-NUMBERED
PROBLEMS AND
REVIEW EXERCISES

Exercises 1.2

1. Yes

3. No

5. Yes

7. Yes

9. Yes

11. 5, 6, 7, 9, 10

13. If government spending is increasing, then taxes are higher than ever.

15. If government spending is increasing, then there is a freeze on wage increases.

17. If taxes are higher than ever and there is a freeze on wage increases, then government spending is increasing.

19. a. True   b. True   c. True   d. True

Exercises 1.3

1. Logic is important and I will pass this course.

3. Logic is not important and I will pass this course.

5. It is not the case that logic is important and I will pass this course.

7. $q \vee \sim p$

9. $(p \wedge q) \vee \sim p$

11. TTFT

13. FFFT

15. FTTT

17. True

19. True

21. True

23. True

25. FFTF

Exercises 1.4

1. TTTF

3. TTFT

5. FTFF

7. TTTT

9. FTTT

11. $p \rightarrow q$

13. $p \rightarrow q$

15. $q \leftrightarrow p$

17. False

19. False

21. True

23. True

25. False

27. Either or, but not both

Exercises 1.5

1. TTTFTFTF

3. TTTFTTTT

5. TTTFTTTT

7. TTTFFTTT

9. #8

11. Tautology

13. Neither

15. Tautology

17. $(p \lor \sim r) \land q$, false

19. $\sim (p \lor q)$, false

Exercises 1.6

7. a. If profits rise, efficiency in management will increase.
  b. If efficiency in management is not increased, profits will not rise.
  c. If profits don't rise, then efficiency in management will not increase.
  d. a. and b.

9. $\sim p \lor q$

11. $\sim p \rightarrow q$

13. $p \rightarrow q$, yes

15. $q \rightarrow p$, no

17. $\sim p \rightarrow \sim q$, no

19. $\sim (p \land \sim q)$, yes

21. $(p \rightarrow q) \land (q \rightarrow p)$

23. True

Exercises 1.7

1. Yes

3. Yes

5. Yes

7. No

9. $p \land q$

11. $p \land \sim q$

13. $p \land q$

15. $\sim (p \land \sim q) \land \sim r$

17. The bank loan is not approved or construction cannot be started.

19. The credit union does not pay 6% interest and the Loan and Trust Company does not pay $6\frac{1}{4}$% interest.

21. Yes

23. Yes

25. No

Exercises 1.8

1. Invalid

3. Valid

5. Valid

7. Invalid

9. Valid

11. Valid

13. Valid

15. Invalid

17. Invalid

19. Invalid

Exercises 1.9

1. Valid

3. Valid

5. Invalid

7. Valid

9. Valid

11. Invalid

13. r

15. No conclusion possible

17. No conclusion possible

19. Sally is not allowed to register for Logic 24.

21. Denver will host the Winter Olympics.

Chapter 1 Review Exercises

1. FTFF

2. TFFF

3. FF

4. TTTF

5. TFFFTFTF

6. TFFF

7. FTFT

8. TTTT

9. TTTT

10. TTTT

11. 8, 9, 10

12. 3

13. 9, 10

14. False

15. False

16. True

17. True

18. Valid

19. Invalid

20. Invalid

21. Invalid

22. Valid

23. r

24. No conclusion possible

25. If profits are not limited, then profits are limited.

Exercises 2.1

1. $\{x|\ x$ is one of the first seven counting numbers$\}$

3. $\{p|p$ is one of the first four presidents of the U.S.A. not named James$\}$

5. $\{x \mid x$ is an integer$\}$ **7.** $\{6, 7, 8, 9, \ldots\}$

9. $\{$Secretary of state, of defense, of the interior, of health, education, and welfare, of the treasury, of labor, of commerce, of housing and transportation, of agriculture; attorney general$\}$

11. $A \subset \mathcal{U}$ **13.** $C \nsubseteq \mathcal{U}$

15. $E = \mathcal{U}$ **17.** $\{$McDonald's, IBM$\}$

19. $\varnothing$

Exercises 2.2

1. $\{1, 2, 4, 6, 8\}$ **3.** $\{8\}$

5. $\{6\}$ **7.** $\{1, 2, 3, 4, 5, 7, 8, 9\}$

9. $\{2, 4, 6, 8\}$ **11.** $\{2, 4, 6\} = A$

13. $\{2, 4, 6, 8\}$ **15.** $\mathcal{U}$

17. $\{3, 5, 7, 9\}$ **19.** $\varnothing$

21. 281   **23.** 654

25. 623   **27.** 794

29. The business administration majors or the economics majors taking data processing

31. The economics majors or the psychology majors

33. All psychology majors or all students enrolled in statistics

35. Those students not taking English   **37.** $A \subseteq B$

39. A and B are disjoint.

Exercises 2.3

1.

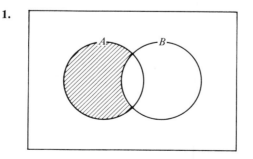

3.

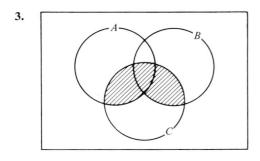

5.

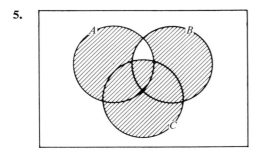

7.

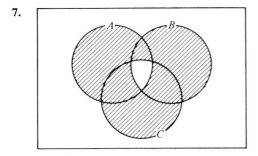

9.

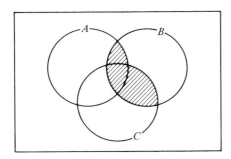

11. $A \cup (B \cap C)$

13. $[(A \cap B) \cup (A \cap C) \cup (B \cap C)]'$ **15.** $(A \cup B \cup C)' \cup (A \cap B \cap C)$

Exercises 2.4

1. Empty set, $\varnothing$

3. Logically false

5. Neither

7. $(A \cup B') \cap (A' \cup B)$

9. *Not* empty

11. Valid

13. Invalid

15. Valid

Exercises 2.5

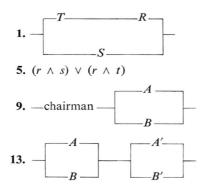

1.

3.

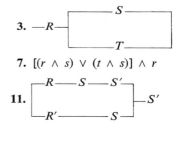

5. $(r \wedge s) \vee (r \wedge t)$

7. $[(r \wedge s) \vee (t \wedge s)] \wedge r$

9. —chairman

11.

13.

Exercises 2.6

1. $\{(1, a), (2, a), (3, a), (1, b), (2, b), (3, b)\}$

3. $\{(1, x), (1, y), (1, z), (2\ x), (2, y), (2, z), (3, x), (3, y), (3, z)\}$

5. $\{(a, a), (a, b), (b, a), (b, b)\}$

7. $\{(1, a, a), (1, a, b), (1, b, a), (1, b, b), (2, a, a), (2, a, b), (2, b, a), (2, b, b),$
$(3, a, a), (3, a, b), (3, b, a), (3, b, b)\}$

9. $\{(BJ, JN), (JS, LT), (FD, AP), (JF, SS)\}$,
$\{(BJ, LT), (JS, JN), (FD, SS), (JF, AP)\}$,
$\{(BJ, SS), (JS, AP), (FD, LT), (JF, JN)\}$,
and so forth. There are 24 such sets of different pairings possible.

11.

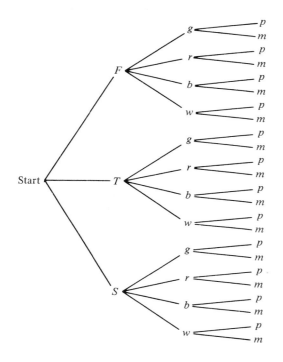

13.

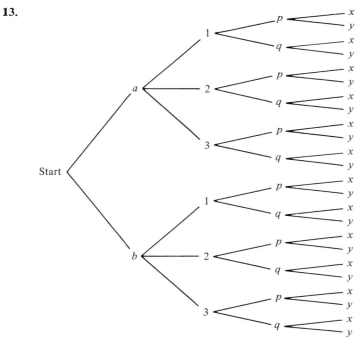

15.

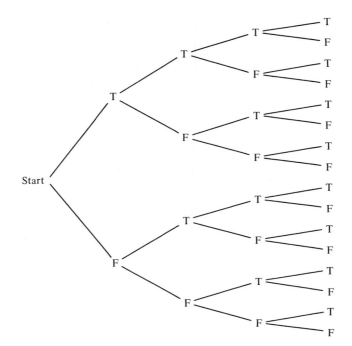

17.

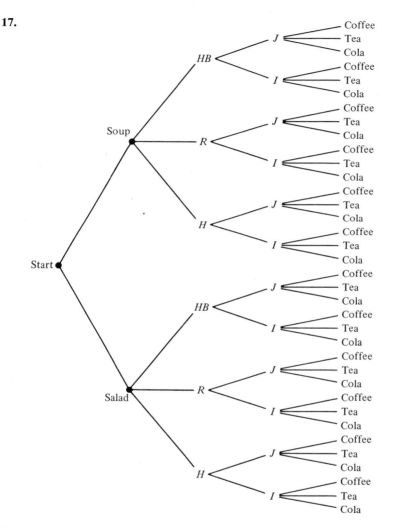

Exercises 2.7

1. {2, 4, 6, 8, 10, 12, 14, 16, 18}

3. $n(A) = 9$, $n(B) = 3$, $n(C) = 4$, $n(D) = 3$

5. 11   **7.** 27

9. 39   **11.** $n(A) \cdot n(B) \cdot n(C)$

13. $A \sim \mathcal{U}$, $B \sim D$ **15.** 16

19. $n(L) = 501$, $n(H) = 253$, $n(N) = 246$

21. 426   **23.** 206

25. a. 225   b. 175   c. 235   d. 65

Chapter 2 Review Exercises

1. $A \subset \mathcal{U}$

2. $B \nsubseteq \mathcal{U}$

3. $\mathcal{U} \subset C$

4. $D = \mathcal{U}$

5. All employees who are not secretaries

6. All female executives

7. Custodians or part-time employees

8. Part-time custodians

9. All male secretaries

10. Male part-time sales persons

11. $\{(1, 2), (5, 2), (7, 2), (1, 4), (5, 4), (7, 4)\}$

12. $\{(2, 1), (2, 5), (2, 7), (4, 1), (4, 5), (4, 7)\}$

13. $\{(1, 2, 1), (5, 2, 1), (7, 2, 1), (1, 4, 1), (5, 4, 1), (7, 4, 1), (1, 2, 4), (5, 2, 4),$
$(7, 2, 4), (1, 4, 4), (5, 4, 4), (7, 4, 4)\}$

14. $\{(1, 2, 1), (1, 2, 5), (1, 2, 7), (1, 4, 1), (1, 4, 5), (1, 4, 7), (4, 2, 1), (4, 2, 5),$
$(4, 2, 7), (4, 4, 1), (4, 4, 5), (4, 4, 7)\}$

15. 6

16. 12

17. 5

18. 9

19. 9

20. 5

21. Team, meat, mate, tame

22. $S = \{4\}$

23. $A \subseteq B$

24.

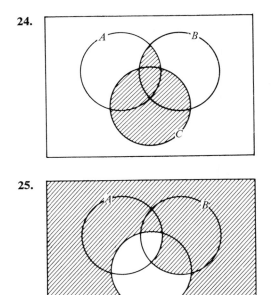

25.

26.

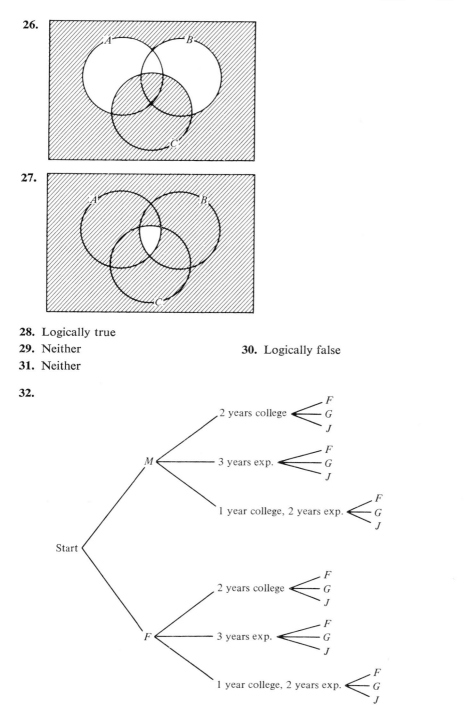

27.

28. Logically true

29. Neither   **30.** Logically false

31. Neither

32.

33. 3

Exercises 3.1

1. $(\{a\}, \{b\}, \{c\}), (\{a, b\}, \{c\})$
3. $(\{a\}, \{b\}, \{c\}), (\{a, b\}, \{c\}), (\{a\}, \{b, c\}), (\{a, c\}, \{b\})$
5. $(\{1\}, \{2\}, \{4\}, \{3, 5\}, \varnothing)$
7. $(P \cap Q, P' \cap Q, P \cap Q', P' \cap Q')$
9. $(\{a\}, \{b\}, \{c\}, \{d\}, \{e\}, \{f\}, \varnothing)$
13. Cross partition is $(P', Q, P \cap Q', \varnothing)$
15. a. 40   b. 32   17. 84
19. 145   21. Not possible—error in data
23. AB—: 25   rh—: 180

Exercises 3.2

1. 2,520   3. 40,320
5. 42   7. 24
9. Not possible   11. $26^3 \cdot 10^3 = 17,576,000$
13. 5,040   15. 6,720
17. $5! = 120, 72$
19. There are $26^3 = 17,576$ different possibilities for three letters (initials) to occur. Therefore, if 250,000 people live in Sacramento, at least two people must have the same three initials.

Exercises 3.3

1. 10   3. 7
5. 84   7. 5
9. 1   11. 1
13. 15 and 15

15. $_nC_r = \dfrac{n!}{(n-r)!\, r!}$

$$_nC_{n-r} = \dfrac{n!}{[n-(n-r)]!\,(n-r)!}$$

$$= \dfrac{n!}{r!\,(n-r)!}$$

$$= {_nC_r}$$

17. $\binom{5}{3} = 10$ 19. $\binom{15}{4} = 1,365$

21. $\binom{10}{3} \cdot \binom{8}{3} = 6,720$ 23. $4\binom{13}{5} = 5,148$

25. 4, 4   27. 15

29. 64

31. Exercise 29 includes six pieces of optional equipment, and the answer is 2^6. With five options the answer is 2^5, as in exercise 30.

Exercises 3.4

1. 35

3. 420

5. 1

7. 200

9. 6

11. Not possible

13. $\begin{pmatrix} & & 5 & & \\ 1 & 1 & 1 & 1 & 1 \end{pmatrix}$

15. $\dfrac{10!}{2!\,4!\,4!}$

17. $\dfrac{7!}{2!\,2!\,3!}$

19. $\dfrac{8!}{1!\,2!\,2!\,2!\,1!}$

Exercises 3.5

3. $x^4 + 16x^3 + 96x^2 + 256x + 256$

5. $32x^5 + 80x^4 + 80x^3 + 40x^2 + 10x + 1$

7. $16x^4 + 96x^3y + 216x^2y^2 + 216xy^3 + 81y^4$

9. $1 - 6y + 15y^2 - 20y^3 + 15y^4 - 6y^5 + y^6$

11. 161,051

13. 96,059,601

15. $-56x^3y^5$

17. 126

19. 9

21. 90,720

Exercises 3.6

1. $a^3 + 3a^2b + 3a^2c + 6abc + b^3 + 3b^2c + 3ab^2 + c^3 + 3ac^2 + 3bc^2$

3. $a^4 + b^4 + c^4 + 4a^3b + 4a^3c + 4ac^3 + 4ab^3 + 4b^3c + 4bc^3 + 6a^2b^2 + 6a^2c^2 + 6b^2c^2 + 12a^2bc + 12ab^2c + 12abc^2$

5. $4x^2 + 9y^2 + 4z^2 - 12xy + 8xz - 12yz$

7. $4w^2 + x^2 + 4y^2 + z^2 - 4wx + 8wy - 4wz - 4xy + 2xz - 4yz$

9. $a^6 + 12a^5b + 60a^4b^2 + 160a^3b^3 + 240a^2b^4 + 192ab^5 + 64b^6$

11. 2,520

13. 80

15. Not possible

17. 21

Chapter 3 Review Exercises

1. $(\{*\}, \{\triangle\}, \{\square\}), (\{*, \triangle\}, \{\square\}), (\{*, \square\}, \{\triangle\}), (\{*\}, \{\triangle, \square\})$

2. $(\{1\}, \{2\}, \{3\}, \{4\}), (\{1, 2\}, \{3, 4\}), (\{1\}, \{2, 3\}, \{4\}), (\{1, 3\}, \{2, 4\}),$
$(\{1, 2, 3\}, \{4\}), (\{1\}, \{2, 3, 4\}), (\{1, 4\}, \{2, 3\}), (\{2\}, \{1, 3, 4\}), (\{3\}, \{1, 2, 4\}),$
$(\{1\}, \{2\}, \{3, 4\}), (\{1\}, \{3\}, \{2, 4\}), (\{2\}, \{3\}, \{1, 4\}), (\{2\}, \{4\}, \{1, 3\}), (\{3\}, \{4\},$
$\{1, 2\})$

3. $(\{1\}, \{2\}, \{3\}, \{4, 5\}, \varnothing)$

4. 60

5. 42

6. 1

7. 6

8. 56

9. Impossible

10. 1

11. Impossible

12. 3

13. 30

14. 1,120

15. 138,600

16. $x^5 - 5x^4y + 10x^3y^2 - 10x^2y^3 + 5xy^4 - y^5$

17. $a^8 + 8a^7b + 28a^6b^2 + 56a^5b^3 + 70a^4b^4 + 56a^3b^5 + 28a^2b^6 + 8ab^7 + b^8$

18. $a^2 + b^2 + c^2 + 2ab + 2ac + 2bc$

19. $a^2 + b^2 + c^2 + d^2 + 2ab + 2ac + 2ad + 2bc + 2bd + 2cd$

20. $8a^3 - b^3 + 27c^3 - 12a^2b + 36a^2c + 6ab^2 - 36abc + 54ac^2 + 9b^2c - 27bc^2$

21. 1,260

22. 576

23. 84

24. 5,040

25. 90

26. 1,287

27. 210

28. 16

29. 3,150

Exercises 4.1

1. {H1, H2, H3, H4, H5, H6, T1, T2, T3, T4, T5, T6}

3. {red, not red}

5. a. {three red, three not red}   b. ∅

7. {BBBB, BBBG, BBGB, BBGG, BGBB, BGBG, BGGB, BGGG, GBBB, GBBG, GBGB, GBGG, GGBB, GGBG, GGGB, GGGG}

9. {(1, 1), (1, 2), (1, 3), (1, 4), (1, 5), (1, 6), (2, 1), (2, 2), (2, 3), (2, 4), (2, 5), (2, 6), (3, 1), (3, 2), (3, 3), (3, 4), (3, 5), (3, 6), (4, 1), (4, 2), (4, 3), (4, 4), (4, 5), (4, 6), (5, 1), (5, 2), (5, 3), (5, 4), (5, 5), (5, 6), (6, 1), (6, 2), (6, 3), (6, 4), (6, 5), (6, 6)}

11. No

Exercises 4.2

1. $\frac{1}{4}$

3. $\frac{1}{2}$

5. $\frac{12}{13}$

7. $\frac{4}{13}$

9. $\frac{25}{658}$

11. $\frac{1}{4}$

13. $\frac{3}{4}$

15. 1

17. $\frac{3}{8}$

19. $\frac{1}{36}$

21. $\frac{4}{9}$

23. 0

25. $\frac{2}{3}$

Exercises 4.3

1. $\frac{17}{18}$

3. 0.92

5. $\frac{2}{13}$

7. $\frac{3}{13}$

9. $\frac{4}{13}$

11. $\frac{1}{16}$

13. $\frac{1}{169}$

15. $\frac{1}{2704}$

17. 0.45, 0.05, 0.95; the events are independent

19. 0.12, 0.58, 0.42

21. $P(A \cap B) = 0$

Exercises 4.4

1. 0.117

3. 0.970

5. 0.99992

7. 23

9. 41

11. 3.07×10^{-7}

13. 0.294

Exercises 4.5

1. $\frac{1}{3}, 0$

3. $\frac{1}{6}, \frac{1}{6}$

5. $\frac{2}{5}$

7. $\frac{3}{10}$

9. $\frac{1}{6}$

11. 0

13. $\frac{4}{5}$

15. 1

17. $\frac{1}{169}, \frac{1}{221}$

19. $\frac{11}{1110}, \frac{899}{1110}$

21. 0.04 or 4%

Exercises 4.6

1. $-\$0.83$; loss of 83 cents

3. 0

5. 0

7. $-\$0.23$

9. 0

11. $-\$0.05$

13. $-\$1.26$

15. $40

17. $2,500

19. A: $59,400   B: $89,175

Exercises 4.7

1. $\frac{3}{8}$

3. $\frac{405}{1024} \approx 0.40$

5. $\frac{1}{4}, \frac{1}{2}, \frac{1}{4}$

7. $\frac{1}{2}$

9. $\dfrac{1440}{(13)^5}$

11. $\frac{169}{512}$

13. 0.00856

15. 0.08146

17. $\left(\frac{3}{4}\right)^{10} + \binom{10}{9}\left(\frac{3}{4}\right)^{9}\left(\frac{1}{4}\right) + \binom{10}{8}\left(\frac{3}{4}\right)^{8}\left(\frac{1}{4}\right)^{2}$

19. 4

Exercises 4.8

1. 0.037

3. 0.194

5. 0.000

7. 0.000

9. 0.147

11. 0.849

13. 0.051

15. 0.011

17. 0.999

19. 2 or more

21. $\binom{n}{x} p^x q^{n-x} = \binom{n}{n-x} q^x p^{n-x}$

Exercises 4.9

1. $\frac{1}{3}$

3. $\frac{3}{8}$

5. $\frac{1}{2}$

7. 0

9. $\frac{21}{32}$

11. $\frac{11}{36}$

13. 0.55

15. 0.413

17. x: 61%, y: $22\frac{1}{2}\%$, z: $16\frac{1}{2}\%$

Chapter 4 Review Exercises

1. $\frac{3}{13}$

2. $\frac{12}{13}$

3. $\frac{4}{13}$

4. $\frac{11}{26}$

5. $\frac{3}{8}, \frac{1}{2}$

6. $\frac{1}{2}$

7. 1

8. $\frac{7}{20}$

9. $\frac{2}{7}$

10. $\frac{2}{5}$

11, 0

12. 0.56, 0.94

13. $1 - 3.07 \times 10^{-7}$

14. $\frac{19}{12475} \approx 0.0015$

15. $+\$0.17$

16. $-\$0.03$

17. $\$10$

18. 0.2048

19. 0.986

20. 9 or more

21. 0.44

22. 0.661

Exercises 5.1

1. $x = -3$

3. Not possible—one is a row vector and one is a column vector

5. x is any real number

7. $[-3, -7, 4, 4]$

9. $\begin{bmatrix} -4 \\ -4 \\ 15 \end{bmatrix}$

11. Not possible

13. $B = [-1, 4, -6]$

15. $C = [x, y]$, where x and y are any real numbers

17. $C = \bar{0}$

19.

$$A = \begin{bmatrix} 1,700 \\ 1,500 \\ 2,000 \\ 2,275 \end{bmatrix} \quad B = \begin{bmatrix} 2,000 \\ 1,800 \\ 2,230 \\ 2,500 \end{bmatrix} \quad C = \begin{bmatrix} 2,250 \\ 1,600 \\ 2,275 \\ 2,600 \end{bmatrix} \quad D = \begin{bmatrix} 1,975 \\ 1,915 \\ 2,540 \\ 2,575 \end{bmatrix}$$

21.

$$B - A = \begin{bmatrix} 300 \\ 300 \\ 230 \\ 225 \end{bmatrix} \quad C - B = \begin{bmatrix} 250 \\ -200 \\ 45 \\ 100 \end{bmatrix} \quad D - C = \begin{bmatrix} -275 \\ 315 \\ 265 \\ -25 \end{bmatrix}$$

Exercises 5.2

1. $[23, -2, 21, -27]$

3. $\begin{bmatrix} 6 \\ 40 \\ 24 \end{bmatrix}$

5. Not possible

7. $[2, -\frac{1}{2}, \frac{3}{2}]$

9. $[-7, -7, -7]$

11. $\begin{bmatrix} 7.5 \\ 31.5 \\ 10.5 \end{bmatrix}$ Total $= 49.5$ hr

13. $\begin{bmatrix} 38.6 \\ 288.2 \\ 171.6 \end{bmatrix}$

15. 4

17. 0

19. 12

21. $64.55

23. $92.55

25. Always true

27. Always true, provided $A \cdot B$ exists

29. Always true if A and B are row vectors of the same dimension or column vectors of the same dimension and if $(A + B) \cdot C$ exists

Exercises 5.3

1. a. 3   b. 5

3. $x = 2, y = 3, z = -7$

5. $x = 5$ or $x = -2$
$y = 2$ or $y = -2$

7. $x = y, z = 3$

9. Never equal since $a_{13} \neq b_{13}$; $-4 \neq -2$

11. $x = 1, y = 3, z = 4$

13. $\begin{bmatrix} -2 & 2 & 10 \\ 12 & -2 & -9 \\ 1 & 4 & -1 \end{bmatrix}$

15. $\begin{bmatrix} -7 & -5 & -3 & -1 \\ 1 & 3 & 5 & 7 \end{bmatrix}$

17. No

19. Yes

Exercises 5.4

1. $\begin{bmatrix} -10 & 9 \\ -8 & 22 \end{bmatrix}$

3. $\begin{bmatrix} -1 & -4 \\ 0 & 36 \end{bmatrix}$

5. $\begin{bmatrix} 1 & 2 \\ 2 & 4 \\ 3 & 6 \end{bmatrix}$

7. $[23]$

9. $\begin{bmatrix} 1 & -2 & -3 & 7 \\ 4 & -8 & -12 & 28 \\ 3 & -6 & -9 & 21 \\ 2 & -4 & -6 & 14 \end{bmatrix}$

Exercises 5.5

3. $\begin{bmatrix} \frac{4}{11} & -\frac{3}{11} \\ \frac{1}{11} & \frac{2}{11} \end{bmatrix}$

5. $\begin{bmatrix} \frac{1}{6} & 0 & \frac{1}{6} \\ \frac{1}{15} & \frac{1}{5} & -\frac{2}{15} \\ -\frac{13}{30} & \frac{1}{5} & \frac{11}{30} \end{bmatrix}$

7. $\begin{bmatrix} \frac{5}{14} & -\frac{1}{14} & \frac{1}{7} \\ -\frac{1}{7} & \frac{3}{7} & \frac{1}{7} \\ \frac{2}{7} & \frac{1}{7} & -\frac{2}{7} \end{bmatrix}$

9. $\begin{bmatrix} 1 & -1 & -1 \\ 1 & -2 & -2 \\ 1 & -2 & -1 \end{bmatrix}$

11. $\begin{bmatrix} \frac{1}{2} & \frac{1}{2} & 0 \\ 1 & 0 & -\frac{1}{2} \\ \frac{1}{4} & -\frac{1}{4} & 0 \end{bmatrix}$

13. $\begin{bmatrix} 2 & -1 & 0 \\ 5 & -4 & 1 \\ -6 & 5 & -1 \end{bmatrix}$

15. $\begin{bmatrix} \frac{4}{13} & -\frac{7}{13} & \frac{1}{13} \\ \frac{11}{13} & -\frac{3}{13} & -\frac{7}{13} \\ -\frac{6}{13} & \frac{4}{13} & \frac{5}{13} \end{bmatrix}$

17. $\begin{bmatrix} \frac{1}{22} & \frac{1}{22} & \frac{5}{22} & \frac{2}{11} \\ \frac{71}{154} & \frac{5}{154} & \frac{3}{154} & -\frac{12}{77} \\ \frac{5}{22} & \frac{5}{22} & \frac{3}{22} & -\frac{1}{11} \\ \frac{20}{77} & \frac{9}{77} & -\frac{10}{77} & \frac{3}{77} \end{bmatrix}$

19. $C = \begin{bmatrix} 1 & 0 \\ 0 & 0 \end{bmatrix}$ $D = \begin{bmatrix} 0 & 0 \\ 0 & 1 \end{bmatrix}$ (answer not unique)

Exercises 5.6

1. $x_1 = -\frac{1}{6}, x_2 = -\frac{5}{6}$

3. No solution

5. $x_1 = \frac{3}{4}, x_2 = \frac{1}{2}, x_3 = -\frac{1}{4}$

7. $x_1 = \frac{11}{7}, x_2 = -\frac{10}{7}, x_3 = -\frac{8}{7}$

9. Dependent system

11. $x_1 = -1, x_2 = 0, x_3 = -\frac{1}{2}$

13. $x_1 = 10, x_2 = 29, x_3 = -35$

15. $x_1 = -\frac{9}{13}, x_2 = -\frac{7}{13}, x_3 = \frac{36}{13}$

17. $x_1 = \frac{11}{49}, x_2 = \frac{5}{7}, x_3 = \frac{4}{7}, x_4 = \frac{20}{49}$

19. $x_1 = 0, x_2 = 0$

Exercises 5.7

1. 5 salesmen in first division, 6 salesmen in second division

3. 20 lb A, 180 lb B, 260 lb E

5. 2 lb B_1, 1 lb B_2, 3 lb B_3

7. $\frac{1}{5}, \frac{3}{5}, \frac{1}{5}$

9. Person (1): \$4,000; person (2): \$6,000; person (3): \$6,000

11. 75 units of A and 75 units of B

13. 60 lb type 1, 40 lb type 2, 20 lb type 3

15. \$10,000 at 5%, \$9,333$\frac{1}{3}$ at 7%, \$10,666$\frac{2}{3}$ at 10%

Exercises 5.8

1. $\begin{bmatrix} a_{11} & a_{12} & a_{13} & a_{14} & a_{15} \\ a_{21} & a_{22} & a_{23} & a_{24} & a_{25} \\ a_{31} & a_{32} & a_{33} & a_{34} & a_{35} \end{bmatrix}$

3. $\begin{bmatrix} 1 & \frac{3}{2} \\ \frac{3}{2} & 2 \end{bmatrix}$

5. $[a_{ij}]_{23}$, where $a_{ij} = j$ for all i and j

7. $[a_{ij}]_{33}$, where $a_{ij} = ij$ for all i and j

9. $\begin{bmatrix} 2 & 4 & 6 \\ 2 & 4 & 6 \\ 2 & 4 & 6 \end{bmatrix}$

11. $\begin{bmatrix} 3 & 1 & -1 \\ 5 & 0 & -5 \\ 7 & -1 & -9 \end{bmatrix}$

13. 55

15. 91

17. 70

19. 10

21. $\begin{bmatrix} 11 & 2 & -7 \\ 14 & 2 & -10 \\ 17 & 2 & -13 \end{bmatrix}$

23. $\begin{bmatrix} 20 & 26 & 32 \\ 40 & 52 & 64 \\ 60 & 78 & 96 \end{bmatrix}$

25. $\begin{bmatrix} 160 & 40 & -80 \\ 208 & 52 & -104 \\ 256 & 64 & -128 \end{bmatrix}$

Chapter 5 Review Exercises

1. $x = 1$ or $x = -2$, $y = 3$, $z = 4$

2. $x = 1$

3. Not possible

4. $x = 0$, $y = 1$

5. -23

6. 26

7. $\begin{bmatrix} -3 & -7 & -5 \\ 6 & 14 & 10 \\ 9 & 21 & 15 \end{bmatrix}$

8. $\begin{bmatrix} 12 & 8 & -28 & 4 \\ 6 & 4 & -14 & 2 \\ 18 & 12 & -42 & 6 \\ 9 & 6 & -21 & 3 \end{bmatrix}$

9. $\begin{bmatrix} -55 \\ 220 \\ 330 \end{bmatrix}$

10. $\begin{bmatrix} -3 \\ 12 \\ 18 \end{bmatrix}$

11. Yes, $n \times q$

12. Yes, $n \times q$

13. Yes

14. $\begin{bmatrix} -\frac{3}{13} & \frac{4}{13} & \frac{3}{13} \\ \frac{1}{13} & \frac{3}{13} & -\frac{1}{13} \\ \frac{11}{13} & -\frac{6}{13} & \frac{2}{13} \end{bmatrix}$

15. $\begin{bmatrix} \frac{3}{4} & -\frac{1}{4} & 0 \\ \frac{7}{8} & -\frac{9}{8} & 1 \\ -\frac{5}{8} & \frac{3}{8} & 0 \end{bmatrix}$

16. $\begin{bmatrix} \frac{3}{7} & -\frac{1}{14} & \frac{1}{7} \\ -\frac{4}{7} & \frac{3}{7} & \frac{1}{7} \\ \frac{1}{7} & \frac{1}{7} & -\frac{2}{7} \end{bmatrix}$

17. $\begin{bmatrix} 1 & 1 & -1 \\ -4 & -3 & 5 \\ -10 & -8 & 13 \end{bmatrix}$

18. $\begin{bmatrix} -\frac{165}{14} & \frac{37}{7} & \frac{12}{7} & -\frac{3}{14} \\ \frac{74}{7} & -\frac{32}{7} & -\frac{10}{7} & \frac{3}{7} \\ -\frac{13}{7} & \frac{6}{7} & \frac{1}{7} & -\frac{1}{7} \\ \frac{75}{7} & -\frac{33}{7} & -\frac{9}{7} & \frac{2}{7} \end{bmatrix}$

19. $\begin{bmatrix} \frac{1}{11} & -\frac{9}{77} & \frac{6}{7} & -\frac{6}{11} \\ -\frac{1}{11} & -\frac{13}{77} & \frac{4}{7} & -\frac{5}{11} \\ -\frac{3}{11} & \frac{16}{77} & \frac{1}{7} & -\frac{4}{11} \\ \frac{3}{11} & \frac{6}{77} & -\frac{4}{7} & \frac{4}{11} \end{bmatrix}$

20. $\begin{bmatrix} \frac{1}{2} & \frac{1}{2} & -1 & 0 \\ \frac{1}{4} & -\frac{1}{4} & -\frac{1}{2} & \frac{1}{2} \\ \frac{1}{4} & -\frac{1}{4} & \frac{1}{2} & -\frac{1}{2} \\ -\frac{1}{4} & -\frac{3}{4} & \frac{3}{2} & \frac{1}{2} \end{bmatrix}$

21. $x_1 = 3, x_2 = 6, x_3 = 5$

22. $x_1 = -\frac{5}{3}, x_2 = \frac{17}{3}, x_3 = \frac{44}{3}$

23. $x_1 = 0, x_2 = -1, x_3 = 0, x_4 = 0$

24. $x_1 = -5, x_2 = \frac{1}{3}, x_3 = \frac{8}{3}, x_4 = 0$

25. $71,000 at 5%, $14,000 at 6%

26. $50,000 at 5%, $60,000 at 6%, $34,000 at 8%

Exercises 6.1

1.

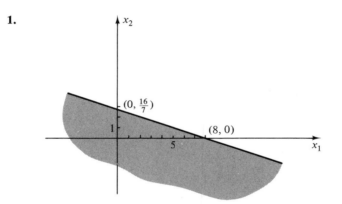

3.

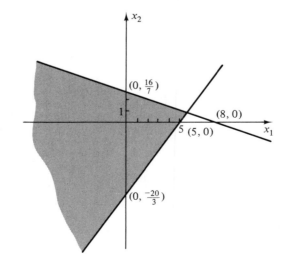

5.

7.

9.

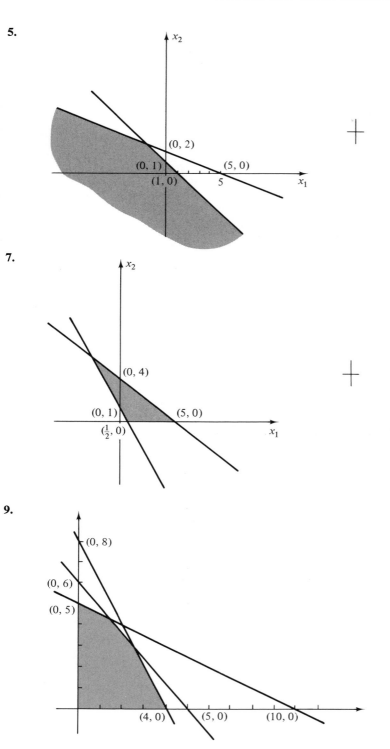

11.

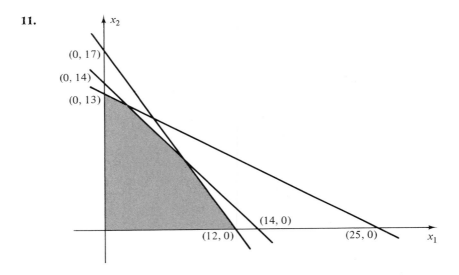

13. No solution

15.

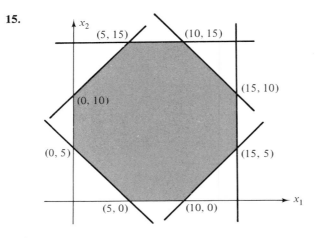

17.

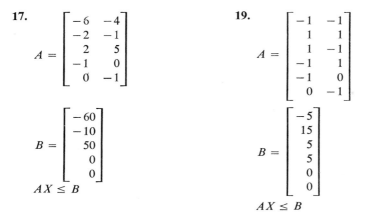

$$A = \begin{bmatrix} -6 & -4 \\ -2 & -1 \\ 2 & 5 \\ -1 & 0 \\ 0 & -1 \end{bmatrix}$$

$$B = \begin{bmatrix} -60 \\ -10 \\ 50 \\ 0 \\ 0 \end{bmatrix}$$

$AX \le B$

19.

$$A = \begin{bmatrix} -1 & -1 \\ 1 & 1 \\ 1 & -1 \\ -1 & 1 \\ -1 & 0 \\ 0 & -1 \end{bmatrix}$$

$$B = \begin{bmatrix} -5 \\ 15 \\ 5 \\ 5 \\ 0 \\ 0 \end{bmatrix}$$

$AX \le B$

Exercises 6.2

1. At $(-\frac{5}{2}, 6)$, $P = 16\frac{1}{2}$
3. At $(0, 13)$, $P = -52$
5. At $(4, 0)$, $P = 64$
7. 44
9. At $(5, 15)$, $P = -90$. At $(10, 0)$, $P = 30$; $30 - (-90) = 120$
11. 25 six-family units for a profit of \$298,000
13. 5 units of the second drug, none of the first, for \$62.50
15. 10 cu yd of chemical A, 5 cu yd of chemical B

Exercises 6.3

1. 60
3. 61
5. 72
7. 30
9. $\frac{166}{3}$
11. 150
13. 2 units of B, none of A
15. $\frac{176}{125}$ kg of A, $\frac{126}{125}$ kg of B

Exercises 6.4

1. 8
3. $133\frac{1}{3}$
5. $\frac{43}{29}$
7. $\frac{191}{35}$
9. $20\frac{8}{9}$
11. 40
13. 2 units of A, none of B or C
15. $16\frac{2}{3}$ units of type II only
17. 6 kg of type I only

Exercises 6.5

1. $\frac{24}{5}$
3. 18
5. 24
7. 57
9. 18
11. $\frac{16}{7}$ oz of A to $\frac{10}{7}$ oz of B
13. 4 gal of type III only
15. $6\frac{2}{5}$ carloads from A, $1\frac{1}{5}$ carloads from C, none from B

Chapter 6 Review Exercises

1. 110

2. 120

3. 168

4. 98

5. 10

6. 16

7. 30

8. $\frac{155}{7}$

9. $\frac{1022}{17}$

10. 135

11. 42

12. 100

13. 120

14. 216

15. $35\frac{1}{3}$

16. $\frac{768}{19}$

17. $\frac{38}{3}$

18. 32

19. 15

20. $\frac{104}{3}$

21. 4 kg barley, no rice

22. $\frac{82}{17}$ kg barley, $\frac{10}{17}$ kg oatmeal, no rice

23. 8 kg oatmeal, no rice

Exercises 7.1

1.
$$\begin{array}{cc} & C_1 \quad C_2 \\ \begin{array}{c} R_1 \\ R_2 \end{array} & \begin{bmatrix} +1 & -1 \\ -1 & +1 \end{bmatrix} \end{array}$$
zero-sum

3.
$$\begin{array}{cc} & C_1 \quad C_2 \\ \begin{array}{c} R_1 \\ R_2 \end{array} & \begin{bmatrix} 0 & +7 \\ +15 & -5 \end{bmatrix} \end{array}$$
zero-sum

5.
$$A = \begin{array}{cc} & C_1 \qquad C_2 \\ \begin{array}{c} R_1 \\ R_2 \end{array} & \begin{bmatrix} -1,000 & +1,500 \\ +1,500 & -1,000 \end{bmatrix} \end{array}$$
not zero-sum
$$\bar{A} = \begin{array}{cc} & C_1 \qquad C_2 \\ \begin{array}{c} R_1 \\ R_2 \end{array} & \begin{bmatrix} -1,000 & -1,000 \\ -1,000 & -1,000 \end{bmatrix} \end{array}$$

7. Max. $R = 1$, max. $C = 1$; no best strategy

9. Max. $R = 5$, max. $C = 5$; no best strategy

11. Max. $R = 6$, max. $C = 6$; best strategy R: row 1, C: column 1

13. Yes, R_1, C_1

15. Yes, R_2, C_1

Exercises 7.2

1. 0

3. 1

5. 2

7. 0

9. $\frac{7}{12}$

11. $p_i \; a_{ij} \; q_j$, where $p_i = q_j = 1$ and all other p's and q's $= 0$

13. $P = \begin{bmatrix} \frac{1}{2} & \frac{1}{2} \end{bmatrix}$
$$Q = \begin{bmatrix} \frac{1}{3} \\ \frac{1}{3} \\ \frac{1}{3} \end{bmatrix}$$

15.
$$A = \begin{array}{cc} & B \qquad RD \\ \begin{array}{c} R \\ S \\ L \end{array} & \begin{bmatrix} 5 & 1 \\ 15 & 3 \\ 0 & -20 \end{bmatrix} \end{array}$$
$$P = \begin{bmatrix} \frac{1}{3} & \frac{1}{3} & \frac{1}{3} \end{bmatrix}$$
$$Q = \begin{bmatrix} \frac{1}{4} \\ \frac{3}{4} \end{bmatrix}$$

Exercises 7.3

1. $v = 1$ **3.** $v = 1$

5. $v = -2$ **7.** $v = 0$

9. $v = \frac{9}{11}$ **11.** No

13. $v = -2$ **15.** $v = 3$

$\bar{P} = \begin{bmatrix} 1 & 0 \end{bmatrix}$ $\bar{P} = \begin{bmatrix} 1 & 0 & 0 & 0 \end{bmatrix}$

$\bar{Q} = \begin{bmatrix} 0 \\ 1 \end{bmatrix}$ $\bar{Q} = \begin{bmatrix} 0 \\ 0 \\ 1 \end{bmatrix}$

17.

$$A' = \begin{bmatrix} 5 & 4 & 3 \\ 1 & 4 & 0 \\ 5 & -2 & -1 \\ 1 & 3 & -2 \end{bmatrix}$$

$v = 3 = 1 + 2$

Exercises 7.4

1. Yes, $\bar{P} = \begin{bmatrix} 1 & 0 \end{bmatrix}$ **3.** No

$\bar{Q} = \begin{bmatrix} 1 \\ 0 \end{bmatrix}$

5. Yes, $\bar{P} = \begin{bmatrix} 1 & 0 & 0 \end{bmatrix}$ **7.** Yes, $\bar{P} = \begin{bmatrix} 0 & 1 \end{bmatrix}$

$\bar{Q} = \begin{bmatrix} 0 \\ 1 \\ 0 \end{bmatrix}$ $\bar{Q} = \begin{bmatrix} 0 \\ 1 \\ 0 \\ 0 \end{bmatrix}$

9. Yes, $\bar{P} = \begin{bmatrix} 1 & 0 & 0 \end{bmatrix}$ **11.** No

$\bar{Q}_1 = \begin{bmatrix} 1 \\ 0 \\ 0 \\ 0 \end{bmatrix}$ or $\bar{Q}_2 = \begin{bmatrix} 0 \\ 1 \\ 0 \\ 0 \end{bmatrix}$

$\bar{Q}_3 = \begin{bmatrix} 0 \\ 0 \\ 1 \\ 0 \end{bmatrix}$ or $\bar{Q}_4 = \begin{bmatrix} 0 \\ 0 \\ 0 \\ 1 \end{bmatrix}$

13. $a \le b$ and $a \ge c$, or $b \le a$ and $b \ge d$, or $c \le d$ and $c \ge a$, or $d \le c$ and $d \ge b$

15. T: support bill 1, C: support bill 1

$\bar{P} = \begin{bmatrix} 1 & 0 \end{bmatrix}, \bar{Q} = \begin{bmatrix} 1 \\ 0 \end{bmatrix}$

Exercises 7.5

1. $\bar{P} = \begin{bmatrix} 1 & 0 \end{bmatrix}, \bar{Q} = \begin{bmatrix} 1 \\ 0 \end{bmatrix},$ **3.** $\bar{P} = \begin{bmatrix} \frac{1}{3} & \frac{2}{3} \end{bmatrix}, \bar{Q} = \begin{bmatrix} \frac{3}{4} \\ \frac{1}{4} \end{bmatrix},$

$v = -1$ $v = 1$

5.
$$\bar{P} = [0 \quad 1], \bar{Q} = \begin{bmatrix} 1 \\ 0 \\ 0 \end{bmatrix},$$
$$v = -2$$

7.
$$\bar{P} = [0 \quad \tfrac{4}{9} \quad \tfrac{5}{9}], \bar{Q} = \begin{bmatrix} \tfrac{2}{3} \\ \tfrac{1}{3} \end{bmatrix},$$
$$v = -\tfrac{13}{3}$$

9. $\bar{P} = [\tfrac{3}{5} \quad \tfrac{2}{5} \quad 0],$
$$\bar{Q} = \begin{bmatrix} \tfrac{1}{2} \\ \tfrac{1}{2} \\ 0 \end{bmatrix},$$
$$v = 3$$

11.
$$\bar{P} = [0 \quad 1], \bar{Q} = \begin{bmatrix} 0 \\ 0 \\ 1 \\ 0 \\ 0 \end{bmatrix},$$
$$v = 1$$

13. $v = -1, \bar{P} = [0 \quad 0 \quad 1 \quad 0]$
$$\bar{Q} = \begin{bmatrix} 0 \\ 0 \\ 1 \\ 0 \end{bmatrix}$$

15. $\tfrac{11}{13}$ of the time buy crop 2, $\tfrac{2}{13}$ of the time buy crop 3, never buy crop 1

Exercises 7.6

3. All entries in $\bar{P}$, A, and $\bar{Q}$ are positive. Therefore, $v = \bar{P}A\bar{Q}$ must be positive.

5. $v = \tfrac{5}{7},$
$$\bar{P} = [\tfrac{2}{7} \quad \tfrac{5}{7}], \bar{Q} = \begin{bmatrix} \tfrac{6}{7} \\ \tfrac{1}{7} \end{bmatrix}$$

7.
$$v = -\tfrac{11}{3}, \bar{Q} = \begin{bmatrix} \tfrac{2}{3} \\ \tfrac{1}{3} \\ 0 \end{bmatrix},$$
$$\bar{P} = [\tfrac{1}{3} \quad \tfrac{2}{3}]$$

9. $v = \tfrac{4}{3}, \bar{Q} = \begin{bmatrix} \tfrac{5}{6} \\ \tfrac{1}{6} \end{bmatrix},$
$$\bar{P} = [0 \quad \tfrac{1}{3} \quad \tfrac{2}{3}]$$

11.
$$\bar{P} = [1 \quad 0 \quad 0], \bar{Q} = \begin{bmatrix} 0 \\ 0 \\ 1 \end{bmatrix},$$
$$v = -3$$

13.
$$v = -\tfrac{5}{3}, \bar{Q} = \begin{bmatrix} \tfrac{1}{3} \\ 0 \\ \tfrac{2}{3} \end{bmatrix},$$
$$\bar{P} = [\tfrac{7}{9} \quad \tfrac{2}{9} \quad 0]$$

15. $v = 0, \bar{P} = [0 \quad 1 \quad 0],$
$$\bar{Q} = \begin{bmatrix} 0 \\ 1 \\ 0 \\ 0 \end{bmatrix}$$

17. $v = -2, \bar{P} = [0 \quad \tfrac{1}{2} \quad \tfrac{1}{2}]$

Chapter 7 Review Exercises

1. 9, 6

2. 8, 6

3. All positive entries for R, and all negative entries for C

4. All negative entries for R, and all positive entries for C

5. $v = 3, \bar{P} = [1 \quad 0],$
$$\bar{Q} = \begin{bmatrix} 0 \\ 1 \end{bmatrix}$$

6. $v = 0, \bar{P} = [1 \quad 0],$
$$\bar{Q} = \begin{bmatrix} 0 \\ 1 \\ 0 \end{bmatrix}$$

7. $v = 1, \bar{P} = [0 \quad 1 \quad 0]$,

$$\bar{Q} = \begin{bmatrix} 0 \\ 1 \\ 0 \end{bmatrix}$$

8. $v = 0, \bar{P} = [0 \quad 1 \quad 0]$,

$$\bar{Q} = \begin{bmatrix} 1 \\ 0 \\ 0 \\ 0 \end{bmatrix}$$

9. $v = 1, \bar{P} = [1 \quad 0]$,

$$\bar{Q} = \begin{bmatrix} 0 \\ 1 \end{bmatrix}$$

10. $v = 1, \bar{P} = [\frac{5}{11} \quad \frac{6}{11}]$,

$$\bar{Q} = \begin{bmatrix} \frac{1}{2} \\ \frac{1}{2} \end{bmatrix}$$

11. $v = -\frac{1}{2}, \bar{P} = [\frac{1}{2} \quad \frac{1}{2}]$,

$$\bar{Q} = \begin{bmatrix} \frac{1}{2} \\ \frac{1}{2} \end{bmatrix}$$

12. $v = 1, \bar{P} = [1 \quad 0]$,

$$\bar{Q} = \begin{bmatrix} 1 \\ 0 \end{bmatrix}$$

13. $v = 3, \bar{P} = [\frac{1}{4} \quad 0 \quad \frac{3}{4}]$,

$$\bar{Q} = \begin{bmatrix} \frac{1}{2} \\ \frac{1}{2} \\ 0 \end{bmatrix}$$

14. $v = \frac{22}{13}, \bar{P} = [\frac{8}{13} \quad 0 \quad \frac{5}{13}]$,

$$\bar{Q} = \begin{bmatrix} \frac{6}{13} \\ \frac{7}{13} \\ 0 \end{bmatrix}$$

15. $v = 4, \bar{P} = [0 \quad 0 \quad 1]$,

$$\bar{Q} = \begin{bmatrix} 0 \\ 0 \\ 1 \end{bmatrix}$$

16. $v = \frac{30}{11}, \bar{P} = [\frac{9}{11} \quad \frac{2}{11} \quad 0]$,

$$\bar{Q} = \begin{bmatrix} \frac{7}{11} \\ 0 \\ \frac{4}{11} \\ 0 \end{bmatrix}$$

17.
$$\bar{P} = [1 \quad 0 \quad 0], \quad \bar{Q} = \begin{bmatrix} 1 \\ 0 \\ 0 \end{bmatrix}$$

18. $\bar{P} = [0 \quad \frac{5}{8} \quad \frac{3}{8}]$; short pass five times out of eight, long pass three times out of eight, never run

Exercises A.1

1. 18

3. 416

5. 35

7. 54

9. 7.5

11. 117.2

13. 3,011

15. 2,050

17. 120,020

19. 1,100,100

21. 200

23. 553

25. 1,859

27. 1,001,011 + 101,001 = 1,110,100

29. 10,001 × 1,010 = 10,101,010

31. 32 × 20 = 1,140

Exercises A.3

1. $1 + 4 + 7 + 10 + 13 + 16 + 19$

3. $\frac{1}{2} + \frac{1}{3} + \frac{1}{4} + \frac{1}{5} + \frac{1}{6} + \frac{1}{7} + \cdots + \frac{1}{16} + \frac{1}{17}$

5. $0 - 1 + 4 - 9 + 16 - 25 + \cdots$

7. $2 + \frac{5}{2} + \frac{10}{3} + \frac{17}{4} + \frac{26}{5} + \frac{37}{6}$

9. $\frac{1}{2} + \frac{1}{5} + \frac{1}{8} + \frac{1}{11} + \frac{1}{14} + \cdots$

11. $\sum_{i=1}^{7} 2i$

13. $\sum\limits_{n=1}^{5} n^2$

15. $\sum\limits_{i=1}^{4} i(i + 1)$

17. $\sum\limits_{i=1}^{12} (-1)^i i$

19. $\sum\limits_{k=0}^{5} 5^{-k}$

21. $\sum\limits_{i=1}^{\infty} \dfrac{n}{(n + 2)(n + 4)}$

23. 55

25. 26

31. $\sqrt{\frac{26}{5}} \approx 2.28$

33

f_i	$(\bar{x} - x_i)$	$(\bar{x} - x_i)^2$	$f_i(\bar{x} - x_i)^2$	
12	14	196	2,352	
18	10	100	1,800	
22	4	16	352	
6	-4	16	96	
16	-8	64	1,024	
$\dfrac{5}{79}$	-16	256	$\dfrac{1,280}{6,904}$	$\sigma^2 = 87.4$

Exercises A.4

1. $(1, 1)$

3. $(c + 3, c)$

5. $(1, 0, 0)$

7. $(\frac{9}{7}, -\frac{3}{7}, -\frac{1}{7})$

9. $\varnothing$

11. $(1, -2, -1, 3)$

13. $(x_1, 2 - x_1, x_1 - 1)$

15. $(\frac{29}{7}, -\frac{3}{7})$

Exercises A.5

1. -14

3. -4

5. $\begin{vmatrix} -2 & 7 & 7 \\ 1 & 0 & 0 \\ 2 & -1 & -3 \end{vmatrix} = 14$

7. $\begin{vmatrix} 1 & 2 & 0 \\ 1 & 4 & 0 \\ 1 & -1 & 2 \end{vmatrix} = 4$

9. $\begin{vmatrix} 1 & 2 & 3 \\ 0 & 2 & 6 \\ 0 & 0 & 6 \end{vmatrix}$

11. $\begin{vmatrix} 1 & 3 & 2 & 2 \\ 0 & -1 & 5 & 5 \\ 0 & 11 & 7 & 0 \\ 0 & 7 & 7 & 0 \end{vmatrix} = 140$

13. $(\frac{13}{7}, \frac{3}{14})$

15. $(1, 6, 3)$

17. $(\frac{3}{4}, -\frac{1}{2}, -\frac{2}{3})$

19. Cannot be solved by determinants

21. $(0, 5, 3, -1)$

Exercises A.6

7. a. No   b. No   c. No   d. Yes

9. Exponentiation   **11.** Not equal to

13. $\dfrac{A}{3} + \dfrac{B}{2} - \dfrac{C}{4}$ **15.** $A^2 + B^2 + 2AB$

17. Let A take on the larger value of C or 8.

19. Go to statement 75 for next instruction.

21. If X does not equal A + B, go to statement 75; otherwise go to next in sequence.

23. Print the result A = 6.2B

25. 70   LET   B = (X1 + X2 + X3)/3

27. 40   GO TO   600   **29.** 45   INPUT   X, Y, Z, A3

Exercises A.7

1. First column starts with 3 and the corresponding values are incorrect, being one line out of phase.

3. 5   PRINT   "N   P(NO MATCH) P(MATCH)"

5. None

7. 70   IF   Y > = .5   THEN   110

9. 10   LET   X = 2   **11.** 25   LET   P2 = 1 − P1
    20   LET   Z = 1   27   LET   Q2 = 1 − Q1
    30   LET   W = 6
    40   LET   Z = (Z * W)/7
    50   LET   Y = 1 − Z
    60   PRINT   X, Z, Y
    70   IF   X > = 7   THEN   110
    80   LET   X = X + 1
    90   LET   W = W − 1
   100   GO TO   40
   110   PRINT   "END OF CALCULATION"
   120   END

13. 108   IF   X = A   THEN   120
   109   IF   X = B   THEN   130
   110   IF   X = C   THEN   140
   111   GO TO   150
   120   PRINT   "SADDLE POINT AT A"
   125   GO TO   5
   130   PRINT   "SADDLE POINT AT B"
   135   GO TO   5
   140   PRINT   "SADDLE POINT AT C"
   145   GO TO   5
   150   PRINT   "SADDLE POINT AT D"

```
155     GO TO  5
160     END
```

15.
```
10     PRINT  "ENTER 10 NUMBERS"
20     INPUT N, N1, N2, N3, N4, N5, N6, N7, N8, N9
30     Z = (N + N1 + N2 + N3 + N4 + N5 + N6 + N7 + N8 + N9)/10
40     PRINT  "THE MEAN IS"; Z
50     GO TO  10
60     END
```

INDEX